Flora Europaea check-list and chromosome index

T0296987

Flora Europaea check-list and chromosome index

D.M. MOORE

CAMBRIDGE UNIVERSITY PRESS
Cambridge
London New York New Rochelle
Melbourne Sydney

CAMBRIDGE UNIVERSITY PRESS
Cambridge, New York, Melbourne, Madrid, Cape Town, Singapore, São Paulo, Delhi

Cambridge University Press
The Edinburgh Building, Cambridge CB2 8RU, UK

Published in the United States of America by Cambridge University Press, New York

www.cambridge.org
Information on this title: www.cambridge.org/9780521105736

First published 1982
This digitally printed version 2009

A catalogue record for this publication is available from the British Library

Library of Congress Catalogue Card Number: 82-1335

ISBN 978-0-521-23759-8 hardback
ISBN 978-0-521-10573-6 paperback

CONTENTS

Introduction

This book has grown out of the requests by many users of Flora Europaea (Tutin et al., 1964, 1968, 1972, 1976, 1980) for information on the sources of the chromosome numbers cited in that work. During discussions about the publication of these data it became apparent that there is also a demand for a basic checklist of the taxa recognised in the Flora, and that these two requirements could readily be met in the same volume.

The Checklist

Apart from the fundamental value of Flora Europaea to all modern taxonomic studies of European plants, the names and sequence of taxa it presents are widely used for arranging herbaria and libraries concerned with these plants. It is intended that the handy format of this checklist will facilitate these latter activities, as well as being useful in recording and assessing local floras within Europe.

The sequence of taxa is exactly the same as that used in Flora Europaea except that the Rubiaceae, which for technical reasons had to be published in volume 4 rather than in volume 3, have been positioned correctly in the Checklist. All species and subspecies 'accepted' in Flora Europaea (i.e. those prefixed by a number or a letter) are included here. Those unnumbered taxa of doubtful status are only included if a chromosome number is given for them in Flora Europaea, as is also the case for the agamospecies ('microspecies') listed, but not described, within the 'species-groups' of apomictic or partially apomictic genera. It has not been possible to include any synonyms in the Checklist but access to these is provided by the Consolidated Index of Flora Europaea (Halliday and Beadle, 1982) as well as the volumes of the Flora itself.

The Chromosome Index

As noted in the Introductions to the five volumes of Flora Europaea, diploid ($2n$) chromosome numbers were given when they were derived from material of known wild origin. In the case of naturalized or cultivated taxa chromosome numbers were included when they were known to be based on material with a status which justified their inclusion in the Flora: aliens had to be common and to have been naturalized for at least 25 years, while crops had to be widely cultivated. As can be seen from the Index, between 98% and 99% of the chromosome numbers included in Flora Europaea fulfil these criteria, whether they were based on a single observation from one locality or on many counts from a number of areas. In many instances the Editors were provided with unpublished chromosome data; some of these were subsequently published, hence references to them often postdate the publication of the relevant volume of the Flora, others are apparently still unpublished. The provenance of the material is indicated by using the abbreviations for the territories recognised in Flora Europaea (see p. iii), except where the data only permit a wider area to be cited. In several cases the chromosome numbers were published under names different from those adopted in the Flora Europaea, but these can readily be traced by reference to the synonymy given in the Flora.

Ideally, of course, a definitive chromosome index would give due weight to the number of observations on which each number is based, but the practical-ities of Flora-writing preclude this. The Index, consequently, cites only one reference for each chromosome number given, even though in many instances far more data were available to the editors of Flora Europaea; this has been done to obviate any impression of completeness in the coverage, which would be illusory. Less than 2% of the chromosome numbers cited in the Flora cannot be substantiated or are shown to be derived from material which does not fulfil the criteria referred to above. The commonest reason for this is that pre-liminary accounts of genera included chromosome numbers taken uncritically from the various worldwide compilations and their credentials were mis-interpreted or inadequately checked during the various stages of the editorial process; in other cases, far fewer, the sources cannot now be traced in the records of the Editorial Committee; occasionally, the numbers included in the Flora are simple misprints, undetected during proof reading.

The Chromosome Index reveals, therefore, the problems of fully documenting multi-author Floras of substantial areas produced within a realistic, or realisable, time-scale. It is felt that the level of accuracy achieved fully justifies the approach adopted by the editors of Flora Europaea and the publication of this documented list of chromosome numbers. The extra effort required to achieve complete accuracy (if such were possible) would not have been worthwhile.

Abbreviations of geographical territories

(For precise definitions of these territories, see maps in Flora Europaea)

Al	Albania	Hs	Spain
Amer.*	America	Hu	Hungary
Au	Austria	Is	Iceland
Az	Açores (Azores	It	Italy
Be	Belgium and Luxembourg	Ju	Jugoslavia
Bl	Islas Baleares (Balearic Islands)	Lu	Portugal
		No	Norway
Br	Britain	Po	Poland
Bu	Bulgaria	Rm	Romania
Co	Corse (Corsica)	Rs	U.S.S.R. (European part), subdivided thus:
Cr	Kriti (Crete)		
Cz	Czechoslovakia		(N) Northern region
Da	Denmark		(B) Baltic region
Fa	Faeröer (Faeroes)		(C) Central region
Fe	Finland		(W) South-western region
Ga	France		(K) Krym (Crimea)
Ge	Germany		(E) South-eastern region
Gr	Greece	Sa	Sardegna (Sardinia)
Hb	Ireland	Sb	Svalbard (Spitsbergen)
He	Switzerland	Scand.*	Scandinavia
Ho	Netherlands	Si	Sicilia (Sicily)

ix

Su   Sweden   Tu   Turkey (European part)

*not used in volumes of Flora Europaea

Acknowledgements

This book is founded on the endeavours of the Editorial Committee of Flora Europaea (T.G. Tutin, V.H. Heywood, N.A. Burges, D.M. Moore, D.H. Valentine, S.M. Walters and D.A. Webb), to which I am the most recent recruit. Professor D.H. Valentine and later Dr. S.M. Walters were primarily responsible for the chromosome numbers cited in volume 1 of Flora Europaea, and Dr. Walters for a significant part of those included in the second volume. They, and my other colleagues on the Editorial Committee, have been a constant source of support and information. Dr. P.W. Ball, Mr. A.O. Chater, Dr. R.A. de Fillips and Dr. I.B.K. Richardson, Research Associates during the various phases of the Flora Europaea project, checked hundreds of chromosome references, while Dr. Áskell Löve freely gave of his wide cytotaxonomical knowledge. Above all, however, the Chromosome Index is a testament to the many authors of Flora Europaea who provided accurate documentation of their chromosomal data; the cooperation of them and other friends and colleagues who made available un- published information is greatly appreciated. The onerous task of typing the camera-ready copy was undertaken by Miss Valerie Norris, whose efforts are grate- fully acknowledged, whilst I am also indebted to the Flora Europaea Trust Fund of the Linnean Society of London for a grant towards the costs of preparing the manuscript.

I LYCOPODIACEAE

1 Huperzia Bernh.
1 selago (L.) Bernh. ex Schrank & Mart.   90   Fe   Sorsa, 1963
                                           264   Br   Manton, 1950
                                         c̲.272   Is   Löve & Löve, 1961a
(a) selago
(b) arctica (Grossh.) Á. & D. Löve
(c) dentata (Herter) Valentine

2 Lepidotis Beauv.
1 cernua (L.) Beauv.
2 inundata (L.) Börner   156   Br Fe   V. Sorsa, 1958

3 Lycopodium L.
1 annotinum L.   68   Br He Su   Manton, 1950
2 dubium Zoega   68   Is   Löve & Löve, 1958
3 clavatum L.   68   Br He   Manton, 1950

4 Diphasium C. Presl.
1 complanatum (L.) Rothm.   46   Scand.   Löve & Löve, 1961a
2 tristachyum (Pursh) Rothm.   46   Scand.   Löve & Löve, 1961a
3 issleri (Rouy) J. Holub
4 alpinum (L.) Rothm.   46   Su   Wilce, 1961
                                         48-50   Cz   Manton, 1950

II SELAGINELLACEAE

1 Selaginella Beauv.
1 selaginoides (L.) Link   18   Br   Manton, 1950
2 helvetica (L.) Spring   18   He   Manton, 1950
3 denticulata (L.) Link   18   It   Manton, 1950
4 apoda (L.) Fernald
5 kraussiana (G. Kunze) A. Braun

III ISOETACEAE

1 Isoetes L.
1 lacustris L.   c̲.110   Is   Löve & Löve, 1961d
2 setacea Lam.   22   Is   Löve & Löve, 1961d
                                           100   Hb   Manton, 1950

3 brochonii Moteley
4 azorica Durieu ex Milde
5 boryana Durieu
6 delilei Rothm.
7 malinverniana Cesati & De Not.
8 heldreichii Wettst.
9 velata A. Braun
10 tenuissima Boreau
11 histrix Bory   20   Br   Manton, 1950
12 durieui Bory

IV EQUISETACEAE

1 Equisetum L.
1 hyemale L.   216   Is   Löve & Löve, 1961c
2 ramosissimum Desf.   c̲.216   It   Manton, 1950
3 variegatum Schleicher ex Weber &
    Mohr   216   Is   Löve & Löve, 1961c

```
 4 scirpoides Michx.                          c.216      No      Manton, 1950
 5 fluviatile L.                              c.216      Ho      Bir, 1960
 6 palustre L.                                c.216      Ho      Bir, 1960
 7 sylvaticum L.                              c.216      Is      Löve & Löve, 1961c
 8 pratense Ehrh.                             c.216      Is      Löve & Löve, 1961c
 9 arvense L.                                 c.216      Ho      Bir, 1960
10 telmateia Ehrh.                            c.216      Br      Manton, 1950
```

V OPHIOGLOSSACEAE

```
1 Ophioglossum L.
  1 lusitanicum                               250-260    Ga      Manton, 1950
  2 azoricum C. Presl.                         c.480
  3 vulgatum L.                                  480      Is      Löve & Löve, 1961c
                                               c.496      Ho      Verma, 1958
                                               500-520    Br      Manton, 1950

2 Botrychium Swartz
  1 simplex E. Hitchc.
  2 lunaria (L.) Swartz.                          90      Br      Manton, 1950
                                                  96      Da      Hagerup, 1941
  3 boreale Milde                                 90      Is      Löve & Löve, 1961c
  4 matricariifolium A. Braun ex Koch
  5 lanceolatum (S.G. Gmelin) Angström.           90      Is      Löve & Löve, 1961c
  6 multifidum (S.G. Gmelin) Rupr.
  7 virginianum (L.) Swartz
```

VI OSMUNDACEAE

```
1 Osmunda L.
  1 regalis L.                                    44      Br      Manton, 1950
```

VII SINOPTERIDACEAE

```
1 Cheilanthes Swartz
  1 marantae (L.) Domin                           58      It      Fabbri, 1957
  2 fragrans (L. fil.) Swartz
  3 hispanica Mett.
  4 catanensis (Cosent.) H.P. Fuchs
  5 persica (Bory) Mett. ex Kuhn

2 Pellaea Link
  1 calomelanos (Swartz) Link
```

VIII ADIANTACEAE

```
1 Adiantum L.
  1 capillus-veneris L.                           60    Hb Hs It  Manton, 1950
```

IX PTERIDACEAE

```
1 Pteris L.
  1 serrulata Forskål
  2 cretica L.                                    58      It      Manton, 1950
  3 vittata L.
```

X CRYPTOGRAMMACEAE

1 Cryptogramma R. Br.
  1 crispa (L.) R. Br. ex Hooker   120   Br   Manton, 1950
  2 stelleri (S.G. Gmelin) Prantl

XI GYMNOGRAMMACEAE

1 Anogramma Link
  1 leptophylla (L.) Link

XII DICKSONIACEAE

1 Culcita C. Presl
  1 macrocarpa C. Presl   132-136   Az   Manton, 1958

XIII HYPOLEPIDACEAE

1 Pteridium Scop.
  1 aquilinum (L.) Kuhn   104   Br   Manton, 1950

XIV DAVALLIACEAE

1 Davallia Sm.
  1 canariensis (L.) Sm.

XV HYMENOPHYLLACEAE

1 Hymenophyllum Sm.
  1 tunbrigense (L.) Sm.   26   Br   Manton, 1950
  2 wilsonii Hooker   36   Br   Manton, 1950

2 Trichomanes L.
  1 speciosum Willd.   144   Hb   Manton, 1950

XVI THELYPTERIDACEAE

1 Thelypteris Schmidel
  1 limbosperma (All.) H.P. Fuchs   68   Br   Manton, 1950
  2 palustris Schott   70   Fe   Sorsa, 1958
  3 phegopteris (L.) Slosson   90   Fe   Sorsa, 1958
  4 pozoi (Lag.) C.V. Morton

2 Cyclosorus Link
  1 dentatus (Forskål) R.-C. Ching

XVII ASPLENIACEAE

1 Asplenium L.
  1 hemionitis L.
  2 marinum L.   72   Br   Manton, 1950
  3 petrarchae (Guerin) DC.   144   Ge   Manton, 1950
  4 monanthes L.
  5 trichomanes L.

3

(a) trichomanes	73	Ge	Meyer, 1957
(b) quadrivalens	144	Ge	Meyer, 1957
6 adulterinum Milde	144	Au Ge	Meyer, 1957
7 viride Hudson	72	Br	Manton, 1950
8 jahandiezii (Litard.) Rouy	72	Ge	Meyer, 1960
9 fontanum (L.) Bernh.	72	He	Manton, 1950
majoricum Litard.	144	Bl	Lovis, 1963
10 forisiense Le Grand	144	Ge	Meyer, 1961
11 obovatum Viv.	72	Sa	Meyer, 1962
12 billotii F.W. Schultz	144	Br	Manton, 1950
13 adiantum-nigrum L.	144	Br	Manton, 1950
14 onopteris L.	72	Hb	Manton, 1955
15 cuneifolium Viv.	72	Ge	Meyer, 1967
16 septentrionale (L.) Hoffm.	144	Fe	Sorsa, 1958
17 seelosii Leybold	72	It	Meyer, 1957
18 ruta-muraria L.	144	Br	Manton, 1950
19 lepidum C. Presl	144	Europe	Meyer, 1960
20 fissum Kit. ex Willd.	72	Ge	Meyer, 1958

2 Ceterach DC.			
1 officinarum DC.	144	Br	Manton, 1950

3 Pleurosorus Fée			
1 hispanicus (Cosson) C.V. Morton	72	Hs	Lovis, 1963

4 Phyllitis Hill			
1 scolopendrium (L.) Newman	72	Br	Manton, 1950
2 sagittata (DC.) Guinea & Heywood	72	Ga	Manton, 1950
3 hybrida (Milde) C. Chr.	144	It	Manton, 1950

XVIII ATHYRIACEAE

1 Athyrium Roth			
1 filix-femina (L.) Roth	80	Br	Manton, 1950
2 distentifolium Tausch ex Opiz	80	Br	Manton, 1950

2 Diplazium Swartz			
1 caudatum (Cav.) Jermy			
2 sibiricum (Turcz. ex G. Kunze) Jermy	82	No	Brøgger, 1960

3 Cystopteris Bernh.			
1 fragilis (L.) Bernh.	168	Br He	Manton, 1950
	252	Br	Manton, 1950
2 dickieana R. Sim	168	Br	Manton, 1950
3 montana (Lam.) Desv.	168	He	Manton, 1950
4 sudetica A. Braun & Milde	168	No	Brøgger, 1960

4 Woodsia R. Br.			
1 ilvensis (L.) R. Br.	82	Br	Manton, 1950
2 alpina (Bolton) S.F. Gray	164	Is	Löve & Löve, 1961c
3 pulchella Bertol.	78	It	Meyer, 1959
4 glabella R. Br.	c.80	Su	Löve & Löve, 1961b
	164	Fe	Sorsa, 1963

5 Matteuccia Tod.			
1 struthiopteris (L.) Tod.	c.80	Fe	Sorsa, 1958

6 Onoclea L.			
1 sensibilis L.			

XIX ASPIDIACEAE

1 Polystichum Roth
  1 lonchitis (L.) Roth   82   Br He   Manton, 1950
  2 aculeatum (L.) Roth   164   Br He   Manton, 1950
  3 setiferum (Forskal) Woynar   82   Br He   Manton, 1950
  4 braunii (Spenner) Fée   164   Ge   Manton &
                                           Reichstein, 1961

2 Dryopteris Adanson
  1 filix-mas (L.) Schott   164   Fe   Sorsa, 1958
  2 borreri Newman   82   Br He   Manton, 1950
                          123   Br He No Manton, 1950
                          130   No   Knaben, 1948
  3 abbreviata (DC.) Newman   82   Br   Manton, 1950
  4 villarii (Bellardi) Woynar ex Schinz
    & Thell.   82   He No Su Manton, 1950
                         164   Br   Manton, 1950
    (a) villarii
    (b) pallida (Boey) Heywood
  5 cristata (L.) A. Gray   164   Br He   Manton, 1950
  6 carthusiana (Vill.) H.P. Fuchs 164   Fe   Sorsa, 1958
  7 dilatata (Hoffm.) A. Gray   164   Br   Manton, 1950
  8 assimilis S. Walker   82   Fe   Sorsa, 1958
  9 aemula (Aiton) O. Kuntze   82   Br   Manton, 1950
 10 fragrans (L.) Schott

3 Gymnocarpium Newman
  1 dryopteris (L.) Newman   160   Fe   Sorsa, 1958
  2 robertianum (Hoffm.) Newman 160-168   Br   Manton, 1950

XX ELAPHOGLOSSACEAE

1 Elaphoglossum Schott ex Sm.
  1 hirtum (Swartz) C. Chr.

XXI BLECHNACEAE

1 Blechnum L.
  1 spicant (L.) Roth   68   Br   Manton, 1950

2 Woodwardia Sm.
  1 radicans (L.) Sm.   68   Br   Manton, 1950

XXII POLYPODIACEAE

1 Polypodium L.
  1 australe Fée   74   Ga He It Manton, 1950
  2 vulgare L.   148   Br He   Manton, 1950
  3 interjectum Shivas   222   Br   Manton, 1950

XXIII MARSILEACEAE

1 Marsilea L.
  1 quadrifolia L.
  2 aegyptiaca Willd.
  3 strigosa Willd.

2 Pilularia L.
  1 globulifera L.
  2 minuta Durieu ex A. Braun

XXIV SALVINIACEAE

1 Salvinia Adanson
  1 natans (L.) All.   8   Rs(W)   Arnoldi, 1910
                                        18   It   D'Amato-Avanzi,
                                                       1957
  2 rotundifolia Willd.

XXV AZOLLACEAE

1 Azolla Lam.
  1 filiculoides Lam.
  2 caroliniana Willd.

SPERMATOPHYTA-GYMNOSPERMAE
CONIFEROPSIDA

XXVI PINACEAE

1 Abies Miller
  1 procera Rehder
  2 grandis (D. Don) Lindley
  3 lasiocarpa (Hooker) Nutt.
  4 sibirica Ledeb.
  5 nordmanniana (Steven) Spach
  6 alba Miller   24   ?Ge   Seitz, 1951
  7 nebrodensis (Lojac.) Mattei
  8 cephalonica Loudon
  9 pinsapo Boiss.

2 Pseudotsuga Carrière
  1 menziesii (Mirbel) Franco

3 Tsuga (Antoine) Carrière
  1 heterophylla (Rafin.) Sarg.
  2 canadensis (L.) Carrière

4 Picea A. Dietr.
  1 abies (L.) Karsten
    (a) abies   22
                                        24   Su   Lövkvist, 1963

    (b) obovata (Ledeb.) Hultén
  2 orientalis (L.) Link
  3 glauca (Moench) Voss
  4 engelmannii Parry ex Engelm.
  5 pungens Engelm.
  6 sitchensis (Bong.) Carrière
  7 omorika (Pančić) Purkyne

5 Larix Miller
  1 kaempferi (Lamb.) Carrière
  2 decidua Miller
  3 russica (Endl.) Sabine ex Trautv.

6

    4 gmelinii (Rupr.) Kuzeneva

6 Cedrus Trew
    1 deodara (D. Don) G. Don fil.
    2 atlantica (Endl.) Carrière

7 Pinus L.
    1 contorta Douglas ex Loudon
    2 pinaster Aiton   24   Ga   Dangeard, 1941
      (a) pinaster
      (b) atlantica H. del Villar
    3 radiata D. Don
    4 banksiana Lamb.
    5 ponderosa Douglas ex P. & C. Lawson
    6 nigra Arnold
      (a) nigra
      (b) salzmannii (Dunal) Franco
      (c) laricio (Poiret) Maire
      (d) dalmatica (Vis.) Franco
      (e) pallasiana (Lamb.) Holmboe
    7 sylvestris L.   24   Su   Lövkvist, 1963
    8 mugo Turra
    9 uncinata Miller ex Mirbel
   10 leucodermis Antoine
   11 heldreichii Christ
   12 halepensis Miller
   13 brutia Ten.
   14 canariensis Sweet ex Sprengel
   15 pinea L.
   16 cembra L.
   17 sibirica Du Tour
   18 wallichiana A.B. Jackson
   19 peuce Griseb.
   20 strobus L.

XXVII TAXODIACEAE

1 Sequoia Endl.
    1 sempervirens (Lamb.) Endl.

2 Sequoiadendron Buchholz
    1 giganteum (Lindley) Buchholz

3 Taxodium L.C.M. Richard
    1 distichum (L.) L.C.M. Richard

4 Cryptomeria D. Don
    1 japonica (L. fil.) D. Don

XXVIII CUPRESSACEAE

1 Cupressus L.
    1 sempervirens L.
    2 macrocarpa Hartweg
    3 lusitanica Miller
    4 arizonica E.L. Greene

2 Chamaecyparis Spach
    1 lawsoniana (A. Murray) Parl.

7

3 Thuja L.
  1 plicata D. Don ex Lambert
    orientalis L.   22 Bot. Gard. Dark, 1932

4 Tetraclinis Masters
  1 articulata (Vahl) Masters

5 Juniperus L.
  1 drupacea Labill.
  2 communis L.
    (a) communis   22   Su   Lövkvist, 1963
    (b) hemisphaerica (J. & C. Presl) Nyman
    (c) nana Syme   22   Is   Löve & Löve, 1956
  3 oxycedrus L.
    (a) oxycedrus
    (b) macrocarpa (Sibth. & Sm.) Ball
    (c) transtagana Franco
  4 brevifolia (Seub.) Antoine
  5 phoenicea L.
  6 thurifera L.
  7 foetidissima Willd.
  8 excelsa Bieb.
  9 sabina L.   22 Bot. Gard. Seitz, 1951
 10 virginiana L.

TAXOPSIDA

XXIX TAXACEAE

1 Taxus L.
  1 baccata L.   24   Su   Lövkvist, 1963

GNETOPSIDA

XXX EPHEDRACEAE

1 Ephedra L.
  1 fragilis Desf.
   (a) fragilis
   (b) campylopoda (C.A. Meyer) Ascherson & Graebner
  2 distachya L.
   (a) distachya   24   Ga   Berudye &
                                                       Jandar, 1907
                                         28   Ga   Delay, 1947
   (b) helvetica (C.A. Meyer) Ascherson & Graebner
  3 major Host
   (a) major   14   Ju   Geitler, 1929
   (b) procera (Fischer & C.A. Meyer) Markgraf

SPERMATOPHYTA-ANGIOSPERMAE

XXXI SALICACEAE

1 Salix L.
  1 pentandra L.

8

```
2 fragilis L.                              79      Br       Wilkinson, 1941
3 alba L.
  (a) alba
  (b) coerulea (Sm.) Rech. fil.
  (c) vitellina (L.) Arcangeli
  (d) micans (N.J. Andersson) Rech. fil.
4 babylonica L.
5 triandra L.
  (a) triandra                             38      Br       Blackburn & Heslop-
                                                            Harrison, 1924
  (b) discolor (Koch) Arcangeli
6 reticulata L.                            38      Rs(N)    Sokolovskaya &
                                                            Strelkova, 1960
7 herbacea L.                              38      Is       Löve & Löve, 1956
8 polaris Wahlenb.
9 retusa L.
10 kitaibeliana Willd.
11 serpyllifolia Scop.
12 rotundifolia Trautv.
13 myrsinites L.                          190      No       Wilkinson, 1944
14 breviserrata B. Flod.
15 alpina Scop.
16 arctica Pallas
17 pulchra Cham.
18 glauca L.
   callicarpaea Trautv.                   190      Is       Löve & Löve, 1956
19 stipulifera B. Flod. ex Hayren         152      Su       Löve & Löve, 1956
20 glaucosericea B. Flod
21 pyrenaica Gouan
22 reptans Rupr.
23 lanata L.
24 glandulifera B. Flod.                   38      Scand.   Marklund, 1931
25 phylicifolia L.                        114      Is       Löve & Löve, 1956
26 bicolor Willd.
27 hegetschweileri Heer
28 hibernica Rech. fil.
29 nigricans Sm.                          114      Su       Håkansson, 1955
30 borealis Fries
31 mielichhoferi Sauter
32 glabra Scop.
33 crataegifolia Bertol.
34 pedicellata Desf.
35 silesiaca Willd.
36 aegyptiaca L.
37 appendiculata Vill.
38 laggeri Wimmer
39 cinerea L.                              76      Br       Blackburn & Heslop-
40 atrocinerea Brot.                       76      Br       Harrison, 1924
41 aurita L.                               76      Br       Blackburn & Heslop-
                                                            Harrison, 1924

42 caprea L.
43 coaetanea (Hartman) B. Flod
44 salvifolia Brot.
45 cantabrica Rech. fil.
46 starkeana Willd.
47 xerophila B. Flod.
48 myrtilloides L.
49 repens L.                               38      Br       Blackburn & Heslop-
                                                            Harrison, 1924
50 rosmarinifolia L.                       38      Su       Lövkvist, 1963
51 arenaria L.                             38      Su       Lövkvist, 1963
```

52 arbuscula L.
53 foetida Schleicher
54 waldsteiniana Willd.
55 hastata L.   c.110   Rs(N)   Sokolovskaya &
                                                                  Strelkova, 1960
  (a) hastata
  (b) vegeta N.J. Andersson
  (c) subintegrifolia (B. Flod.) B. Flod.
56 pyrolifolia Ledeb.
57 helvetica Vill.
58 marrubifolia Tausch ex N.J. Andersson
59 lapponum L.
60 viminalis L.
61 rossica Nasarov
62 elaeagnos Scop.
  (a) elaeagnos   38   Europe   Håkansson, 1955
  (b) angustifolia (Cariot) Rech. fil.
63 purpurea L.
  (a) purpurea
  (b) lambertiana (Sm.) A. Neumann ex Rech. fil.
64 amplexicaulis Bory
65 caspica Pallas
66 wilhelmsiana Bieb.
67 caesia Vill.
68 tarraconensis Paul
69 daphnoides Vill.   38   Scand.   Marklund, 1931
70 acutifolia Willd.   38   Scand.   Marklund, 1931

2 Populus L.
  1 alba L.   38   Ge   Wettstein-
                                                                  Westerskeim, 1933
  2 canescens (Aiton) Sm.   38   Au   Seitz, 1954
  3 grandidentata Michx
  4 tremula L.
  5 simonii Carrière
  6 gileadensis Rouleau
  7 x berolinensis C. Koch
  8 nigra L.   38   Br   Blackburn & Heslop-
                                                                  Harrison, 1924
  9 x canadensis Moench
  10 deltoides Marshall
  11 euphratica Olivier

XXXII MYRICACEAE

1 Myrica L.
  1 gale L.   48   Su   Hagerup, 1941
  2 faya Aiton
  3 caroliniensis Miller

XXXIII JUGLANDACEAE

1 Juglans L.
  1 regia L.
  2 nigra L.
  3 cinerea L.

2 Carya Nutt.

1 cordiformis (Wangenh.) C. Koch
2 glabra (Miller) Sweet
3 alba (L.) Nutt.

3 Pterocarya Kunth
  1 fraxinifolia (Poiret) Spach

XXXIV BETULACEAE

1 Betula L.

1 pendula Roth	28	Da	Helms & Jörgensen, 1925
2 pubescens Ehrh.			
(a) pubescens	56	Da	Helms & Jörgensen, 1925
(b) carpatica (Willd.)	56	Po	Skalińska, 1959
(c) tortuosa (Ledeb.) Nyman	56	Is	Löve & Löve, 1956
3 humilis Schrank	28	Ge	Natho, 1959
4 nana L.	28	Sb	Flovik, 1940

2 Alnus Miller

1 viridis (Chaix) DC.			
(a) viridis			
(b) fruticosa (Rupr.) Nyman			
(c) suaveolens (Req.) P.W. Ball			
2 glutinosa (L.) Gaertner	28	Su	Eklundh-Ehrenburg, 1946
3 rugosa (Duroi) Sprengel			
4 incana (L.) Moench			
(a) incana	28	Su	Eklundh-Ehrenburg, 1946
(b) kolaensis (Orlova) Á. & D. Löve			
5 cordata (Loisel.) Loisel.			

XXXV CORYLACEAE

1 Carpinus L.

1 betulus L.	64	Su	Johnsson, 1942
2 orientalis Miller			

2 Ostrya Scop.
  1 carpinifolia Scop.

3 Corylus L.

1 avellana L.	22	Su	Lövkvist, 1963
2 colurna L.			
3 maxima Miller			

XXXVI FAGACEAE

1 Fagus L.

1 sylvatica L.	24	Su	Lövkvist, 1963
2 orientalis Lipsky			

2 Castanea Miller
  1 sativa Miller
  2 crenata Siebold & Zucc.

11

3 Quercus L.
  1 rubra L.
  2 palustris Muenchh.

3 coccifera L.	24	Lu	Natividade, 1937
4 ilex L.	24	Lu	Natividade, 1937
5 rotundifolia Lam.	24	Lu	Natividade, 1937
6 suber L.	24	Lu	Natividade, 1937
7 trojana Webb			
8 macrolepis Kotschy			
9 cerris L.			
10 mas Thore			
11 polycarpa Schur			
12 petraea (Mattuschka) Liebl.	24	Br	Jones, 1959
13 dalechampii Ten.			
14 hartwissiana Steven			
15 robur L.			
(a) robur	24	Br	Jones, 1959
(b) brutia (Ten.) O. Schwarz			
16 pedunculiflora C. Koch			
17 sicula Borzi			
18 frainetto Ten.			
19 pyrenaica Willd.	24	Lu	Natividade, 1937
20 congesta C. Presl			
21 brachyphylla Kotschy			
22 virgiliana (Ten.) Ten.			
23 pubescens Willd.			
(a) pubescens	24	It	Vignoli, 1933
(b) anatolica O. Schwarz			
(c) palensis (Palassou) O. Schwarz			
24 canariensis Willd			
25 faginea Lam.	24	Lu	Natividade, 1937
26 fruticosa Brot.	24	Lu	Natividade, 1937
27 infectoria Olivier			

XXXVII ULMACEAE

1 Ulmus L.

1 glabra Hudson	28	Su	Løvkvist, 1963
2 procera Salisb.			
3 minor Miller	28	Br	Maude, 1939
4 canescens Melville			
5 laevis Pallas	28	Su	Löve & Löve, 1942

2 Zelkova Spach
  1 abelicea (Lam.) Boiss.

3 Celtis L.
  1 australis L.
  2 caucasica Willd.
  3 glabrata Steven ex Planchon
  4 tournefortii Lam.

XXXVIII MORACEAE

1 Broussonetia L.'Hér. ex Vent.
  1 papyrifera (L.) Vent.

2 Morus L.
  1 nigra L.

2 alba L.

3 Maclura Nutt.
  1 pomifera (Rafin.) C.K. Schneider

4 Ficus L.
  1 carica L.

XXXIX CANNABACEAE

1 Humulus L.
  1 lupulus L.   20   Hu   Pólya, 1949

2 Cannabis L.
  1 sativa L.   20   Hu   Pólya, 1949

XL URTICACEAE

1 Urtica L.
  1 atrovirens Req. ex Loisel.   26   Co   Contandriopoulos, 1962
  2 dioica L.   48   Is   Löve & Löve, 1956
                                  52   Br   Fothergill, 1936
  3 kioviensis Rogow.   22   Hu   Baksay, 1956
  4 cannabina L.
  5 rupestris Guss
  6 urens L.   24   Is   Löve & Löve, 1956
                                  26   Su   Lövkvist, 1963
                                  52   Su or Is Löve & Löve, 1942

  7 dubia Forskål
  8 pilulifera L.

2 Pilea Lindley
  1 microphylla (L.) Liebm.

3 Parietaria L.
  1 officinalis L.
  2 diffusa Mert. & Koch
  3 mauritanica Durieu
  4 lusitanica L.
   (a) lusitanica
   (b) serbica (Pančić) P.W. Ball
  5 debilis Forster fil.
  6 cretica L.

4 Soleirolia Gaud.-Beaup.
  1 soleirolii (Req.) Dandy   20   Co   Contandriopoulos, 1962

5 Forsskalea L.
  1 tenacissima L.

XLI PROTEACEAE

1 Hakea Schrader
  1 sericea Schrader
  2 salicifolia (Vent.) B.L. Burtt

XLII SANTALACEAE

1 Comandra Nutt.
  1 elegans (Rochel ex Reichenb.) Reichenb. fil

2 Osyris L.
  1 alba L.
  2 quadripartita Salzm. ex Decne

3 Thesium L.
  1 alpinum L.   12   Au   Mattick, 1950
  2 pyrenaicum Pourret
   (a) pyrenaicum
   (b) alpestre O. Schwarz
  3 ebracteatum Hayne
  4 rostratum Mert. & Koch   26   Au Ge   Rutishauser, 1936
  5 kerneranum Simonkai
  6 auriculatum Vandas
  7 arvense Horvatovszky
  8 parnassi A. DC.
  9 italicum A. DC.
 10 dollineri Murb.
   (a) dollineri
   (b) simplex (Velen.) Stoj. & Stefanov.
 11 bavarum Schrank   24   Ge   Schulle, 1933
 12 linophyllon L.   14   Hu   Baksay, 1961
 13 divaricatum Jan ex Mert. & Koch
 14 humifusum DC.
 15 bergeri Zucc.
 16 humile Vahl
 17 procumbens C.A. Meyer
 18 brachyphullum Boiss.

XLIII LORANTHACEAE

1 Loranthus L.
  1 europaeus Jacq.

2 Viscum L.
  1 cruciatum Sieber ex Boiss.
  2 album L.
   (a) album
   (b) abietis (Wiesb.) Abromeit

3 Arceuthobium Bieb.
  1 oxycedri (DC.) Bieb.

XLIV ARISTOLOCHIACEAE

1 Asarum L.
  1 europaeum L.   26   Po   Skalińska et al.,
                                                              1959
                                              40   Su   Ehrenberg, 1945
2 Aristolochia L.
  1 sempervirens L.
  2 baetica L.
  3 cretica Lam.
  4 pontica Lam.

```
 5 bodamae Dingler
 6 clematitis L.
 7 pistolochia L.
 8 rotunda L.
 9 pallida Willd.
10 sicula Tineo
11 longa L.
12 bianorii Sennen & Pau
13 microstoma Boiss. & Spruner
```

XLV RAFFLESIACEAE

```
1 Cytinus L.
  1 hypocistis (L.) L.
   (a) hypocistis
   (b) orientalis Wettst.
  2 ruber (Fourr.) Komarov
```

XLVI BALANOPHORACEAE

```
1 Cynomorium L.
  1 coccineum L.
```

XLVII POLYGONACEAE

1 Koenigia L.			
1 islandica L.	28	Sb	Löve & Löve, 1961b
2 Polygonum L.			
1 scoparium Req. ex Loisel.	20	Co	Contandriopoulos, 1962
2 equisetiforme Sibth. & Sm.			
3 icaricum Rech. fil.			
4 romanum Jacq.			
5 maritimum L.	20	Ge Lu	Styles, 1962
6 idaeum Hayek			
7 cognatum Meissner			
8 floribundum Schlecht. ex Sprengel			
9 salsugineum Bieb.			
10 aschersonianum H. Gross			
11 patulum Bieb.	20	Rm	Tarnavschi, 1948
12 arenarium Waldst. & Kit.			
(a) arenarium			
(b) pulchellum (Loisel.) D.A. Webb & Chater	20	Lu	Löve, 1954
13 oxyspermum Meyer & Bunge ex Ledeb.			
(a) oxyspermum	40	Da	Styles, 1962
(b) raii (Bab.) D.A. Webb & Chater	40	Br	Styles, 1962
14 graminifolium Wierzb. ex Heuffel			
15 aviculare L.	60	Br	Styles, 1962
16 boreale (Lange) Small	40	Br	Styles, 1962
17 rurivagum Jordan ex Boreau	60	Br	Styles, 1962
18 arenastrum Boreau			
19 acetosum Bieb.			
20 minus Hudson			
21 foliosum H. Lindb.	20	Fe	Löve, ined.
22 mite Schrank	40	Be	Pauwels, 1959

23 hydropiper L.	20	Is Su	Löve & Löve, 1942
24 salicifolium Brouss. ex Willd.			
25 persicaria L.	44	Be	Pauwels, 1959
26 lapathifolium L.	22	Be	Pauwels, 1959
27 amphibium L.	66	Is	Löve & Löve, 1956
28 hydropiperoides Michx			
29 orientale L.			
30 sagittatum L.			
31 tinctorium Aiton			
32 bistorta L.	44	Ge	Jaretsky, 1928b
33 amplexicaule D. Don			
34 viviparum L.	c.100	Sb	Flovik, 1940
	c.132	Po	Skalińska, 1950
35 alpinum All.			
36 polystachyum Wall. ex Meissner			
3 Bilderdykia Dumort.			
1 convolvulus (L.) Dumort.	40	Be	Pauwels, 1959
2 dumetorum (L.) Dumort.	20	Be	Pauwels, 1959
3 aubertii (Louis Henry) Moldenke			
4 Reynoutria Houtt.			
1 japonica Houtt.			
2 sachalinensis (Friedrich Schmidt Petrop.) Nakai			
5 Fagopyrum Miller			
1 esculentum Moench	16	Rs( )	Mansurova, 1948
2 tataricum (L.) Gaertner	16	Rs( )	Mansurova, 1948
6 Oxyria Hill			
1 digyna (L.) Hill	14	Rs(N)	Sokolovskaya & Strelkova, 1960
7 Rheum L.			
1 tataricum L. fil.			
2 rhaponticum L.			
8 Rumex L.			
1 angiocarpus Murb.	14	Lu	Löve, 1941
2 tenuifolius (Wallr.) Á. Löve	28	Su	Löve, 1940
3 acetosella L.	42	Su	Löve, 1940
4 graminifolius Rudolph ex Lamb	56	Rs(N)	Löve, 1940
5 lunaria L.			
6 tingitanus L.			
7 suffruticosus Gay ex Meissner			
8 scutatus L.	20	Au	Mattick, 1950
9 tuberosus L.			
(a) tuberosus			
(b) creticus (Boiss.) Rech. fil.			
10 nivalis Hegetschw.	♀ 14	Alps	Löve, 1954b
	♂ 15	Alps	Löve, 1954b
11 arifolius All.	♀ 14	Is	Löve & Löve, 1956
	♂ 15	Is	Löve & Löve, 1956
12 amplexicaulis Lapeyr.			
13 gussonei Arcangeli			
14 acetosa L.	♀ 14	Is	Löve, 1940
	♂ 15	Is	Löve, 1940
15 thyrsiflorus Fingerh.	14	Su	Löve, 1940
16 rugosus Campd.			
17 papillaris Boiss. & Reuter			
18 intermedius DC.			

19 thyrsoides Desf.			
20 vesicarius L.			
21 triangulivalvis (Danser) Rech. fil.			
22 frutescens Thouars			
23 alpinus L.	20	Ga	Larsen, 1954
24 aquaticus L.	c.200	Su	Löve, 1942
25 azoricus Rech. fil.			
26 arcticus Trautv.	c.200	Su	Löve, 1942
27 balcanicus Rech. fil.			
28 pseudonatronatus Borbás	40	Au Su	Löve, 1961b
29 cantabricus Rech. fil.			
30 longifolius DC.	60	Su	Löve, 1942
31 confertus Willd.	40	C. Europe	Löve, 1961b
32 hydrolapathum Hudson	200	Su	Löve, 1942
33 cristatus DC.			
34 kerneri Borbás			
35 patientia L.			
(a) patientia	60	C. Europe	Löve, 1961b
(b) orientalis (Bernh.) Danser	60	C. Europe	Löve, 1961b
(c) recurvatus (Rech.) Rech. fil.	60	C. Europe	Löve, 1961b
36 crispus L.	60	Su	Löve, 1942
37 stenophyllus Ledeb.	60	Hu	Pólya, 1950
38 conglomeratus Murray	20	Su	Löve, 1942
39 sanguineus L.	20	Su	Löve, 1942
40 rupestris Le Gall			
41 nepalensis Sprengel	60		
42 pulcher L.			
(a) pulcher	20	C. Europe	Löve, 1961b
(b) divaricatus (L.) Murb.	20	C. Europe	Löve, 1961b
(c) anodontus (Hausskn.) Rech. fil.			
(d) raulinii (Boiss.) Rech. fil.			
43 obtusifolius L.			
(a) obtusifolius	40	Su	Löve, 1961b
(b) subalpinus (Schur) Čelak			
(c) transiens (Simonkai) Rech. fil.			
(d) sylvestris (Wallr.) Rech.	40	C. Europe	Löve, 1961b
44 dentatus L.	40	Su	Löve, 1967
(a) halacsyi (Rech.) Rech. fil.			
(b) reticulatus (Besser) Rech. fil.			
45 palustris Sm.	40	Su	Löve, 1942
46 maritimus L.	40	Su	Löve, 1942
47 rossicus Murb.	40	Rs(C)	Löve & Löve, 1961b
48 marschallianus Reichenb.			
49 ucranicus Besser ex Sprengel	40	Rs(W)	Löve & Löve, 1961b
50 bucephalophorus L.			
(a) bucephalophorus	16	It	Löve, 1967
(b) hispanicus (Steinh.) Rech. fil.			
(c) graecus (Steinh.) Rech. fil.			
(d) aegaeus Rech. fil.			

9 Emex Campd.
1 spinosa (L.) Campd.

10 Muehlenbeckia Meissner
1 sagittifolia (Ortega) Meissner

11 Calligonum L.
1 aphyllum (Pallas) Gurke

12 Atraphaxis L.
1 spinosa L.

2 replicata Lam.
3 frutescens (L.) C. Koch
4 billardieri Jaub. & Spach

XLVIII CHENOPODIACEAE

1 Polycnemum L.
  1 majus A. Braun
  2 arvense L.
  3 verrucosum A.F. Láng
  4 heuffelii A.F. Láng

2 Beta L.
  1 vulgaris L.
    (a) vulgaris   18   Ge   Wulff, 1936
    (b) maritima (L.) Arcangeli   18   Rm   Tarnavschi, 1948
  2 macrocarpa Guss.
  3 patellaris Moq.
  4 trigyna Waldst. & Kit.
  5 nana Boiss. & Heldr.

3 Chenopodium L.
  1 botrys L.
  2 schraderanum Schultes   18   Su   Kjellmark, 1934
  3 aristatum L.
  4 ambrosioides L.   32   Su   Kjellmark, 1934
  5 multifidum L.
  6 bonus-henricus L.   36   Su   Lövkvist, 1963
  7 foliosum Ascherson
  8 capitatum (L.) Ascherson   18   Su   Kjellmark, 1934
  9 glaucum L.   18   Su   Lövkvist, 1963
 10 rubrum L.   36   Su   Lövkvist, 1963
 11 botryodes Sm.
 12 hybridum L.   18   Su   Lövkvist, 1963
 13 polyspermum L.   18   Su   Lövkvist, 1963
 14 vulvaria L.   18   Su   Lövkvist, 1963
 15 urbicum L.   36   Br   Cole, 1962
 16 murale L.   18   Br   Cole, 1962
 17 ficifolium Sm.   18   Su   Kjellmark, 1934
 18 acerifolium Andrz.
 19 jenissejense Aellen & Iljin
 20 opulifolium Schrader ex Koch & Ziz 54   Br   Cole, 1962
 21 album L.
    (a) album   18   Su   Lövkvist, 1963
                                        54   Su   Lövkvist, 1963
    (b) striatum (Krasan.) J. Murr
 22 suecicum J. Murr   18   Su   Lövkvist, 1963
 23 giganteum D. Don

4 Cycloloma Moq.
  1 atriplicifolia (Sprengel) Coulter

5 Spinacia L.
  1 oleracea L.

6 Atriplex L.
  1 halimus L.   18   Lu   Castro &
                                                        Fontes, 1946
  2 glauca L.
  3 cana Ledeb.

18

```
 4 hortensis L.
 5 nitens Schkuhr
 6 aucheri Moq.
 7 oblongifolia Waldst. & Kit.
 8 heterosperma Bunge
 9 sphaeromorpha Iljin
10 rosea L.
11 laciniata L.                        18      Ge        Wulff, 1937a
12 tatarica L.                         18   Ge Hu Rm     Wulff, 1937a
13 littoralis L.                       18      Rm        Tarnavschi, 1948
14 patens (Litv.) Iljin
15 patula L.                           36      Br        Hulme, ined.
16 calotheca (Rafn) Fries              18      Ge        Wulff, 1937a
17 hastata L.                          18      Br        Hulme, ined.
18 glabriuscula Edmondston             18      Br        Hulme, ined.
19 longipes Drejer                     18      Br        Hulme, ined.


 7 Halimione Aellen
   1 portulacoides (L.) Aellen         36      Lu Rm     Tarnavschi, 1948
   2 verrucifera (Bieb.) Aellen
   3 pedunculata (L.) Aellen           18      Ge        Wulff, 1937a


 8 Krascheninnikovia Gueldenst.
   1 ceratoides (L.) Gueldenst.


 9 Ceratocarpus L.
   1 arenarius L.


10 Axyris L.
   1 amaranthoides L.


11 Camphorosma L.
   1 monspeliaca L.
   2 lessingii Litv.
   3 annua Pallas
   4 songorica Bunge


12 Bassia All.
   1 hyssopifolia (Pallas) Volk.
   2 hirsuta (L.) Ascherson             18      Rm        Tarnavschi, 1947
   3 sedoides (Pallas) Ascherson


13 Kochia Roth
   1 prostrata (L.) Schrader
   2 saxicola Guss.
   3 laniflora (S.G. Gmelin) Borbás
   4 scoparia (L.) Schrader


14 Corispermum L.
   1 uralense (Iljin) Aellen
   2 marschallii Steven
   3 algidum Iljin
   4 canescens Kit.
   5 aralocaspicum Iljin
   6 nitidum Kit.
   7 filifolium C.A. Meyer
   8 declinatum Stephan ex Steven
   9 hyssopifolium L.
  10 intermedium Schweigger
  11 leptopterum (Ascherson) Iljin
  12 orientale Lam.
```

15 Agriophyllum Bieb. ex C.A. Meyer
   1 squarrosum (L.) Moq.

16 Kalidium Moq.
   1 foliatum (Pallas) Moq.
   2 caspicum (L.) Ung.-Sternb.

17 Halopeplis Bunge ex Ung.-Sternb.
   1 amplexicaulis (Vahl) Ung.-Sternb.
   2 pygmaea (Pallas) Bunge ex Ung.-Sternb.

18 Halostachys C.A. Meyer
   1 belangerana (Moq.) Botsch.

19 Halocnemum Bieb.
   1 strobilaceum (Pallas) Bieb.

20 Arthrocnemum Moq.
   1 perenne (Miller) Moss   18   Br   Maude, 1940
   2 fruticosum (L.) Moq.   54   Ga   Contandriopoulos, 1968
   3 glaucum (Delile) Ung.-Sternb.   36   Lu   Castro & Fontes, 1946

21 Salicornia L.
   1 europaea L.   18   Br   Ball & Tutin, 1959
   2 ramosissima J. Woods   18   Br   Ball & Tutin, 1959
   3 prostrata Pallas   18   Rm   Tarnavschi, 1947
   4 pusilla J. Woods   18   Br   Ball & Tutin, 1959
   5 nitens P.W. Ball & Tutin   36   Br   Ball & Tutin, 1959
   6 fragilis P.W. Ball & Tutin   36   Br   Ball & Tutin, 1959
   7 dolichostachya Moss   36   Br   Ball & Tutin, 1959
    (a) dolichostachya
    (b) strictissima (K. Gram) P.W. Ball

22 Microcnemum Ung.-Sternb.
   1 coralloides (Loscos & Pardo) Font Quer

23 Suaeda Forskål ex Scop.
   1 vera J.F. Gmelin
   2 pruinosa Lange
   3 physophora Pallas
   4 dendroides (C.A. Meyer) Moq.
   5 confusa Iljin
   6 baccifera Pallas
   7 maritima (L.) Dumort.
    (a) maritima   36   Ge   Wulff, 1937a
    (b) salsa (L.) Soó   36   Hu   Pólya, 1948
    (c) pannonica (G. Beck) Soó ex P.W. Ball
   8 corniculata (C.A. Meyer) Bunge
   9 splendens (Pourret) Gren. & Godron   18   Lu   Castro &
                                            Fontes, 1946
  10 altissima (L.) Pallas
  11 linifolia Pallas
  12 eltonica Iljin
  13 heterophylla (Kar. & Kit.) Bunge ex Boiss.
  14 kossingskyi Iljin

24 Bienertia Bunge ex Boiss.
   1 cycloptera Bunge ex Boiss.

25 Salsola L.
   1 foliosa (L.) Schrader

2 brachiata Pallas
3 paulsenii Litv.
4 pellucida Litv.
5 soda L.
6 acutifolia (Bunge) Botsch.
7 tamariscina Pallas
8 kali L.
  (a) kali
  (b) tragus (L.) Nyman
  (c) ruthenica (Iljin) Soó
9 collina Pallas
10 lanata Pallas
11 turcomanica Litv.
12 crassa Bieb.
13 nitraria Pallas
14 affinis C.A. Meyer
15 nodulosa (Moq.) Iljin
16 carpatha P.H. Davis
17 arbuscula Pallas
18 genistoides Juss. ex Poiret
19 webbii Moq.
20 verticillata Schousboe
21 papillosa (Cosson) Willk.
22 vermiculata L.
23 aegaea Rech. fil.
24 dendroides Pallas
25 laricina Pallas

26 Noaea Moq.
  1 mucronata (Forskål) Ascherson & Schweinf.

27 Ofaiston Rafin.
  1 monandrum (Pallas) Moq.

28 Girgensohnia Bunge
  1 oppositifolia (Pallas) Fenzl in Ledeb.

29 Anabasis L.
  1 salsa (C.A. Meyer) Paulsen
  2 aphylla L.
  3 cretacea Pallas
  4 articulata (Forskål) Moq.

30 Haloxylon Bunge
  1 articulatum (Moq.) Bunge

31 Nanophyton Less.
  1 erinaceum (Pallas) Bunge

32 Petrosimonia Bunge
  1 brachiata (Pallas) Bunge
  2 triandra (Pallas) Simonkai
  3 litwinowii Korsh.
  4 monandra (Pallas) Bunge
  5 oppositifolia (Pallas) Litv.
  6 brachyphylla (Bunge) Iljin
  7 glaucescens (Bunge) Iljin

33 Halimocnemis C.A. Meyer
  1 sclerosperma (Pallas) C.A. Meyer

21

34 Halogeton C.A. Meyer
   1 sativus (L.) Moq.

XLIX AMARANTHACEAE

1 Celosia L.
   1 argentea L.

2 Amaranthus L.
   1 hybridus L.   32   Lu   Grant, 1959
   2 cruentus L.   34   non-Europe   Grant, 1959
   3 bouchonii Thell.
   4 retroflexus L.
   5 muricatus (Moq.) Gillies ex Hicken
   6 blitoides S. Watson
   7 crispus (Lesp. & Thev.) N. Terracc.
   8 standleyanus Parodi ex Covas
   9 albus L.
  10 graecizans L.
  11 deflexus L.
  12 lividus L.

3 Achyranthes L.
   1 aspera L.
   2 sicula (L.) All.

4 Alternanthera Forskål
   1 pungens Kunch
   2 peploides (Humb. & Bonpl.) Urban
   3 nodiflora R.Br.

L NYCTAGINACEAE

1 Commicarpus Standley
   1 plumbagineus (Cav.) Standley

2 Mirabilis L.
   1 jalapa L.
   2 longiflora L.

3 Oxybaphus L'Hér. ex Willd.
   1 nyctagineus (Michx) Sweet

LI PHYTOLACCACEAE

1 Phytolacca L.
   1 americana L.
   2 dioica L.

LII AIZOACEAE

1 Aizoon L.
   1 hispanicum L.

2 Sesuvium L.
   1 portulacastrum L.

3 Carpobrotus N.E. Br.
  1 acinaciformis (L.) L. Bolus
  2 edulis (L.) N.E. Br.   18   Lu   Rodrigues, 1953
  3 chilensis (Molina) N.E. Br.

4 Mesembryanthemum L.
  1 nodiflorum L.
  2 crystallinum L.

5 Aptenia N.E. Br.
  1 cordifolia (L. fil.) N.E. Br.

6 Disphyma N.E. Br.
  1 crassifolium (L.) L. Bolus

7 Lampranthus N.E. Br.
  1 glaucus (L.) N.E. Br.

LIII MOLLUGINACEAE

1 Mollugo L.
  1 verticillata L.
  2 cerviana (L.) Ser.

2 Glinus L.
  1 lotoides L.

LIV TETRAGONIACEAE

1 Tetragonia L.
  1 tetragonoides (Pallas) O. Kuntze   32   Lu   Rodrigues, 1953

LV PORTULACACEAE

1 Portulaca L.
  1 oleracea L.
    (a) oleracea   54   Africa Hagerup, 1932
    (b) sativa (Haw.) Čelak.   54

2 Montia L.
  1 fontana L.   18   No   Hagerup, 1941
                                                20   Ge   Scheerer, 1940
    (a) fontana   18   Is   Löve & Löve, 1956
                                                20   Su   Lövkvist, 1963
    (b) variabilis Walters
    (c) amporitana Sennen
    (d) chondrosperma (Fenzl) Walters   18   Ge   Scheerer, 1940
                                                20   Su   Lövkvist, 1963
  2 perfoliata (Donn ex Willd.) Howell   36   Br   Rutland, 1941
  3 sibirica (L.) Howell

LVI BASELLACEAE

1 Boussingaultia Humb., Bonpl. & Kunth
  1 cordifolia Ten.

23

LVII CARYOPHYLLACEAE

1 Arenaria L.
  1 rigida Bieb.
  2 cephalotes Bieb.
  3 gypsophiloides L.
  4 procera Sprengel
    (a) procera
    (b) glabra (F.N. Williams) J. Holub
  5 longifolia Bieb.
  6 purpurascens Ramond ex DC.
  7 lithops Heywood ex McNeill
  8 tetraquetra L.
  9 tomentosa Willk.
  10 armerina Bory
  11 aggregata (L.) Loisel.
    (a) aggregata
    (b) erinacea (Boiss.) Font Quer
  12 pungens Clemente ex Lag.
  13 valentina Boiss.
  14 grandiflora L.   44   He   Favarger, 1959
  15 bertolonii Fiori   30   Co   Favarger, 1962
  16 huteri Kerner
  17 montana L.
    (a) montana   28   Ga   Favarger, 1962
    (b) intricata (Dufour) Pau
  18 balearica L.
  19 biflora L.   22   Au   Favarger, 1962
  20 humifusa Wahlenb.   40   No   Horn, 1948
  21 norvegica Gunnerus
    (a) norvegica   80   Is   Löve & Löve, 1956
    (b) anglica Halliday   80   Br   Halliday, 1960
  22 ciliata L.
    (a) ciliata   80   He   Favarger, 1962
                                                        120   Au   Favarger, 1965
                                                        240   Au   Favarger, 1965
          ssp. pseudofrigida Ostenf. & O. Dahl 40   No   Horn, 1948
    (b) moehringioides (J. Murr) Br.-Bl.   40   Ga   Favarger, 1969
  23 gothica Fries   100   Su   Halliday, 1960
  24 gracilis Waldst. & Kit.
  25 filicaulis Fenzl
    (a) filicaulis
    (b) graeca (Boiss.) McNeill
  26 deflexa Decne   44   Ju   Favarger, 1962
  27 fragillima Rech. fil.
  28 ligericina Lecoq & Lamotte
  29 cretica Sprengel
  30 halacsyi Bald.
  31 cinerea DC.
  32 hispida L.   40   Ga   Favarger, 1962
  33 serpyllifolia L.   40   Su   Lövkvist, 1963
  34 leptoclados (Reichenb.) Guss   20   Su   Lövkvist, 1963
  35 conferta Boiss.
  36 serpentini A.K. Jackson
  37 nevadensis Boiss. & Reuter
  38 muralis (Link) Sieber ex Sprengel
  39 saponarioides Boiss. & Balansa
  40 guicciardii Heldr. ex Boiss.
  41 cerastioides Poiret
  42 emarginata Brot.
  43 algarbiensis Welw. ex Willk.

24

```
44 conimbricensis Brot.                        22    Ga    Favarger, 1962
45 obtusiflora G. Kunze
   (a) obtusiflora
   (b) ciliaris (Loscos) Font Quer
46 controversa Boiss.
47 conica Boiss.
48 modesta Dufour                              26    Ga    Favarger, 1962
49 retusa Boiss.
50 capillipes Boiss.
51 provincialis Chater & Halliday             40    Ga    Favarger, 1962

2 Moehringia L.
  1 trinervia (L.) Clairv.                     24    Br Lu Blackburn &
                                                           Morton, 1957
  2 pentandra Gay                              48    Lu    Blackburn &
                                                           Morton, 1957
  3 lateriflora (L.) Fenzl
  4 minutiflora Bornm.
  5 intricata Willk.
  6 fontqueri Pau
  7 tejedensis Huter, Porta & Rigo ex Willk.
  8 diversifolia Dolliner ex Koch             24    Au    Blackburn, 1950
  9 pendula (Waldst. & Kit.) Fenzl
 10 villosa (Wulfen) Fenzl
 11 jankae Griseb. ex Janka
 12 grisebachii Janka
 13 dielsiana Mattf.
 14 papulosa Bertol.
 15 tommasinii Marchesetti
 16 markgrafii Merxm.& Guterm.
 17 bavarica (L.) Gren.
    (a) bavarica                               24    Au    Mattick, 1950
    (b) insubrica (Degen) Sauer
 18 sedifolia  Willd.
 19 glaucovirens Bertol.
 20 muscosa L.                                 24    Au    Mattick, 1950
 21 ciliata (Scop.) Dalla Torre                24    Au    Mattick, 1950

3 Minuartia L.
  1 thymifolia (Sibth. & Sm.) Bornm.
  2 mesogitana (Boiss.) Hand.-Mazz.
  3 viscosa (Schreber) Schinz & Thell.        46    Ju    Favarger, 1962
  4 bilykiana Klokov
  5 hybrida (Vill.) Schischkin
    (a) hybrida                                46    Ga    Favarger, 1962
                                               70    Ga    Favarger, 1962
    (b) lydia L.(Boiss.) Rech. fil.
  6 velenovskyi (Rohlena) Hayek
  7 mediterranea (Link) K. Maly                24    Lu    Favarger, 1962
  8 regeliana (Trautv.) Mattf.
  9 hamata (Hausskn.) Mattf.
 10 dichotoma L.
 11 montana L.
 12 campestris L.
 13 globulosa (Labill.) Schinz & Thell.
 14 glomerata (Bieb.) Degen                    28    Rs(K) Favarger, 1962
    (a) glomerata
    (b) velutina (Boiss. & Orph.) Mattf.
 15 rubra (Scop.) McNeill                      30    He    Favarger, 1962
 16 funkii (Jordan) Graebner
 17 mutabilis Schinz & Thell. ex Becherer      28    He    Favarger, 1959
```

18 setacea (Thuill.) Hayek	30	Cz	Favarger, 1962
(a) setacea			
(b) banatica (Reichenb.) Prodan			
19 bosniaca (G. Beck) K. Maly			
20 krascheninnikovii Schischkin			
21 adenotricha Schischkin			
22 anatolica (Boiss.) Graebner			
23 trichocalycina (Ten. & Guss.) Grande			
24 hirsuta (Bieb.) Hand.-Mazz.			
(a) hirsuta			
(b) falcata (Griseb.) Mattf.			
(c) frutescens (Kit.) Hand.-Mazz.	30	Europe	Favarger, 1961
	32	Hu	Baksay, 1958
25 eurytanica (Boiss. & Heldr.) Hand.-Mazz.			
26 recurva (All.) Schinz & Thell.			
(a) recurva	30	He	Favarger, 1962
(b) juressi (Willd. ex Schlecht.) Mattf.			
27 bulgarica (Velen.) Graebner			
28 saxifraga (Friv.) Graebner			
29 graminifolia (Ard.) Jáv.			
(a) graminifolia			
(b) clandestina (Portenschl.) Mattf.			
30 stellata (E.D. Clarke) Maire & Petitmengin			
31 cerastiifolia (Lam. & DC.) Graebner			
32 rupestris (Scop.) Schinz & Thell.	c.72	Ga	Favarger, 1962
33 lanceolata (All.) Mattf.	36	He	Favarger, 1959
34 grignensis (Reichenb.) Mattf.	36	It	Favarger, 1959
35 cherlerioides (Hoppe) Becherer			
(a) cherlerioides			
(b) rionii (Gremli) Friedrich			
36 austriaca (Jacq.) Hayek	26	Au	Favarger, 1959
37 helmii (Ser.) Schischkin	26	Rs(C)	Favarger, 1962
38 villarii (Balbis) Chenevard	26	Ga	Favarger, 1962
39 taurica (Steven) Graebner			
40 juniperina (L.) Maire & Petitmengin			
41 pichleri (Boiss.) Maire & Petitmengin			
42 verna (L.) Hiern			
(a) verna	24	Br Ga He	Halliday, 1961
(b) collina (Neilr.) Halliday	48	Hu	Favarger, 1962
(c) valentina (Pau) Font Quer			
(d) oxypetala (Woloszczak) Halliday			
(e) attica (Boiss. & Spruner) Hayek			
(f) idaea (Halácsy) Hayek			
43 rubella (Wahlenb.) Hiern	24	Br	Blackburn & Morton, 1957
	26	Is	Löve & Löve, 1956
44 stricta (Swartz) Hiern	22	Br	Blackburn & Morton, 1957
	26	Is	Löve & Löve, 1956
45 rossii (R. Br.) Graebner			
46 wettsteinii Mattf.			
47 capillacea (All.) Graebner	26	Ga	Favarger, 1959
48 baldaccii (Halácsy) Mattf.			
(a) baldaccii			
(b) doerfleri (Hayek) Hayek			
(c) skutariensis Hayek			
49 garckeana (Ascherson & Sint.) Mattf.			
50 handelii Mattf.			
51 laricifolia (L.) Schinz & Thell.			
(a) laricifolia			
(b) kitaibelii (Nyman) Mattf.	26	Po	Favarger, 1962
52 macrocarpa (Pursh) Ostenf.			

53 arctica (Ser.) Graebner
54 biflora (L.) Schinz & Thell.
55 olonensis (Bonnier) P. Fourn.
56 sedoides (L.) Hiern   26   He   Favarger, 1959
57 geniculata (Poiret) Thell.

4 Honkenya Ehrh.
  1 peploides (L.) Ehrh.   66   Sb   Flovik, 1940
                                 68-70   Rs(N)   Sokolovskaya &
                                                    Strelkova, 1960

5 Bufonia L.
  1 paniculata F. Dubois
  2 tenuifolia L.
  3 tuberculata Loscos
  4 macropetala Willk.
  5 willkommiana Boiss.
  6 perennis Pourret
  7 stricta (Sibth. & Sm.) Gurke

6 Stellaria L.
  1 nemorum L.
    (a) nemorum
    (b) glochidisperma Murb.
  2 bungeana Fenzl
  3 media (L.) Vill.
    (a) media   40 44   Br   Blackburn &
                                        Morton, 1957
                             42   Is   Löve & Löve, 1956
    (b) postii Holmboe
    (c) cupaniana (Jordan & Fourr.) Nyman
  4 neglecta Weihe   22   Su   Peterson, 1933
  5 pallida (Dumort.) Pire   22   Su   Peterson, 1935
  6 holostea L.   26   Br Lu   Blackburn &
                                        Morton, 1957
  7 alsine Grimm   24   Br Lu   Blackburn &
                                        Morton, 1957
  8 palustris Retz.
  9 hebecalyx Fenzl
 10 graminea L.   26   Br   Blackburn &
                                        Morton, 1957
 11 longipes Goldie
 12 crassipes Hultén   104   No   Knaben, 1950
 13 ciliatisepala Trautv.
 14 longifolia Muhl. ex Willd.
 15 calycantha (Ledeb.) Bong. 52   Is   Löve & Löve, 1956
 16 crassifolia Ehrh.   26   No   Knaben, 1950
 17 humifusa Rottb.   26   Sb   Flovik, 1940

7 Pseudostellaria Pax
  1 europaea Schaeftlein   32   Au   Favarger, 1961

8 Holosteum L.
  1 umbellatum L.
    (a) umbellatum   20   Su   Lövkvist, 1963
    (b) glutinosum (Bieb.) Nyman

9 Cerastium L.
  1 cerastoides (L.) Britton   36   Au   Mattick, 1950
                                 38   He   Söllner, 1954
  2 dubium (Bast.) O. Schwarz 38   Ga   Söllner, 1954

```
 3 maximum L.
 4 dahuricum Fischer ex Sprengel
 5 nemorale Bieb.
 6 perfoliatum L.                        c.37      Hs      Söllner,1954
 7 dichotomum L.
 8 grandiflorum Waldst. & Kit.
 9 candidissimum Correns
10 biebersteinii DC.
11 tomentosum L.                         c.108     It      Söllner, 1954
12 moesiacum Friv.
13 decalvans Schlosser & Vuk.
14 boissieri Gren.                       72        Co      Söllner, 1954
15 lineare All.
16 gibraltaricum Boiss.
17 banaticum (Rochel) Heuffel
   (a) banaticum
   (b) alpinum (Boiss.) Buschm.
18 julicum Schellm.                      36        It      Söllner,1954
19 soleirolii Ser. ex Duby              72         Co      Söllner,1954
20 scaranii Ten.
21 vagans Lowe
22 arvense L.
   (a) thomasii (Ten.) Rouy & Fouc.
   (b) strictum (Haenke) Gaudin          36        He      Söllner,1954
   (c) allasii (Vest) Walters
   (d) suffruticosum (L.) Hegi           36        Hs      Favarger, 1969c
                                         72        Ga      Küpfer, 1969
   (e) ciliatum (Waldst. & Kit.) Reichenb.c.72
   (f) arvense                           72        Br He   Söllner,1954
   (g) lerchenfeldianum (Schur) Ascherson & Graebner
   (h) glandulosum (Kit.) Soó
23 alsinifolium Tausch
24 jenisejense Hultén
25 regelii Ostenf.                       72        Sb      Flovik, 1940
26 alpinum L.                            c.54      Rs(N)   Sokolovskaya &
                                                           Strelkova, 196
                                         72        Br      Söllner,1954
                                         144       He      Brett, 1955
   (a) squalidum (Lam.) Hultén
   (b) lanatum (Lam.) Ascherson & Graebner
   (c) alpinum
   (d) glabratum (Hartman) Â. & D. Löve
27 arcticum Lange
   (a) arcticum                          108       Br      Brett, 1955
   (b) edmondstonii (H.C. Watson) Â. &
       D. Löve                           108       Br      Söllner,1954
28 uniflorum Clairv.                     36        Au Ile  Söllner,1954
29 latifolium L.                         36        He      Söllner,1954
30 pyrenaicum Gay
31 runemarkii Möschl & Rech. fil.
32 pedunculatum Gaudin                   36        Au      Söllner,1954
33 dinaricum G. Beck & Szysz.
34 carinthiacum Vest
   (a) carinthiacum                      36        Au      Söllner,1954
   (b) austroalpinum (H. Kunz) H. Kunz   36        Au      Sollner,1954
35 transsilvanicum Schur
36 subtriflorum (Reichenb.) Pacher       36        Ju      Söllner,1954
37 sylvaticum Waldst. & Kit.             36        Au      Söllner,1954
38 fontanum Baumg.
   (a) macrocarpum (Schur) Jalas         144       Au      Söllner,1954
   (b) scandicum H. Gartner              144       Is      Löve & Löve, 1956
```

(c) fontanum	144	Is	Löve & Löve, 1956
(d) hispanicum H. Gartner			
(e) triviale (Link) Jalas	144	Co	Söllner, 1954
	134-152	It	Söllner, 1954

39 pauciflorum Steven ex Ser.
40 illyricum Ard.
  (a) comatum (Desv.) P.D. Sell &
    Whitehead   34 (not 68) Co Gr   Söllner, 1954
  (b) decrescens (Lonsing) P.D. Sell &
    Whitehead
  (c) prolixum (Lonsing) P.D. Sell &
    Whitehead
  (d) illyricum
  (e) crinitum (Lonsing) P.D. Sell &
    Whitehead
41 pedunculare Bory & Chaub.
42 scaposum Boiss. & Heldr.
43 brachypetalum Pers.
  (a) pindigenum (Lonsing) P.D. Sell &
    Whitehead
  (b) corcyrense (Möschl) P.D. Sell &
    Whitehead
  (c) doerfleri (Halácsy ex Hayek) P.D.
    Sell & Whitehead

(d) brachypetalum	c.90	He	Söllner, 1954
(e) tenoreanum (Ser.) Soó	c.52	Au He	Söllner, 1954

  (f) atheniense (Lonsing) P.D. Sell &
    Whitehead

(g) tauricum (Sprengel) Murb.	c.90	Ga	Söllner, 1954

  (h) roeseri (Boiss. & Heldr.) Nyman

44 glomeratum Thuill.	72	Br Da Ga He	Söllner, 1954

45 rectum Friv.
  (a) rectum
  (b) petricola (Pančić) H. Gartner
46 ligusticum Viv.

(a) ligusticum	34	It	Söllner, 1954

  (b) palustre (Moris) P.D. Sell &
    Whitehead
  (c) granulatum (Huter, Porta & Rigo)
    P.D. Sell & Whitehead
  (d) trichogynum (Möschl) P.D. Sell
    & Whitehead
47 semidecandrum L.

(a) semidecandrum	36	Hs	Söllner, 1954

  (b) macilentum (Aspegren) Möschl

48 pumilum Curtis	90-100	Au Br He	Söllner, 1954

  (a) pumilum
  (b) pallens (F.W. Schultz) Schinz &
    Thell.
  (c) litigiosum (De Lens) P.D. Sell
    & Whitehead
49 diffusum Pers.
  (a) gussonei (Tod. ex Lojac.) P.D.
    Sell & Whitehead

(b) diffusum	36	Ge	Wulff, 1937
	72	Br	Brett, 1955

  (c) subtetrandrum (Lange) P.D. Sell

& Whitehead	72	Da	Hagerup, 1941

50 siculum Guss.

51 gracile Dufour	88-92	Lu	Söllner, 1954

10 Moenchia Ehrh.
   1 erecta (L.) P. Gaertner, B. Meyer   36   Br Lu   Blackburn &
     & Scherb.   Morton, 1957
     (a) erecta
     (b) octandra (Ziz) Coutinho
   2 graeca Boiss. & Heldr.
   3 mantica (L.) Bartl.
     (a) mantica
     (b) caerulea (Boiss.) Clapham

11 Myosoton Moench
   1 aquaticum (L.) Moench   28   Hs Lu   Blackburn &
                                                                    Morton, 1957

12 Sagina L.
   1 nodosa (L.) Fenzl   22-24   Ge   Wulff, 1937b
                                                 44   Is   Löve & Löve, 1956
                                                 56   Br   Blackburn &
                                                                    Morton, 1957
   2 intermedia Fenzl   88   Is   Löve & Löve, 1956
   3 caespitosa (J. Vahl) Lange   88   Is   Löve & Löve, 1956
   4 glabra (Willd.) Fenzl
   5 pilifera (DC.) Fenzl
   6 subulata (Swartz) C. Presl   22   Is   Löve & Löve, 1956
   7 sabuletorum (Gay) Lange
   8 nevadensis Boiss. & Reuter
   9 saginoides (L.) Karsten   22   Is   Löve & Löve, 1956
  10 x normaniana Lagerh.
  11 procumbens L.   22   Is   Löve & Löve, 1956
     (a) procumbens
     (b) muscosa (Jordan) Nyman
         boydii Buchanan-White   22   Br   Blackburn &
                                                                    Morton, 1957
  12 apetala Ard.   12   Br Lu   Blackburn &
                                                                    Morton, 1957
     (a) apetala
     (b) erecta (Hornem.) F. Hermann
  13 maritima G. Don.   22-24   Ge   Wulff, 1937a
                                                 28   Br   Blackburn &
                                                                    Morton, 1957

13 Scleranthus L.
   1 perennis L.
     (a) perennis   22   Su   Lövkvist, 1963
     (b) prostratus P.D. Sell
     (c) burnatii (Briq.) P.D. Sell
     (d) polycnemoides (Willk. & Costa)
         Font Quer
     (e) vulcanicus (Strobl) Béguinot
     (f) marginatus (Guss.) Arcangeli
     (g) dichotomus (Schur) Stoj. &
         Stefanov
   2 annuus L.   44   Br Lu   Blackburn &
                                                                    Morton, 1957
     (a) annuus
     (b) polycarpos (L.) Thell.
     (c) verticillatus (Tausch) Arcangeli
     (d) ruscinonensis (Gillot & Coste)
         P.D. Sell
   3 uncinatus Schur

14 Corrigiola L.
  1 litoralis L.   18   Br Lu   Blackburn &
                                                   Morton, 1957
  2 telephiifolia Pourret   18   Lu   Blackburn &
                                                   Morton, 1957

15 Paronychia Miller
  1 cymosa (L.) DC.
  2 echinulata Chater
  3 rouyana Coincy
  4 suffruticosa (L.) Lam.
  5 argentea Lam.
  6 polygonifolia (Vill.) DC.
  7 kapela (Hacq.) Kerner
    (a) kapela
    (b) serpyllifolia (Chaix) Graebner
    (c) chionaea (Boiss.) Borhidi
  8 taurica Borhidi & Sikura
  9 cephalotes (Bieb.) Besser   36   Hu   Baksay, 1956
10 aretioides DC.
11 capitata (L.) Lam.

16 Herniaria L.
  1 alpina Chaix   18   He   Favarger, 1959a
  2 boissieri Gay
  3 latifolia Lapeyr.
  4 parnassica Heldr. & Sart. ex Boiss.
  5 glabra L.   18   Br   Blackburn &
                                                   Morton, 1957
  6 ciliolata Melderis   72   Br   Blackburn &
                                                   Morton, 1957
                              108   ?Br   Blackburn &
                              126   Lu   Adams, 1955
  7 maritima Link
  8 scabrida Boiss.
  9 baetica Boiss. & Reuter
10 incana Lam.
11 hirsuta L.   36   Lu   Blackburn &
                                                     Morton, 1957
12 polygama Gay
13 nigrimontium F. Hermann
14 fontanesii Gay
    (a) fontanesii
    (b) almeriana Brummitt & Heywood
15 fruticosa L.
    (a) fruticosa
    (b) erecta (Willk.) Batt.

17 Illecebrum L.
  1 verticillatum L.   10   Br Lu   Blackburn &
                                                     Morton, 1957

18 Pteranthus Forskål
  1 dichotomus Forskål

19 Polycarpon Loefl. ex L.
  1 tetraphyllum (L.) L.
  2 diphyllum Cav.
  3 alsinifolium (Biv.) DC.
  4 polycarpoides (Biv.) Zodda

20 Ortegia L.
    1 hispanica L.

21 Loeflingia L.
    1 hispanica L.
    2 baetica Lag.
    3 tavaresiana Samp.

22 Spergula L.
    1 arvensis L.   18   Br Lu   Blackburn &
                                                                         Morton, 1957
    2 morisonii Boreau   18   Ge   Rohweder, 1939
    3 pentandra L.
    4 viscosa Lag.

23 Spergularia (Pers.) J. & C. Presl
    1 azorica (Kindb.) Lebel
    2 fimbriata Boiss.
    3 rupicola Lebel ex Le Jolis   36   Br   Blackburn &
                                                                         Morton, 1957
    4 macrorhiza (Req.) Heynh.   36   ?Co   Monnier, ined.
    5 australis Samp.   40   Lu   Monnier, ined.
    6 media (L.) C. Presl   18   Br   Ratter, 1964
                                                   36   Br   Blackburn &
                                                                         Morton, 1957
    7 marina (L.) Griseb.   36   Is   Löve & Löve, 1956
    8 tangerina P. Monnier   18   S. Europe   Ratter, 1976
    9 segetalis (L.) G. Don fil.   18   Lu   Blackburn &
                                                                         Morton, 1957
   10 diandra (Guss.) Boiss.   18   Lu   Blackburn &
                                                                         Morton, 1957
   11 purpurea (Pers.) G. Don fil.   18   Lu   Blackburn &
                                                                         Morton, 1957
                                                   36   Hs Lu   Ratter, 1964
   12 rubra (L.) J. & C. Presl   36   Br Fe Lu   Ratter, 1964
                                                   54   Ga   Ratter, 1964
   13 nicaeensis Sarato ex Burnat   36   Ga Hs   Ratter, 1964
   14 capillacea (Kindb. & Lange) Willk.   18   Lu   Blackburn &
                                                                         Morton, 1957
   15 bocconii (Scheele) Ascherson &   36   Lu   Blackburn &
      Graebner   Morton, 1957
   16 heldreichii Fouc. ex E. Simon
      secundus & P. Monnier   36   N. Africa   Ratter, 1964
   17 echinosperma Celak.

24 Telephium L.
    1 imperati L.
    (a) imperati
    (b) orientale (Boiss.) Nyman

25 Lychnis L.
    1 chalcedonica L.
    2 coronaria (L.) Desr.
    3 flos-jovis (L.) Desr.
    4 flos-cuculi L.
    (a) flos-cuculi   24   He   Favarger, 1946
    (b) subintegra Hayek
    5 sibirica L.
    6 viscaria L.
    (a) viscaria   24   Ge Su   Rohweder, 1939
    (b) atropurpurea (Griseb.) Chater
    7 alpina L.   24   Is   Löve & Löve, 1956
    8 nivalis Kit.

32

26 Agrostemma L.
   1 githago L.   48   Su   D. Löve, 1942
   2 linicola Terechov
   3 gracilis Boiss.

27 Petrocoptis A. Braun
   1 pyrenaica (J.P. Bergeret) A. Braun
   2 viscosa Rothm.
   3 glaucifolia (Lag.) Boiss.
   4 grandiflora Rothm.
   5 hispanica (Willk.) Pau
   6 crassifolia Rouy
   7 pardoi Pau

28 Silene L.
   1 italica (L.) Pers.

(a) italica	24	S. Europe	D. Löve, 1942
(b) nemoralis (Waldst. & Kit.) Nyman	24	Au	Mattick, 1950
2 patula Desf.	24	Lu	Blackburn & Morton, 1957

   3 hifacensis Rouy ex Willk.
   4 mollissima (L.) Pers.
   5 velutina Pourret ex Loisel.
   6 pseudovelutina Rothm.
   7 mellifera Boiss. & Reuter
   8 paradoxa L.
   9 nodulosa Viv.
  10 cythnia (Halácsy) Walters
  11 dictaea Rech. fil.
  12 spinescens Sibth. & Sm.
  13 fruticosa L.
  14 gigantea L.
  15 nutans L.

(a) nutans	24	He	Favarger, 1946
(b) dubia (Herbich) Zapal.	24	Br	Blackburn & Morton, 1957

  16 brachypoda Rouy
  17 viridiflora L.
  18 catholica (L.) Aiton fil.
  19 viscariopsis Bornm.
  20 longipetala Vent.
  21 niederi Heldr. ex Boiss.
  22 guicciardii Boiss. & Heldr.
  23 bupleuroides L.

(a) bupleuroides	24	Cz	Májovský et al., 1974
(b) staticifolia (Sibth. & Sm.) Chowdhuri			

  24 chlorifolia Sm.
  25 chlorantha (Willd.) Ehrh.
  26 frivaldszkyana Hampe

27 multiflora (Waldst. & Kit.) Pers.	24	Au	Larsen, 1956a
28 viscosa (L.) Pers.	24	Su	Löve & Löve, 1942

  29 radicosa Boiss. & Heldr.
  30 reichenbachii Vis.
  31 skorpilii Velen.
  32 tatarica (L.) Pers.
  33 paucifolia Ledeb.
  34 graminifolia Otth

35 wahlbergella Chowdhuri	24	Su	Nygren, 1949
36 furcata Rafin.	48	No Su	Nygren, 1951

  37 sibirica (L.) Pers.
  38 roemeri Friv.

39 sendtneri Boiss.
40 ventricosa Adamović
41 borysthenica (Gruner) Walters   24   Hu   Baksay, 1958
42 media (Litv.) Kleopow
43 hellmannii Claus
44 otites (L.) Wibel   24   Rm   Vladesco, 1940
45 densiflora D'Urv.
46 exaltata Friv.
47 wolgensis (Willd.) Besser ex Sprengel   24   Rs( ) F. Wrigley, ined.
48 baschkirorum Janisch.
49 auriculata Sibth. & Sm.
50 zawadzkii Herbich
51 elisabetha Jan
52 requienii Otth
53 cordifolia All.
54 foetida Link ex Sprengel   24   Lu   Blackburn &
                                                         Morton, 1957

55 macrorhiza Gay & Durieu ex Lacaita
56 vulgaris (Moench) Garcke
   (a) vulgaris   24   Au   Griesinger, 1937
   (b) angustifolia (Miller) Hayek
   (c) commutata (Guss.) Hayek   48   It Si K. Larsen, ined.
   (d) macrocarpa Turrill
   (e) glareosa (Jordan) Marsden-Jones &   24   Ga   Marsden-Jones &
       Turrill   Turrill, 1957
   (f) prostrata (Gaudin) Chater &
       Walters   24   Au   Griesinger, 1937
   (g) maritima (With.) A. & D. Löve   24   Su   Griesinger, 1937
   (h) thorei (Duf.) Chater & Walters
57 csereii Baumg.
58 fabaria (L.) Sibth. & Sm.
59 caesia Sibth. & Sm.
60 variegata (Desf.) Boiss. & Heldr.
61 fabarioides Hausskn.
62 procumbens Murray
63 flavescens Waldst. & Kit.
64 congesta Sibth. & Sm.
65 vallesia L.
   (a) vallesia
   (b) graminea (Vis. ex Reichenb.)
       Ascherson & Graebner
66 boryi Boiss.
67 repens Patrin
68 succulenta Forskål
69 supina Bieb.   24   Rm   Vladesco, 1940
70 thymifolia Sibth. & Sm.   48   Rm   Vladesco, 1940
71 altaica Pers.
72 cretacea Fischer ex Sprengel
73 suffrutescens Bieb.
74 taliewii Kleopow
75 linifolia Sibth. & Sm.
76 schwarzenbergeri Halácsy
77 campanula Pers.
78 saxifraga L.
79 schmuckeri Wettst.
80 multicaulis Guss.
81 waldsteinii Griseb.
82 pindicola Hausskn.
83 orphanidis Boiss.
84 barbeyana Heldr. ex Boiss.
85 falcata Sibth. & Sm.

```
86 cephallenia Heldr.
87 ciliata Pourret              24      Hs      Blackburn &
                                                Morton, 1957
                                36      Hs      Blackburn &
                                                Morton, 1957
                                48      Hs      Blackburn &
                                                Morton, 1957
                                72
                                84
                                c.96
                                c.120   Bot. Card. Favarger, 1946
88 legionensis Lag.
89 borderi Jordan
90 acaulis (L.) Jacq.           24      Au Su   Jörgensen et al.,
                                                1958
   (a) longiscapa (Kerner ex Vierh.)
       Hayek
   (b) exscapa (All.) J. Braun
91 dinarica Sprengel
92 rupestris L.                 24      It      Griesinger, 1937
93 lerchenfeldiana Baumg.
94 pusilla Waldst. & Kit.       24      It      Rohweder, 1939
95 veselskyi (Janka) Béguinot
96 macrantha (Pančić) Neumayer
97 tommasinii Vis.
98 retzdorffiana (K. Maly) Walters
99 chromodonta Boiss. & Reuter
100 alpestris Jacq.
101 armeria L.
102 compacta Fischer
103 asterias Griseb.
104 noctiflora L.                24      Su      Löve & Löve, 1942
105 alba (Miller) E.H.L. Krause
    (a) alba                     24      Br      Blackburn, 1923
    (b) divaricata (Reichenb.) Walters  24  Lu  Blackburn &
                                                Morton, 1957
    (c) eriocalycina (Boiss.) Walters
106 dioica (L.) Clairv.          24      Br      Baker, 1948
107 heuffelii Soó
108 diclinis (Lag.) M. Laínz
109 portensis L.
110 echinosperma Boiss. & Heldr.
111 pinetorum Boiss. & Heldr.
112 inaperta L.
113 fuscata Link
114 pseudatocion Desf.
115 rubella L.
    (a) rubella
    (b) turbinata (Guss.) Chater & Walters
116 bergiana Lindman
117 insularis W. Barbey
118 divaricata Clemente
119 integripetala Bory & Chaub.
120 sedoides Poiret
121 pentelica Boiss.
122 haussknechtii Heldr. ex Hausskn.
123 laconica Boiss. & Orph.
124 cretica L.
125 ungeri Fenzl
126 graeca Boiss. & Spruner
127 muscipula L.
```

```
128 stricta L.
129 behen L.
130 holzmannii Heldr. ex Boiss.
131 linicola C.C. Gmelin
132 echinata Otth
133 squamigera Boiss.
134 trinervia Sebastiani & Mauri
135 laeta (Aiton) Godron
136 coeli-rosa (L.) Godron
137 pendula L.
138 psammitis Link ex Sprengel
139 littorea Brot.
140 adscendens Lag.
141 boissieri Gay
142 almolae Gay
143 dichotoma Ehrh.                    24      Rm      Vladesco, 1940
144(a) dichotoma
    (b) racemosa (Otth) Graebner
144 remotiflora Vis.
145 nicaeensis All.
146 scabriflora Brot.
147 micropetala Lag.
148 discolor Sibth. & Sm.
149 obtusifolia Willd.
150 sericea All.
151 nocturna L.
    (a) nocturna
    (b) neglecta (Ten.) Arcangeli
152 ramosissima Desf.
153 gallica L.                         24      Br Lu   Blackburn &
                                                       Morton, 1957

154 giraldii Guss.
155 bellidifolia Juss. ex Jacq.
156 cerastoides L.
157 tridentata Desf.
158 disticha Willd.
159 colorata Poiret
160 secundiflora Otth
161 apetala Willd.
162 longicaulis Pourret ex Lag.
163 ammophila Boiss. & Heldr.
    (a) ammophila
    (b) carpathae Chowdhuri
164 conica L.
    (a) conica                         20      Su      Lövkvist, 1963
    (b) subconica (Friv.) Gavioli
    (c) sartorii (Boiss. & Heldr.) Chater
        & Walters
165 lydia Boiss.
166 conoidea L.


29 Cucubalus L.
   1 baccifer L.                        24      Lu      Blackburn &
                                                        Morton, 1957


30 Drypis L.
   1 spinosa L.                         60      Bot. Gard.  Favarger, 1946
    (a) spinosa
    (b) jacquiniana Murb. & Wettst. ex Murb.
```

31 Gypsophila L.
  1 nana Bory & Chaub.
  2 achaia Bornm.
  3 spergulifolia Griseb.
  4 repens L.   34   Po   Skalińska, 1950
  5 altissima L.
  6 litwinowii Kos.-Pol.
  7 fastigiata L.   34   Hu   Baksay, 1956
  8 papillosa Porta
  9 hispanica Willk.
 10 struthium Loefl.
 11 collina Steven ex Ser.
 12 patrinii Ser.
 13 uralensis Less.
 14 petraea (Baumg.) Reichenb.
 15 glomerata Pallas ex Bieb.
 16 globulosa Steven ex Boiss.
 17 paniculata L.
 18 arrostii Guss.
 19 belorossica Barkoudah
 20 acutifolia Steven ex Sprengel
 21 perfoliata L.
 22 scorzonerifolia Ser.
 23 tomentosa L.
 24 linearifolia (Fischer & C.A. Meyer) Boiss.
 25 elegans Bieb.
 26 muralis L.
 27 macedonica Vandas
 28 pilosa Hudson

32 Bolanthus (Ser.) Reichenb.
  1 laconicus (Boiss.) Barkoudah
  2 fruticulosus (Bory & Chaub.) Barkoudah
  3 thessalus (Jaub. & Spach) Barkoudah
  4 graecus (Schreber) Barkoudah

33 Saponaria L.
  1 bellidifolia Sm.
  2 lutea L.
  3 caespitosa DC.
  4 pumilio (L.) Fenzl ex A. Braun
  5 glutinosa Bieb.
  6 sicula Rafin.
   (a) sicula
   (b) intermedia (Simmler) Chater
   (c) stranjensis (Jordanov) Chater
  7 calabrica Guss.
  8 ocymoides L.   28   He   Favarger, 1946
  9 officinalis L.   28   He   Favarger, 1946
 10 orientalis L.

34 Vaccaria Medicus
  1 pyramidata Medicus   30   He   Favarger, 1946

35 Petrorhagia (Ser. ex DC.) Link
  1 illyrica (L.) P.W. Ball & Heywood
   (a) illyrica
   (b) haynaldiana (Janka) P.W. Ball &
       Heywood
   (c) taygetea (Boiss.) P.W. Ball &
       Heywood
  2 ochroleuca (Sibth. & Sm.) P.W. Ball & Heywood

```
 3 armerioides (Ser.) P.W. Ball & Heywood
 4 candica P.W. Ball & Heywood
 5 cretica (L.) P.W. Ball & Heywood
 6 alpina (Habl.) P.W. Ball & Heywood
 7 phthiotica (Boiss. & Heldr.) P.W. Ball
   & Heywood
 8 fasciculata (Margot & Reuter) P.W. &
   Heywood
 9 saxifraga (L.) Link                 60      Au      Larsen, 1954a
10 graminea (Sibth. & Sm.) P.W. Ball &
   Heywood
11 thessala (Boiss.) P.W. Ball & Heywood
12 dianthoides (Sibth. & Sm.) P.W. Ball
   & Heywood
13 prolifera (L.) P.W. Ball & Heywood   30      Ga      Carolin, 1957
14 nanteuilii (Burnat) P.W. Ball &
   Heywood                              60      Hs      Böcher et al., 1953
15 velutina (Guss.) P.W. Ball & Heywood 30      Lu      Blackburn &
                                                        Morton, 1957
16 glumacea (Chaub. & Bory) P.W. Ball
   & Heywood


36 Dianthus L.
 1 seguieri Vill.
   (a) seguieri                        90      Alps    Carolin, 1957
   (b) guatieri (Sennen) Tutin
   (c) glaber Čelak.
   (d) italicus Tutin
 2 collinus Waldst. & Kit.
   (a) collinus
   (b) glabriusculus (Kit.) Soó
 3 fischeri Sprengel
 4 pratensis Bieb.
   (a) pratensis
   (b) racovitzae (Prodan) Tutin
 5 versicolor Fischer ex Link
 6 pseudoversicolor Klokov
 7 furcatus Balbis
   (a) furcatus
   (b) tener (Balbis) Tutin
   (c) geminiflorus (Loisel.) Tutin     30      Hs      Carolin, 1957
   (d) gyspergerae (Rouy) Burnat ex Briq.
 8 viridescens Vis.
 9 trifasciculatus Kit.
   (a) trifasciculatus
   (b) deserti (Prodan) Tutin
   (c) euponticus (Zapal.) Kleopow
10 urumoffii Stoj. & Acht.
11 eugeniae Kleopow
12 tesquicola Klokov
13 guttatus Bieb.
14 knappii (Pant.) Ascherson & Kanitz
   ex Borbás
15 membranaceus Borbás
16 dobrogensis Prodan
17 barbatus L.
   (a) barbatus
   (b) compactus (Kit.) Heuffel
18 monspessulanus L.                    60      It      Carolin, 1957
   (a) monspessulanus
```

(b) marsicus (Ten.) Novák
(c) sternbergii Hegi
19 repens Willd.
20 alpinus L.   30   Alps   Carolin, 1957
21 nitidus Waldst. & Kit.
22 scardicus Wettst.
23 callizonus Schott & Kotschy
24 glacialis Haenke
   (a) glacialis   30   Alps   Carolin, 1957
   (b) gelidus (Schott, Nyman & Kotschy
       Tutin
25 freynii Vandas
26 microlepis Boiss.
27 pavonius Tausch   30   Alps   Carolin, 1957
28 gratianopolitanus Vill.   60   Ga   Carolin, 1957
29 xylorrhizus Boiss. & Heldr.
30 sylvestris Wulfen   30   Alps   Carolin, 1957
                                              c.60   Ga   K. Jones, ined.
   (a) sylvestris
   (b) siculus (C. Presl) Tutin
   (c) nodosus (Tausch) Hayek
   (d) bertisceus Rech. fil.
   (e) tergestinus (Reichenb.) Hayek
31 caryophyllus L.
32 subacaulis Vill. var rusconinensis   60   Hs   Carolin, 1957
   (a) subacaulis
   (b) brachyanthus (Boiss.) P. Fourn.   30   Hs   Carolin, 1957
33 minutiflorus (Borbás) Halácsy
34 pungens L.
35 hispanicus Asso
36 costae Willk.
37 planellae Willk.
38 langeanus Willk.
39 laricifolius Boiss. & Reuter
40 serratifolius Sibth. & Sm.
41 pyrenaicus Pourret   30   Hs   Carolin, 1957
                                                 60   Ga   K. Jones, 1964
   (a) pyrenaicus
   (b) catalaunicus (Willk. & Costa) Tutin
42 lusitanus Brot.   30   Hs   Carolin, 1957
43 malacitanus Haenseler ex Boiss.   60   Hs   Carolin, 1957
44 scaber Chaix
   (a) scaber
   (b) cutandae (Pau) Tutin   30   Hs   K. Jones,ined.
   (c) toletanus (Boiss. & Reuter) Tutin
45 crassipes R. de Roemer
46 graniticus Jordan   30   Ga   K. Jones,ined.
47 cintranus Boiss. & Reuter
   (a) cintranus   60   Lu   K. Jones,ined.
   (b) multiceps (Costa ex Willk.) Tutin
   (c) charidemii (Pau) Tutin
48 anticarius Boiss. & Reuter
49 petraeus Waldst. & Kit.
   (a) petraeus
   (b) integer (Vis.) Tutin
   (c) simonkaianus (Péterfi) Tutin
   (d) noeanus (Boiss.) Tutin
50 spiculifolius Schur
51 stefanoffii Eig
52 plumarius L.
53 serotinus Waldst. & Kit.

54 acicularis Fischer ex Ledeb.
55 rigidus Bieb.
56 uralensis Korsh.
57 squarrosus Bieb.
58 arenarius L.
  (a) arenarius   60   Su   Lövkvist, 1963
  (b) borussicus (Vierh.) Kleopow
  (c) pseudoserotinus (Blocki) Tutin
  (d) pseudosquarrosus (Novák) Kleopow
59 krylovianus Juz.
60 volgicus Juz.
61 gallicus Pers.
62 superbus L.
  (a) superbus
  (b) speciosus (Reichenb.) Pawl.   30   Ju   Carolin, 1957
  (c) stenocalyx (Trautv.) Kleopow
63 deltoides L.
64 degenii Bald.
65 myrtinervius Griseb.
66 sphacioticus Boiss. & Heldr.
67 pallidiflorus Ser.
68 campestris Bieb.
  (a) campestris
  (b) laevigatus (Gruner) Klokov
  (c) steppaceus Širj.
69 hypanicus Andrz.
70 carbonatus Klokov
71 aridus Griseb. ex Janka
72 marschallii Schischkin
73 cinnamomeus Sibth. & Sm.
74 pallens Sibth. & Sm.
75 lanceolatus Steven ex Reichenb.
76 leptopetalus Willd.
77 roseoluteus Velen.
78 mercurii Heldr.
79 gracilis Sibth. & Sm.   60   Ju   Carolin, 1957
  (a) gracilis
  (b) xanthianus (Davidov) Tutin
  (c) armerioides (Griseb.) Tutin
  (d) friwaldskyanus (Boiss.) Tutin
80 drenowskianus Rech. fil.
81 haematocalyx Boiss. & Heldr.
  (a) haematocalyx
  (b) pruinosus (Boiss. & Orph.) Hayek
  (c) sibthorpii (Vierh.) Hayek
  (d) pindicola (Vierh.) Hayek
82 biflorus Sibth. & Sm.
83 strictus Banks & Solander   30   Balkans   Carolin, 1957
84 tripunctatus Sibth. & Sm.
85 corymbosus Sibth. & Sm.   30   Bu   Carolin, 1957
86 viscidus Bory & Chaub.
87 diffusus Sibth. & Sm.
88 formanekii Borbás ex Form.
89 armeria L.
  (a) armeria   30   Br   Carolin, 1957
  (b) armeriastrum (Wolfner) Velen. 30   Ju   Carolin, 1957
90 pseudarmeria Bieb.
91 humilis Willd. ex Ledeb.
92 ciliatus Guss.
  (a) ciliatus
  (b) dalmaticus (Čelak.) Hayek

```
 93 arpadianus Ade & Bornm.
 94 nardiformis Janka                         30      Bu     Carolin, 1957
 95 ingoldbyi Turrill
 96 juniperinus Sm.                           30      Cr     Carolin, 1957
 97 rupicola Biv.
 98 arboreus L.                               30      Gr     Carolin, 1957
 99 fruticosus L.
100 ferrugineus Miller
    (a) ferrugineus                           30 Bot. Gard.  Carolin, 1957
    (b) liburnicus (Bartl.) Tutin
    (c) vulturius (Guss. & Ten.) Tutin
101 capitatus Balbis ex DC.
    (a) capitatus
    (b) andrzejowskianus Zapal.
102 pinifolius Sibth. & Sm.
    (a) pinifolius                            30      Bu     Carolin, 1957
    (b) serbicus Wettst.
    (c) lilacinus (Boiss. & Heldr.) Wettst.
103 giganteus D'Urv.
    (a) giganteus
    (b) croaticus (Borbás) Tutin
    (c) banaticus (Heuffel) Tutin
    (d) haynaldianus (Borbás) Tutin
    (e) italicus Tutin
    (f) leucophoeniceus (Dörfler & Hayek) Tutin
104 pontederae Kerner
    (a) pontederae                            30      Au     Carolin, 1957
    (b) giganteiformis (Borbás) Soó
    (c) kladovanus (Degen) Stoj. & Acht.
105 diutinus Kit.
106 platyodon Klokov
107 bessarabicus (Kleopow) Klokov
108 carthusianorum L.                         30      Alps   Carolin, 1957
109 puberulus (Simonkai) Kerner
110 tenuifolius Schur
111 borbasii Vandas
    (a) borbasii
    (b) capitellatus (Klokov) Tutin
112 henteri Heuffel ex Griseb
113 cruentus Griseb.                          30      Bu     Carolin, 1957
    (a) cruentus
    (b) turcicus (Velen.) Stoj. & Acht.
114 quadrangulus Velen.                       30      Bu     Carolin, 1957
115 tristis Velen.                            30 Bot. Gard.  Rohweder, 1934
116 stribrnyi Velen.
117 brachyzonus Borbás & Form.                30      Al     Carolin, 1957
118 pelviformis Heuffel
119 stenopetalus Griseb.
120 burgasensis Tutin
121 moesiacus Vis. & Pančić


37 Velezia L.
    1 rigida L.                               28      Lu     Blackburn &
                                                             Morton, 1957

    2 quadridentata Sibth. & Sm.


LVIII NYMPHAEACEAE

1 Nymphaea L.
    1 alba L.                                 84      Br     Y. Heslop-
                                                             Harrison, 1955
```

| | c.105 | Su | Ehrenberg, 1945 |
| | 112 | Scand. | Langlet & Söderberg, 1927 |

2 candida C. Presl   c.160   Scand.   Lohammon, 1942
3 tetragona Georgi
4 lotus L.

2 Nuphar Sm.
  1 lutea (L.) Sibth. & Sm.   34   Br   Y. Heslop-Harrison,
  2 pumila (Timm) DC.   34   Br   1953b

LIX NELUMBONACEAE

1 Nelumbo Adanson
  1 nucifera Gaertner

LX CERATOPHYLLACEAE

1 Ceratophyllum L.
  1 demersum L.   34   Scand.   Langlet & Söderberg, 1927

    (a) demersum
    (b) platyacanthum (Cham.) Nyman
  2 submersum L.   24   Ge   Strasburger, 1902
                          40   Ge   Wulff, 1938
  3 tanaiticum Sapjegin
  4 muricatum Cham.

LXI RANUNCULACEAE

1 Helleborus L.
  1 foetidus L.
  2 lividus Aiton
    (a) lividus
    (b) corsicus (Willd.)
  3 cyclophyllus Boiss.
  4 orientalis Lam.
  5 viridis L.
    (a) viridis
    (b) occidentalis (Reuter) Schiffner
  6 odorus Waldst. & Kit.
    (a) odorus
    (b) laxus (Host) Merxm. & Podl.
  7 multifidus Vis.
    (a) multifidus
    (b) serbicus (Adamović) Merxm. & Podl.
    (c) istriacus (Schiffner) Merxm. & Podl.
  8 bocconei Ten.
    (a) bocconei
    (b) siculus (Schiffner) Merxm. & Podl.
  9 dumetorum Waldst. & Kit.
    (a) dumetorum
    (b) atrorubens (Waldst. & Kit.) Merxm. & Podl.
  10 purpurascens Waldst. & Kit.
  11 niger L.
    (a) niger
    (b) macranthus (Freyn) Schiffner

2 Eranthis Salisb.
  1 hyemalis (L.) Salisb.

3 Callianthemum C.A. Meyer
  1 anemonoides (J. Zahlbr.) Endl. ex
    Heynh   32   Au   Mattick, 1950
  2 kerneranum Freyn ex Kerner
  3 coriandrifolium Reichenb.   16   Po   Skalińska et al.,
                                                   1959

4 Nigella L.
  1 arvensis L.
    (a) arvensis
    (b) aristata (Sibth. & Sm.) Nyman
    (c) rechingeri (Tutin) Tutin
  2 hispanica L.
    (a) hispanica
    (b) atlantica Murb.
  3 gallica Jordan
  4 segetalis Bieb.
  5 degenii Vierh.
  6 cretica Miller
  7 fumariifolia Kotschy
  8 doerfleri Vierh.
  9 sativa L.   12   Bot. Gard. Pereira, 1942
 10 elata Boiss.
 11 damascena L.
 12 orientalis L.

5 Garidella L.
  1 nigellastrum L.

6 Trollius L.
  1 europaeus L.
    (a) europaeus   16   Rs(N) Sokolovskaya, 1958
    (b) transsilvanicus (Schur) Jáv.   16   Po   Skalińska et al.,
                                                    1959
  2 asiaticus L.

7 Isopyrum L.
  1 thalictroides L.   14   Po   Skalińska et al.,
                                                    1959
8 Actaea L.
  1 spicata L.   16   Po   Skalińska et al.,
                                                    1959
  2 erythrocarpa Fischer

9 Cimicifuga L.
  1 europaea Schipcz.   16   Hu   Baksay, 1957b

10 Caltha L.
  1 palustris L.   32   Po   Skalińska et al.,
                                                    1959
                                     53-62   Ge   Reese, 1954

11 Aconitum L.
  1 septentrionale Koelle   16   Su   Schafer & La
                                                    Cour, 1934
  2 lasiostomum Reichenb.
  3 moldavicum Hacq. ex Reichenb.
  4 vulparia Reichenb.   16   Alps   Schafer & La
                                                    Cour, 1934

43

5 lamarckii Reichenb.
6 anthora L.
7 variegatum L.   16   Au   Schafer & La
                                                         Cour, 1934
8 toxicum Reichenb.
9 paniculatum Lam.   16   Au   Mattick, 1950
10 firmum Reichenb.
11 napellus L.   32   Br   Schafer & La
                                                         Cour, 1934
12 tauricum Wulfen
13 compactum Reichenb.
14 nevadense Uechtr. ex Gáyer

12 Delphinium L.
  1 montanum DC.
  2 dubium (Rouy & Fouc.) Pawl.
  3 oxysepalum Borbas & Pax   32   Po   Skalińska et al.,
                                                         1959
  4 elatum L.
   (a) elatum   32   Rs(N)   Sokolovskaya &
                                                         Strelkova, 1960
   (b) helveticum Pawl.
   (c) austriacum Pawl.
  5 simonkaianum Pawl.
  6 cuneatum Steven ex DC.
  7 rossicum Litv.
  8 dictyocarpum DC.
   (a) dictyocarpum
   (b) uralense (Nevski) Pawl.
  9 pentagynum Lam.
  10 nevadense G. Kunze
  11 emarginatum C. Presl
  12 fissum Waldst. & Kit.
  13 albiflorum DC.
  14 puniceum Pallas
  15 schmalhausenii Albov
  16 peregrinum L.
  17 hirschfeldianum Heldr. & Holzm.
  18 obcordatum DC.
  19 halteratum Sibth. & Sm.
  20 balcanicum Pawl.
  21 verdunense Balbis
  22 gracile DC
  23 hellenicum Pawl.
  24 staphisagria L.
  25 pictum Willd.
  26 requienii DC.

15 Consolida (DC.) S.F. Gray
  1 aconiti (L.) Lindley
  2 thirkeana (Boiss.) Bornm.
  3 hellespontica (Boiss.) Chater
  4 orientalis (Gay) Schrödinger
   (a) orientalis
   (b) phrygia (Boiss.) Chater
  5 ambigua (L.) P.W. Ball & Heywood
  6 brevicornis (Vis.) Soó
  7 uechtritziana (Pančić) Soó
  8 tuntasiana (Halácsy) Soó
  9 regalis S.F. Gray   16   Löve & Löve, 1942
   (a) regalis

```
    (b) paniculata (Host) Soó
10 tenuissima (Sibth. & Sm.) Soó
11 pubescens (DC.) Soó
12 rigida (DC.) Hayek

14 Anemone L.
    1 nemorosa L.                              28-32   Su    Bernström, 1946
                                                 37    Su    Bernström, 1946
                                                 42    Su    Lövkvist, 1963
                                                 45    Su    Lövkvist, 1963
                                                 46    Su    Lövkvist, 1963
    2 altaica Fischer ex C.A. Meyer
    3 trifolia L.                                32    Au    Mattick, 1950
       (a) trifolia
       (b) albida (Mariz) Tutin
    4 ranunculoides L.
       (a) ranunculoides                        32    Su    Bernström, 1946
       (b) wockeana (Ascherson & Graebner) Hegi
    5 uralensis Fischer ex DC.
    6 apennina L.
    7 blanda Schott & Kotschy
    8 reflexa Stephan
    9 narcissiflora L.                           14    He    Larsen, 1954a
   10 dichotoma L.
   11 sylvestris L.                              16    Po    Gajewski, 1947
   12 baldensis L.
   13 pavoniana Boiss.
   14 palmata L.
   15 coronaria L.
   16 hortensis L.
   17 pavonina Lam.

15 Hepatica Miller
    1 nobilis Miller                            14    Po    Skalińska et al.,
                                                               1959
    2 transsilvanica Fuss                       28    He    Baumberger, 1970

16 Pulsatilla Miller
    1 alpina (L.) Delarbre                      16    Au    Rosenthal, 1936
       (a) alpina
       (b) apiifolia (Scop.) Nyman
    2 alba Reichenb.
    3 vernalis (L.) Miller                      16    Europe Bücher, 1932
    4 pratensis (L.) Miller                     16    Ge    Rosenthal, 1936
                                                 32    Da    Bücher, 1954
    5 montana (Hoppe) Reichenb.                 16    It    Rosenthal, 1936
    6 rubra (Lam.) Delarbre
    7 vulgaris Miller
       (a) vulgaris                             32    Da    Bücher, 1954
       (b) grandis (Wenderoth) Zamels
    8 halleri (All.) Willd.                     32    Rs(K) Rosenthal, 1936
       (a) rhodopaea K. Krause
       (b) taurica (Juz.) K. Krause
       (c) slavica (G. Reuss) Zamels
       (d) halleri
       (e) styriaca (G.A. Pritzel) Zamels
    9 patens (L.) Miller
       (a) patens                               16    Rm    Tarnavschi, 1948
       (b) teklae (Zamels) Zamels
       (c) multifida (Pritzel) Zamels
       (d) flavescens (Zucc.) Zamels
```

45

17 Clematis L.
  1 flammula L.
  2 vitalba L.
  3 orientalis L.
  4 recta L.
  5 pseudoflammula Schmalh. ex Lipsky
  6 cirrhosa L.
  7 viticella L.
  8 campaniflora Brot.
  9 integrifolia L.
 10 alpina (L.) Miller
    (a) alpina   16   Po   Skalińska et al.,
                                                             1961

    (b) sibirica (L.) O. Kuntze

18 Adonis L.
  1 vernalis L.   16   Po   Skalińska et al.,
                                                             1961

  2 volgensis Steven
  3 sibirica Patrin ex Ledeb.
  4 pyrenaica DC.
  5 distorta Ten.
  6 cyllenea Boiss., Heldr. & Orph.
  7 annua L.
    (a) annua
    (b) carinata Vierh.
  8 flammea Jacq.
  9 aestivalis L.
 10 microcarpa DC.

19 Ranunculus L.
  1 velutinus Ten.
  2 constantinopolitanus (DC.) D'Urv.
  3 polyanthemos L.
    (a) polyanthemos   16   Hu   Baksay, 1956
    (b) thomasii (Ten.) Tutin
    (c) polyanthemoides (Boreau) Ahlfvengren
  4 nemorosus DC.   16   He   Langlet, 1932
    (a) nemorosus
    (b) polyanthemophyllus (W. Koch & H. Hess)
        Tutin
    (c) serpens (Schrank) Tutin
  5 caucasicus Bieb.
  6 repens L.   32   Fe   Vaarama, 1948
  7 lanuginosus L.   28   Hu   Baksay, 1956
  8 acris L.
    (a) acris   14   Fa   Bücher, 1938a
    (b) borealis (Trautv.) Nyman   14   Rs(N)   Sokolovskaya, 1958
    (c) friesianus (Jordan) Rouy & Fouc.
    (d) granatensis (Boiss.) Nyman
    (e) strigulosus (Schur) Hyl.
  9 serbicus Vis.
 10 brutius Ten.
 11 gouanii Willd.   16   Ga/Hs   Landolt, 1954
 12 ruscinonensis Landolt   16   Hs   Landolt, 1956
 13 sartorianus Boiss. & Heldr.
 14 carinthiacus Hoppe   16   Alps   Landolt, 1954
 15 clethraphilus Litard.
 16 montanus Willd.   32   He   Landolt, 1954
 17 pseudomontanus Schur
 18 croaticus Schott

19 venetus Huter ex Landolt	32	It	Landolt, 1954
20 grenieranus Jordan	16	He	Landolt, 1954
21 oreophilus Bieb.	16	It	Landolt, 1954
22 aduncus Gren.	16	Alps	Landolt, 1954
23 carpaticus Herbich	28	Hu	Baksay, 1956
24 macrophyllus Desf.			
25 demissus DC.			
26 hayekii Dörfler			
27 marschlinsii Steudel			
28 dissectus Bieb.			
29 neapolitanus Ten.	16	Gr	Gregson, ined.
30 pratensis C. Presl			
31 bulbosus L.			
(a) bulbosus	16	Da	Böcher, 1938a
(b) castellanus (Boiss. & Reuter ex Freyn) P.W. Ball & Heywood			
(c) bulbifer (Jordan) Neves	16	Lu	Neves, 1950
(d) aleae (Willk.) Rouy & Fouc.			
(e) adscendens (Brot.) Neves	16	Lu	Neves, 1950
(f) gallecicus (Freyn ex Willk.) P.W. Ball & Heywood			
32 sardous Crantz	16	Ge	Reese, 1961
33 cordiger Viv.			
34 marginatus D'Urv.			
35 trilobus Desf.			
36 muricatus L.	48	Lu	Neves, 1950
	64	Lu	Neves, 1950
37 cornutus DC.			
38 arvensis L.	32	Lu	Neves, 1950
39 parviflorus L.			
40 chius DC.			
41 gracilis E.D. Clarke			
42 oxyspermus Willd.			
43 psilostachys Griseb.	32	?Br	Larter, 1932
44 rumelicus Griseb.			
45 monspeliacus L.			
46 miliarakesii Halácsy			
47 isthmicus Boiss.			
48 illyricus L.			
49 pedatus Waldst. & Kit.			
50 platyspermus Fischer ex DC.			
51 paludosus Poiret	32	Ga	Tutin, 1961
52 cupreus Boiss. & Heldr.			
53 millefoliatus Vahl			
54 garganicus Ten.			
55 millii Boiss. & Heldr.			
56 nigrescens Freyn			
57 cortusifolius Willd.			
58 rupestris Guss.			
59 creticus L.			
60 thasius Halácsy			
61 subhomophyllus (Halácsy) Vierh.			
62 spruneranus Boiss.			
63 gregarius Brot.			
64 asiaticus L.			
65 cassubicus L.	24	Po	Skalińska et al., 1961
	32	Po	Skalińska et al., 1961
	40	Fe	Rousi, 1956
	44	Po	Skalińska et al., 1961

	64	Po	Skalińska et al., 1961
66 fallax (Wimmer & Grab.) Kerner	32	Fe	Rousi, 1956
67 monophyllus Ovcz.	32	Fe	Rousi, 1956
68 auricomus L.	16	He	Häfliger, 1943
	32	Fe	Rousi, 1956
	40	Fe	Rousi, 1956
	48	Fe	Rousi, 1956
69 affinis R. Br.			
70 degenii Kümmerie & Jáv.			
71 flabellifolius Heuffel ex Reichenb.			
72 polyrhizos Stephan ex Willd.			
73 pedatifidus Sm.			
74 pygmaeus Wahlenb.	16	Rs(N)	Sokolovskaya, 1958
	32		
75 nivalis L.	40	Scand.	Nygren, 1948
	48	Scand.	Nygren, 1948
	56	Scand.	Nygren, 1948
76 sulphureus C.J. Phipps	96	Rs(N)	Sokolovskaya & Strelkova, 1960
77 hyperboreus Rottb.			
(a) hyperboreus	32	Rs(N)	Sokolovskaya, 1958
(b) samojedorum (Rupr.) Hultén			
78 sceleratus L.			
(a) sceleratus	32	Po	Skalińska et al., 1961
(b) reptabundus (Rupr.) Hultén			
79 gmelinii DC.			
80 lapponicus L	16	Su	Langlet, 1936
81 pallasii Schlecht.			
82 cymbalaria Pursh			
83 ficaria L.			
(a) calthifolius (Reichenb.) Arcangeli			
(b) bulbifer Lawalrée	32	Br	Heywood & Walker, 1961
(c) ficaria	16	Br	Heywood & Walker, 1961
(d) ficariiformis Rouy & Fouc.			
84 ficarioides Bory & Chaub.			
85 bullatus L.			
86 thora L.			
87 scutatus Waldst. & Kit.			
88 hybridus Biria			
89 brevifolius Ten.			
(a) brevifolius			
(b) pindicus (Hausskn.) E. Mayer			
90 alpestris L.	16	He	Larsen, 1954
91 traunfellneri Hoppe			
92 crenatus Waldst. & Kit.			
93 bilobus Bertol.			
94 aconitifolius L.	14	He	Larsen, 1954
	16	Au	Mattick, 1950
95 platanifolius L.	16	Po	Skalińska, 1950
96 seguieri Vill.			
(a) seguieri			
(b) montenegrinus (Halácsy) Tutin			
97 acetosellifolius Boiss.			
98 glacialis L.	16	Ga He	Larsen, 1954
99 cymbalarifolius Balbis ex Moris			
100 weyleri Marès			
101 flammula L.			

(a) flammula	32	Po	Skalińska et al., 1959
(b) minimus (Ar. Benn.) Padmore	32	Br	Padmore, 1957
(c) scoticus (E.S. Marshall) Clapham			
102 reptans L.	32	Rs(N)	Sokolovskaya, 1958
103 batrachioides Pomel			
104 lingua L.	128	Su	Lövkvist, 1963
105 ophioglossifolius Vill.	16	Lu	Neves, 1944
106 fontanus C. Presl			
107 longipes Lange ex Cutanda			
108 revelieri Boreau			
(a) revelieri			
(b) rodiei (Litard.) Tutin			
109 lateriflorus DC.			
110 nodiflorus L.			
111 polyphyllus Waldst. & Kit. ex Willd.			
112 pyrenaeus L.	32	He	Langlet, 1932
(a) pyrenaeus			
(b) alismoides (Bory) O. Bolós & Font Quer			
113 bupleuroides Brot.	16	Lu	Neves, 1950
114 gramineus L.			
115 abnormis Cutanda & Willk.			
116 parnassifolius L.			
117 wettsteinii Dörfler			
118 amplexicaulis L.			
119 hederaceus L.	16	Br Ge Lu	Cook, 1962
120 omiophyllus Ten.	16	Si	Cook, 1962
	32	Br	Cook, 1962
121 tripartitus DC.	48	Br	Cook, 1962
122 ololeucos Lloyd	16	He	Cook, 1962
123 baudotii Godron	32	Au Br	Cook, 1962
124 peltatus Schrank	32	Da Ge	Cook, 1962
125 pseudofluitans (Syme) Newbould ex Baker & Foggitt	24	Ge	Cook, 1962
	32	Ge	Cook, 1962
	40	Br	Cook, 1962
	48	Br Ge	Cook, 1962
126 sphaerospermus Boiss. & Blanche			
127 aquatilis L.	48	Br Ge	Cook, 1962
128 trichophyllus Chaix			
(a) trichophyllus	32	Br Ge	Cook, 1962
(b) lutulentus (Perr. & Song.) Vierh.	32	Ge	Cook, 1962
129 rionii Lagger	16	Au	Cook, 1962
130 circinatus Sibth.	16	Br Ge	Cook, 1962
131 fluitans Lam.	16	Ge	Cook, 1962
	24	Ge	Cook, 1962
	32	Br	Cook, 1962

20 Ceratocephalus
  1 falcatus (L.) Pers.
  2 testiculatus (Crantz) Roth

21 Myosurus L.			
1 minimus L.	16	Rm	Tarnavschi, 1948
	28	Su	Ehrenberg, 1945
2 heldreichi Léveillé			

22 Aquilegia L.
  1 litardierei Briq.
  2 paui Font Quer

```
  3 transsilvanica Schur
  4 vulgaris L.                            14      Po      Skalińska, 1950
  5 viscosa Gouan
  6 hirsutissima (Lapeyr.) Timb.-Lagr.
    ex Gariod
  7 amaliae Heldr. ex Boiss.
  8 ottonis Orph. ex Boiss.
  9 dinarica G. Beck
 10 dichroa Freyn
 11 pancicii Degen
 12 atrata Koch                            14      Au      Mattick, 1950
 13 nigricans Baumg.
    (a) nigricans
    (b) subscaposa (Borbás) Soó
 14 aurea Janka
 15 kitaibelii Schott
 16 einseleana F.W. Schultz
 17 thalictrifolia Schott & Kotschy
 18 bertolonii Schott
 19 pyrenaica DC.
 20 aragonensis Willk.
 21 discolor Levier & Leresche
 22 guarensis Losa
 23 cazorlensis Heywood
 24 bernardii Gren. & Godron
 25 nevadensis Boiss. & Reuter
 26 grata F. Maly ex Zimmeter
 27 alpina L.

23 Thalictrum L.
  1 aquilegifolium L.                      14      Po      Skalińska et al.,
                                                             1959
  2 calabricum Sprengel                    42      It/Si   Langlet, 1927
  3 uncinatum Rehman
  4 macrocarpum Gren.
  5 tuberosum L.
  6 orientale Boiss.
  7 alpinum L.                             14      Br      Hedberg &
                                                             Hedberg, 1961
  8 foetidum L.                            14      Au      Mattick, 1950
  9 minus L.
    (a) minus                              42      Hu      Baksay, 1956
    (b) majus (Crantz) Rouy & Fouc.
    (c) saxatile Schinz & Keller
    (d) pubescens (Schleicher ex DC.) Rouy
        & Fouc.
    (e) pseudominus (Borbás) Soó
    (f) kemense (Fries) Tutin
 10 simplex L.
    (a) simplex                            56      Su      Lövkvist, 1963
    (b) boreale (F. Nyl.) Tutin
    (c) gallicum (Rouy & Fouc.) Tutin
    (d) bauhinii (Crantz) Tutin
    (e) galioides (Nestler) Borza
 11 lucidum L.
 12 morisonii C.C. Gmelin
    (a) morisonii
    (b) mediterraneum (Jordan) P.W. Ball
 13 flavum L.
    (a) flavum                             84      Ge      Kuhn, 1928
    (b) glaucum (Desf.) Batt.
```

LXII PAEONIACEAE

1 Paeonia L.
  1 tenuifolia L.
  2 anomala L.
  3 peregrina Miller
  4 officinalis L.
    (a) officinalis
    (b) banatica (Rochel) Soó
    (c) humilis (Retz.) Cullen & Heywood
    (d) villosa (Huth) Cullen & Heywood
  5 clusii F.C. Stern   10   Cr   Stern, 1946
  6 broteroi Boiss. & Reuter   10   Lu   Stern, 1946
  7 mascula (L.) Miller
    (a) mascula
    (b) russii (Biv.) Cullen & Heywood
    (c) arietina (G. Anderson) Cullen &
        Heywood
  8 daurica Andrews
  9 coriacea Boiss.
 10 cambessedesii (Willk.) Willk.   10   Bl   Stern, 1946

LXIII BERBERIDACEAE

1 Leontice L.
  1 leontopetalum L.

2 Gymnospermium Spach
  1 altaicum (Pallas) Spach

3 Bongardia C.A. Meyer
  1 chrysogonum (L.) Griseb.

4 Epimedium L.
  1 alpinum L.
  2 pubigerum (DC.) Morren & Decne

5 Berberis L.
  1 vulgaris L.   28   Au   Mattick, 1950
  2 aetnensis C. Presl   28   Co   Contandriopoulos, 1962
  3 hispanica Boiss. & Reuter
  4 cretica L.

6 Mahonia Nutt.
  1 aquifolium (Pursh) Nutt.

LXIV MAGNOLIACEAE

1 Liriodendron L.
  1 tulipifera L.

LXV LAURACEAE

1 Laurus L.
  1 nobilis L.   42   It   Bambacioni-
                                                                  Mezzetti, 1936

  2 azorica (Seub.) J. Franco

2 Persea Miller
  1 indica (L.) Sprengel

LXVI PAPAVERACEAE

1 Papaver L.
  1 somniferum L.
    (a) somniferum
    (b) setigerum (DC.) Corb.
    (c) songaricum Basil.
  2 rhoeas L.   14   Su   Lövkvist, 1963
  3 dubium L.   42   Su   Lövkvist, 1963
  4 lecoqii Lamotte   28   Br   McNaughton, 1960
  5 laevigatum Bieb.
  6 arenarium Bieb.
  7 pinnatifidum Moris
  8 stipitatum Fedde
  9 maeoticum Klokov
 10 argemone L.   42   Su   Lövkvist, 1963
 11 nigrotinctum Fedde
 12 apulum Ten.
 13 hybridum L.   14   Br   McNaughton, 1960
 14 macrostomum Boiss. & Huet
 15 rupifragum Boiss. & Reuter
 16 orientale L.
 17 suaveolens Lapeyr.
    (a) suaveolens
    (b) endressii Ascherson
 18 rhaeticum Leresche   14   Ga   Fabergé, 1943
 19 sendtneri Kerner ex Hayek   14   Au   Fabergé, 1943
 20 burseri Crantz
 21 kerneri Hayek   14   Au   Fabergé, 1943
 22 corona-sancti-stephani Zapal.
 23 lapponicum (Tolm.) Nordh.   56   Scand. Knaben, 1958
 24 dahlianum Nordh.   70   Su   Knaben, 1958
 25 radicatum Rottb.   70   No   Knaben, 1958
 26 laestadianum (Nordh.) Nordh.   56   Scand. Knaben, 1959

2 Meconopsis Vig.
  1 cambrica (L.) Vig.   22   Br   Maude, 1940

3 Argemone L.
  1 mexicana L.

4 Roemeria Medicus
  1 hybrida (L.) DC.

5 Glaucium Miller
  1 flavum Crantz   12   Ga It   Larsen, 1954
  2 leiocarpum Boiss.
  3 corniculatum (L.) J.H. Rudolph

6 Chelidonium L.
  1 majus L.   12   Su   Lövkvist, 1963

7 Eschscholzia Cham.
  1 californica Cham.

8 Hypecoum L.
  1 procumbens L.

```
  2 imberbe Sibth. & Sm.
  3 pendulum L.
  4 ponticum Velen.

9 Dicentra Bernh.
  1 spectabilis (L.) Lemaire

10 Corydalis Vent.
  1 claviculata (L.) DC.
    (a) claviculata                       32      Ge      Reese, 1952
    (b) picta (Samp.) P. Silva & J. Franco
  2 capnoides (L.) Pers.
  3 lutea (L.) DC.
  4 acaulis (Wulfen) Pers.
  5 ochroleuca Koch
    (a) ochroleuca
    (b) leiosperma (Conr.) Hayek
  6 nobilis (L.) Pers.
  7 sempervirens (L.) Pers.
  8 rutifolia (Sibth. & Sm.) DC.
  9 bulbosa (L.) DC.
    (a) bulbosa
    (b) blanda (Schott) Chater
    (c) marschalliana (Pallas ) Chater
  10 paczoskii N. Busch
  11 angustifolia (Bieb.) DC.
  12 intermedia (L.) Mérat                 20      Au      Mattick, 1950
  13 pumila (Host) Reichenb.
  14 solida (L.) Swartz
    (a) solida
    (b) laxa (Fries) Nordst.
    (c) slivenensis (Velen.) Hayek
    (d) wettsteinii (Adamović) Hayek

11 Sarcocapnos DC.
  1 enneaphylla (L.) DC.                   c.32    Hs      Ryberg, 1960
  2 crassifolia (Desf.) DC.                c.32    Hs      Ryberg, 1960
  3 integrifolia (Boiss.) Cuatrec.
  4 baetica (Boiss. & Reuter) Nyman

12 Fumaria L.
  1 agraria Lag.
  2 occidentalis Pugsley
  3 rupestris Boiss. & Reuter
  4 bella P.D. Sell
  5 gaillardotii Boiss.
  6 judaica Boiss.
  7 amarysia Boiss. & Heldr.
  8 macrocarpa Parl.
  9 capreolata L.
    (a) capreolata                        56      Ga      Ryberg, 1960
    (b) babingtonii (Pugsley) P.D. Sell
  10 flabellata Gaspar.
  11 purpurea Pugsley
  12 macrosepala Boiss.
  13 bicolor Sommier ex Nicotra
  14 bastardii Boreau                      c.48    Lu      Ryberg, 1960
  15 martinii Clavaud
  16 sepium Boiss. & Reuter
  17 muralis Sonder ex Koch
    (a) muralis
```

```
    (b) boraei (Jordan) Pugsley
    (c) neglecta Pugsley
18 reuteri Boiss.
19 petteri Reichenb.
    (a) petteri
    (b) thuretii (Boiss.) Pugsley
20 transiens P.D. Sell
21 kralikii Jordan
22 mirabilis Pugsley
23 densiflora DC.
24 rostellata Knaf
25 officinalis L.
    (a) officinalis                       28        Ge       Wulff, 1937b
                                           32  Da Fe Ge Su   Vaarama, 1949
    (b) wirtgenii (Koch) Arcangeli         48        Br       Paker, 1965
26 ragusina (Pugsley) Pugsley
27 jankae Hausskn.
28 caroliana Pugsley
29 schleicheri Soyer-Willimet
30 microcarpa (Hausskn.) Pugsley
31 vaillantii Loisel. in Desv.            32        Su       Ryberg, 1960
32 schrammii (Ascherson) Velen.
33 parviflora Lam.

13 Platycapnos (DC.) Bernh.
   1 spicata (L.) Bernh.
    (a) spicata
    (b) echeandiae (Pau) Heywood
   2 saxicola Willk.

14 Rupicapnos Pomel
   1 africana (Lam.) Pomel

15 Ceratocapnos Durieu
   1 heterocarpa Durieu


LXVII CAPPARIDACEAE

1 Capparis L.
   1 spinosa L.
   2 ovata Desf.

2 Cleome L.
   1 ornithopodioides L.
   2 violacea L.


LXVIII CRUCIFERAE

1 Sisymbrium L.
   1 supinum L.                           42        Hs       Baez-Mayor, 1934
   2 strictissimum L.                     28        Su       Löve & Löve, 1961d
   3 irio L.                              42        Hs       Khoshoo, 1959
   4 loeselii L.
   5 assoanum Loscos & Pardo
   6 polymorphum (Murray) Roth
   7 volgense Bieb. ex E. Fourn.
   8 austriacum Jacq.
    (a) austriacum
    (b) chrysanthum (Jordan) Rouy & Fouc.
```

(c) contortum (Cav.) Rouy & Fouc.
(d) hispanicum (Jacq.) P.W. Ball &
    Heywood
9 crassifolium Cav.
10 laxiflorum Boiss.
11 arundanum Boiss.
12 altissimum L.   14   Su   Lövkvist, 1963
13 orientale L.   14   Hs   Baez-Mayor, 1934
14 polyceratium L.
15 confertum Steven ex Turcz.
16 runcinatum Lag. ex DC.   56   Bot. Gard. Manton, 1932
17 erysimoides Desf.
18 officinale (L.) Scop.   14   Su   Lövkvist, 1963
19 matritense P.W. Ball & Heywood   14   Hs   Baez-Mayor, 1934

2 Lycocarpus O.E. Schulz
  1 fugax (Lag.) O.E. Schulz

3 Murbeckiella Rothm.
  1 pinnatifida (Lam.) Rothm.   16   He   Favarger, 1949a
  2 boryi (Boiss.) Rothm.
  3 zanonii (Ball) Rothm.
  4 sousae Rothm.

4 Descurainia Webb & Berth.
  1 sophia (L.) Webb ex Prantl   28   Su   Lövkvist, 1963

5 Hugueninia Reichenb.
  1 tanacetifolia (L.) Reichenb.
    (a) tanacetifolia
    (b) suffruticosa (Coste & Soulié) P.W. Ball

6 Alliaria Scop.
  1 petiolata (Bieb.) Cavara & Grande   36   Su   Lövkvist, 1963
                                         42   Hu   Pólya, 1949

7 Eutrema R.Br.
  1 edwardsii R. Br.   42   Rs(N)   Sokolovskaya &
                                                        Strelkova, 1960

8 Arabidopsis (DC.) Heynh
  1 thaliana (L.) Heynh   10   Su   Böcher & Larsen,
                                                        1958
  2 suecica (Fries) Norrlin   26   Su   Hylander, 1957
  3 parvula (Schrenk) O.E. Schulz
  4 toxophylla (Bieb.) N. Busch   24   Rs(E)   Titova, 1935
  5 pumila (Stephan) N. Busch   32   Rs(E)   Titova, 1935

9 Thellungiella O.E. Schulz
  1 salsuginea (Pallas) O.E. Schulz

10 Braya Sternb. & Hoppe
   1 alpina Sternb. & Hoppe
   2 linearis Rouy   42   No   Jørgensen et al.,
                                                        1958
   3 purpurascens (R. Br.) Bunge   56   Is   Löve & Löve, 1956

11 Chrysochamela (Fenzl) Boiss.
   1 draboides Woronow

12 Sobolewskia Bieb.
   1 sibirica (Willd.) P.W. Ball

13 Myagrum L.
   1 perfoliatum L.

14 Isatis L.
   1 allionii P.W. Ball
   2 platyloba Link ex Steudel
   3 sabulosa Steven ex Ledeb.
   4 laevigata Trautv.
   5 costata C.A. Meyer
   6 praecox Kit. ex Tratt.
   7 tinctoria L.   28   Su   Lövkvist, 1963
   8 athoa Boiss.
   9 arenaria Aznav.
   10 lusitanica L.

15 Tauscheria Fischer ex DC.
   1 lasiocarpa Fischer ex DC.

16 Bunias L.
   1 erucago L.
   2 orientalis L.   14   Su   Lövkvist, 1963
   3 cochlearioides Murray

17 Goldbachia DC.
   1 laevigata DC.

18 Erysimum L.
   1 sylvestre (Crantz) Scop.
   2 comatum Pančić
   3 helveticum (Jacq.) DC.   56   He   Favarger, 1965c
   4 suffruticosum Sprengel
   5 grandiflorum Desf.   14   Ga   Favarger, 1965c
   6 raulinii Boiss.
   7 myriophyllum Lange
   8 linariifolium Tausch
   9 decumbens (Schleicher ex Willd.) Dennst.
   10 linifolium (Pers.) Gay
      (a) linifolium
      (b) baeticum Heywood
      (c) cazorlense Heywood
   11 pusillum Bory & Chaub.
      (a) pusillum
      (b) parnassii (Boiss. & Heldr.) Hayek
      (c) hayekii Jav. & Rech. fil.
   12 rechingeri Jav.
   13 mutabile Boiss. & Heldr.
   14 diffusum Ehrh.   c.32   Hu   Baksay, 1961
   15 olympicum Boiss.
   16 calycinum Griseb.
   17 graecum Boiss. & Heldr.
   18 smyrnaeum Boiss. & Balansa
   19 crepidifolium Reichenb.
   20 leptostylum DC.
   21 ucranicum Gay
   22 odoratum Ehrh.   24   Hu   Baksay, 1961
   23 witmannii Zawadzki
      (a) witmannii
      (b) transsilvanicum (Schur) P.W. Ball

```
   (c) pallidiflorum (Jáv.) Jáv.
24 carniolicum Dolliner
25 pectinatum Bory & Chaub.
26 pulchellum (Willd.) Gay
27 aureum Bieb.
28 leucanthemum (Stephan) B. Fedtsch.
29 hieracifolium L.                          32      Is      Löve & Löve, 1956
30 durum J. & C. Presl
31 virgatum Roth
32 wahlenbergii (Ascherson & Engler)
     Borbás
33 hungaricum Zapal.
34 pieninicum (Zapal.) Pawl.
35 cuspidatum (Bieb.) DC.
36 repandum L.                               16      Hu      Baksay, 1961
37 incanum G. Kunze
38 cheiranthoides L.
   (a) cheiranthoides                        16      Su      Löve & Löve, 1956
   (b) altum Ahti                            16      Fe      Ahti, 1962


19 Syrenia Andrz.
   1 cana (Piller & Mitterp.) Neilr.
   2 talijevii Klokov
   3 siliculosa (Bieb.) Andrz. ex C.A. Meyer
   4 montana (Pallas) Klokov


20 Hesperis L.
   1 tristis L.
   2 laciniata All.
   (a) laciniata
   (b) secundiflora (Boiss. & Spruner)
       Breistr.
   3 matronalis L.                           24      Su      Löve & Löve, 1956
   (a) matronalis
   (b) cladotricha (Borbás) Hayek
   (c) candida (Kit.) Hegi & E. Schmid
   (d) voronovii (N. Busch) P.W. Ball
   4 elata Hornem.
   5 sylvestris Crantz
   (a) sylvestris
   (b) velenovskyi (Fritsch) Borza
   6 theophrasti Borbás
   7 oblongifolia Schur
   8 steveniana DC.
   9 macedonica Adamović
  10 pycnotricha Borbás & Degen
  11 dinarica G. Beck
  12 vrabelyiana (Schur) Borbás               24      Hu      Baksay, 1957b
  13 nivea Baumg.
  14 inodora L.


21 Malcolmia R. Br.
   1 littorea (L.) R. Br.                     20      Lu      Rodrigues, 1953
   2 lacera (L.) DC.
   3 ramosissima (Desf.) Thell.               14      Lu      Rodrigues, 1953
   4 africana (L.) R. Br.
   5 taraxacifolia Balbis
   6 maritima (L.) R. Br.
   7 flexuosa (Sibth. & Sm.) Sibth. & Sm.     16      Lu      Rodrigues, 1953
```

8 graeca Boiss. & Spruner
9 illyrica Hayek
10 orsiniana (Ten.) Ten.
11 angulifolia Boiss. & Orph.
12 macrocalyx (Halácsy) Rech. fil.
  (a) macrocalyx
  (b) scyria (Rech. fil.) P.W. Ball
13 chia (L.) DC.
14 hydraea (Halácsy) Heldr. & Halácsy
15 bicolor Boiss. & Heldr.

22 Torularia (Cosson) O.E. Schulz
  1 torulosa (Desf.) O.E. Schulz
  2 contortuplicata (Stephan ex Willd.)
    O.E. Schulz

23 Maresia Pomel
  1 nana (DC.) Batt.

24 Leptaleum DC.
  1 filifolium (Willd.) DC.

25 Sterigmostemum Bieb.
  1 tomentosum (Willd.) Bieb.

26 Cheiranthus L.
  1 cheiri L.   14   It   Manton, 1932

27 Clausia Trotzky
  1 aprica (Stephan ex Willd.) Trotzky

28 Parrya R. Br.
  1 nudicaulis (L.) Boiss.

29 Matthiola R. Br.
  1 incana (L.) R. Br.
  (a) incana   14   Bot. Gard. Manton, 1932
  (b) rupestris (Rafin.) Nyman
  2 sinuata (L.) R. Br.
  3 odoratissima (Bieb.) R. Br.
  4 tatarica (Pallas) DC.
  5 fragrans (Fischer) Bunge
  6 fruticulosa (L.) Maire
  (a) fruticulosa
  (b) valesiaca (Gay ex Gaudin) P.W. Ball
  (c) perennis (P. Conti) P.W. Ball
  7 longipetala (Vent.) DC.
  (a) longipetala
  (b) bicornis (Sibth. & Sm.) P.W. Ball
  (c) pumilio (Sibth. & Sm.) P.W. Ball
  8 lunata DC.
  9 parviflora (Schousboe) R. Br.
  10 tricuspidata (L.) R. Br.

30 Notoceras R. Br.
  1 bicorne (Aiton) Amo

31 Tetracme Bunge
  1 quadricornis (Willd.) Bunge

32 Diptychocarpus Trautv.

1 strictus (Fischer) Trautv.

33 Chorispora R. Br. ex DC.
   1 tenella (Pallas) DC.

34 Euclidium R. Br.
   1 syriacum (L.) R. Br.

35 Litwinowia Woronow
   1 tenuissima (Pallas) Woronow ex Pavlov

36 Barbarea R. Br.
   1 vulgaris R. Br.   16   Au   Mattick, 1950
   2 stricta Andrz.
   3 verna (Miller) Ascherson
   4 intermedia Boreau
   5 sicula C. Presl
   6 bosniaca Murb.
   7 bracteosa Guss.
   8 rupicola Moris   16   Co   Contandriopoulos,
   9 conferta Boiss. & Heldr.   1957
   10 balcana Pančić
   11 longirostris Velen.

37 Sisymbrella Spach
   1 dentata (L.) O.E. Schulz
   2 aspera (L.) Spach
     (a) aspera
     (b) praeterita Heywood
     (c) boissieri (Cosson) Heywood
     (d) pseudoboissieri (Degen) Heywood

38 Rorippa Scop.
   1 austriaca (Crantz) Besser   16   Br   Howard, 1947
   2 amphibia (L.) Besser   16   Br   Howard, 1947
                                                32   Ge   Wulff, 1939a
   3 prostrata (J.P. Bergeret) Schinz & Thell.
   4 sylvestris (L.) Besser
     (a) sylvestris   48   Br   Howard, 1946
     (b) kerneri (Menyh.) Soó
   5 islandica (Oeder) Borbás   16   Is   Löve & Löve, 1956
                                                32   Br   Howard, 1947
   6 prolifera (Heuffel) Neilr.
   7 brachycarpa (C.A. Meyer) Hayek
   8 pyrenaica (Lam.) Reichenb.   16   Ga   Manton, 1932
   9 lippizensis (Wulfen) Reichenb.   32   Hu   Pólya, 1948
   10 thracica (Griseb.) Fritsch

39 Armoracia Gilib.
   1 rusticana P. Gaertner, B. Meyer &
     Scherb.   32   Su   Lövkvist, 1963
   2 macrocarpa (Waldst. & Kit.) Kit. ex
     Baumg.

40 Nasturtium R. Br.
   1 officinale R. Br.   32   Br   Howard, 1947
   2 microphyllum (Boenn.) Reichenb..   64   Br   Howard, 1947
                                                48

41 Cardamine L.
   1 bulbifera (L.) Crantz   96   Su   Lövkvist, 1963

2 heptaphylla (Vill.) O.E. Schulz	48	Ge	Schwarzenbach, 1922
3 pentaphyllos (L.) Crantz	48	Ge	Schwarzenbach, 1922
4 quinquefolia (Bieb.) Schmalh.			
5 kitaibelii Becherer	48	He	Schwarzenbach, 1922
6 enneaphyllos (L.) Crantz	52-54	Au	Mattick, 1950
	80	Po	Banach-Pogan, 1954
7 glanduligera O. Schwarz	48	Po	Banach-Pogan, 1954
8 waldsteinii Dyer			
9 macrophylla Willd.			
10 trifida (Lam. ex Poiret) B.M.G. Jones			
11 trifolia L.	16	Po	Banach-Pogan, 1955
12 asarifolia L.			
13 amara L.	16	Au	Mattick, 1950
	32	Au	Mattick, 1950
14 barbaraeoides Halácsy			
15 opizii J. & C. Presl	16	Po	Banach-Pogan, 1954
16 raphanifolia Pourret			
(a) raphanifolia	44		Lövkvist, 1956
	46	?Br	Ellis & Jones, 1970
(b) acris (Griseb.) O.E. Schulz			
17 crassifolia Pourret			
18 nymanii Gand.	60	Su	Lövkvist, 1957
	62		
	64	Su	Lövkvist, 1957
	68	Su	Lövkvist, 1956
	72 74	Su	Lövkvist, 1956
	80	No	Lövkvist, 1956
	90	Su	Lövkvist, 1957
19 matthiolii Moretti	16	Ga	Lövkvist, 1956
20 pratensis L.	28-34	Su	Lövkvist, 1956
	38-44	Scand.	Lövkvist, 1956
	48	Da	Lövkvist, 1956
21 rivularis Schur	16	Alps	Lövkvist, 1956
22 palustris (Wimmer & Grab.) Peterm.	56	Scand.	Lövkvist, 1957
	64	Scand.	Lövkvist, 1957
	72	Scand.	Lövkvist, 1957
	76	Scand.	Lövkvist, 1957
	80	Scand.	Lövkvist, 1957
	84	Da	Lövkvist, 1956
	c.96	Scand.	Lövkvist, 1957
23 tenera J.G. Gmelin ex C.A. Meyer			
24 plumieri Vill.			
25 glauca Sprengel			
26 resedifolia L.	16	Au	Mattick, 1950
27 bellidifolia L.			
(a) bellidifolia	16	Rs(N)	Sokolovskaya & Strelkova, 1960
(b) alpina (Willd.) B.M.G. Jones	16	Au	Mattick, 1950
28 maritima Portenschl. ex DC.			
29 carnosa Waldst. & Kit.			
30 graeca L.	16	?Ju	Ball, ined.
31 chelidonia L.			
32 parviflora L.			
33 impatiens L.	16	Po	Banach-Pogan, 1955
34 caldeirarum Guthnick			
35 flexuosa With.	32	Ho	Gadella & Kliphuis 1955
	c.50		
	16	Su	Lövkvist, 1963

42 Cardaminopsis (C.A. Meyer) Hayek

```
 1 arenosa (L.) Hayek                      16    Cz       Jones, 1963
                                           28    Au       Mattick, 1950
                                           32    Su       Hylander, 1957
 2 petraea (L.) Hiitonen                   16    Br       Jones, 1963
 3 neglecta (Schultes) Hayek
 4 croatica (Schott, Nyman & Kotschy) Jáv.
 5 halleri (L.) Hayek                      16    Po       Jones, 1963
   (a) halleri
   (b) ovirensis (Wulfen) Hegi & E. Schmid

43 Arabis L.
 1 glabra (L.) Bernh.                      12    Su       Lövkvist, 1963
                                           16    Au       Mattick, 1950
                                           32    Ge       Jaretsky, 1928a
 2 pseudoturritis Boiss. & Heldr.
 3 laxa Sibth. & Sm.
 4 pauciflora (Grimm) Garcke
 5 planisiliqua (Pers.) Reichenb.          16    Cz       Novotná, 1962
 6 sagittata (Bertol.) DC.                 16    Au       Titz, 1964
 7 borealis Andrz. ex Ledeb.
 8 hirsuta (L.) Scop.                      32    Br No Su  Jones, 1963
   brownii Jordan                          32    Hb       Jones, 1963
 9 allionii DC.
10 corymbiflora Vest.                      16    Au       Mattick, 1950
11 serpillifolia Vill.
   (a) serpillifolia
   (b) nivalis (Guss.) B.M.G. Jones
   (c) cretica (Boiss. & Heldr.) B.M.G.
       Jones
12 muralis Bertol.
13 collina Ten.
14 turrita L.                              16    Hu       Baksay, 1957b
15 mollis Steven
16 pendula L.
17 recta Vill.
18 nova Vill.
19 parvula Dufour in DC.
20 verna (L.) R. Br.
21 procurrens Waldst. & Kit.
22 vochinensis Sprengel
23 ferdinandi-coburgii J. Kellerer & Sund.
24 scopoliana Boiss.
25 bryoides Boiss.
26 longistyla Rech. fil.
27 stricta Hudson                          16    Br       Jones, 1963
28 subflava B.M.G. Jones
29 caerulea (All.) Haenke                  16    He       Favarger, 1953
30 pumila Jacq.
31 soyeri Reuter & Huet
   (a) soyeri
   (b) jacquinii (G. Beck) B.M.G. Jones    16    Au       Mattick, 1950
32 alpina L.                               16    Au Br Ga  Jones, 1963
                                           32    Au       Mattick, 1950
33 caucasica Schlecht.                     16    Gr Ju    Bordet, 1967
34 cebennensis DC.
35 pedemontana Boiss.

44 Aubrieta Adanson
 1 deltoidea (L.) DC.
 2 columnae Guss.
   (a) columnae
```

(b) italica (Boiss.) Mattf.
(c) croatica (Schott, Nyman & Kotschy) Mattf.
3 scyria Halácsy
4 erubescens Griseb.
5 intermedia Heldr. & Orph. ex Boiss.
6 gracilis Spruner ex Boiss.

45 Ricotia L.
1 cretica Boiss. & Heldr.
2 isatoides (W. Barbey) B.L. Burtt

46 Lunaria L.
1 rediviva L.   30   Au   Mattick, 1950
2 telekiana Jav.
3 annua L.
  (a) annua
  (b) pachyrhiza (Borbás) Hayek

47 Peltaria Jacq.
1 alliacea Jacq.
2 emarginata (Boiss.) Hausskn.

48 Alyssoides Miller
1 utriculata (L.) Medicus
2 cretica (L.) Medicus
3 sinuata (L.) Medicus

49 Degenia Hayek
1 velebetica (Degen) Hayek   16   Ju   Lovka et al.,
                                                           1971
50 Alyssum L.
1 corymbosum (Griseb.) Boiss.
2 leucadeum Guss.
3 petraeum Ard.
4 saxatile L.
  (a) saxatile   48
  (b) orientale (Ard.) Rech. fil.
  (c) megalocarpum (Hausskn.) Rech. fil.
5 linifolium Stephan ex Willd.
6 dasycarpum Stephan ex Willd.
7 alyssoides (L.) L.   32   Da   Bücher & Larsen,
                                                           1958
8 granatense Boiss. & Reuter
9 desertorum Stapf
10 foliosum Bory & Chaub.
11 smyrnaeum C.A. Meyer
12 minutum Schlecht. ex DC.
13 umbellatum Desv.   16   Gr   P.W. Ball, ined.
14 minus (L.) Rothm.   16   Ju   Bücher & Larsen,
                                                           1958
15 strigosum Banks & Solander
16 hirsutum Bieb.
17 rostratum Steven
18 repens Baumg.
  (a) repens   16   Bot. Gard. Manton, 1932
  (b) trichostachyum (Rupr.) Hayek
19 lenense Adams
20 fischeranum DC.
21 scardicum Wettst.
22 wulfenianum Bernh.
23 ovirense Kerner

24 wierzbickii Heuffel
25 calycocarpum Rupr.
26 pulvinare Velen.
27 montanum L.
  (a) montanum   16   Hu   Bücher &
                                                Larsen, 1958
  (b) gmelinii (Jordan) Hegi & E. Schmid
28 fastigiatum Heywood
29 diffusum Ten.
30 cuneifolium Ten.
31 arenarium Loisel.
32 atlanticum Desf.
33 moellendorfianum Ascherson ex G. Beck
34 stribrnyi Velen.
35 sphacioticum Boiss. & Heldr.
36 lassiticum Halácsy
37 idaeum Boiss. & Heldr.
38 handelii Hayek
39 doerfleri Degen
40 taygeteum Heldr.
41 densistellatum T.R. Dudley
42 murale Waldst. & Kit.
43 tenium Halácsy
44 argenteum All.
45 fallacinum Hausskn.
46 corsicum Duby   16   Co   Contandriopoulos, 1957
47 heldreichii Hausskn.
48 robertianum Bernard ex Gren. & Godron 16   Co   Contandriopoulos, 1957
49 smolikanum E.I. Nyárády
50 bertolonii Desv.
  (a) bertolonii
  (b) scutarinum E.I. Nyárády
51 markgrafii O.E. Schulz
52 corymbosoides Form.
53 euboeum Halácsy
54 sibiricum Willd.
55 caliacrae E.I. Nyárády
56 borzaeanum E.I. Nyárády
57 obtusifolium Steven ex DC.
  (a) obtusifolium
  (b) helioscopioides E.I. Nyárády
58 obovatum (C.A. Meyer) Turcz.
59 tortuosum Willd.
60 longistylum (Sommier & Levier) Grossh.
61 alpestre L.
62 serpyllifolium Desf.
63 nebrodense Tineo
64 fragillimum (Bald.) Rech. fil.

51 Fibigia Medicus
  1 clypeata (L.) Medicus
  2 lunarioides (Willd.) Sibth. & Sm.
  3 triquetra (DC.) Boiss. ex Prantl

52 Berteroa DC.
  1 obliqua (Sibth. & Sm.) DC.
  2 mutabilis (Vent.) DC.
  3 orbiculata DC.
  4 incana (L.) DC.   16   Ge   Wulff, 1939b
  5 gintlii Rohlena

53 Lepidotrichum Velen. & Bornm.
1 uechtritzianum (Bornm.) Velen.

54 Ptilotrichum C.A. Meyer
1 pyrenaicum (Lapeyr.) Boiss.
2 reverchonii Degen & Hervier
3 lapeyrousianum (Jordan) Jordan
4 halimifolium Boiss.
5 cyclocarpum Boiss.
6 longicaule (Boiss.) Boiss.
7 macrocarpum (DC.) Boiss.
8 spinosum (L.) Boiss.
9 purpureum (Lag. & Rodr.) Boiss.

55 Bornmuellera Hausskn.
1 dieckii Degen
2 baldaccii (Degen) Heywood
3 tymphaea (Hausskn.) Hausskn.

56 Lobularia Desv.
1 maritima (L.) Desv.   24   It   Larsen, 1955b
2 libyca (Viv.) Webb & Berth.

57 Clypeola L.
1 jonthlaspi L.
2 eriocarpa Cav.

58 Schivereckia Andrz.
1 podolica (Besser) Andrz.
2 doerfleri (Wettst.) Bornm.

59 Draba L.
1 aizoides L.   16   Ga He   Duckert &
                                                            Favarger, 1960
2 hoppeana Reichenb.
3 aspera Bertol.
4 scardica (Griseb.) Degen & Dörfler
5 cuspidata Bieb.
6 athoa (Griseb.) Boiss.   16   Gr   Duckert &
                                                            Favarger, 1960
7 lasiocarpa Rochel   16   Hu   Baksay, 1951

8 compacta Schott, Nyman & Kotschy
9 lacaitae Boiss.
10 haynaldii Stur
11 hispanica Boiss.
12 parnassica Boiss. & Heldr.
13 loiseleurii Boiss.   16   Co   Contandriopoulos, 1957
14 cretica Boiss. & Heldr.
15 dedeana Boiss. & Reuter
16 sauteri Hoppe
17 heterocoma Fenzl
18 alpina L.   62-64   No   Heilborn, 1927
                                           80   Sb Su   Flovik, 1940
19 gredinii Elis. Ekman
20 bellii Holm
21 kjellmanii Lid ex Elis. Ekman
22 oblongata R. Br. ex DC.
23 glacialis Adams
24 ladina Br.-Bl.
25 sibirica (Pallas) Thell.

26 stellata Jacq.
27 nivalis Liljeblad   16   No   Heilborn, 1927
28 norvegica Gunnerus   48   No Su   Heilborn, 1927
29 subcapitata Simmons
30 cacuminum Elis. Ekman   64   No   Knaben &
                                                             Engelskjön, 1967

31 carinthiaca Hoppe
32 dubia Suter
33 kotschyi Stur
34 tomentosa Clairv.
35 fladnizensis Wulfen   16   No   Heilborn, 1927
36 daurica DC.   64   Su   Heilborn, 1927
   norica   64   Au   Heilborn, 1941
37 cinerea Adams   48   Rs(N)   Heilborn, 1927
38 incana L.   32   No Su   Heilborn, 1927
39 muralis L.   32   Ge   Reese, 1952
40 nemorosa L.
41 lutescens Cosson
42 crassifolia R.C. Graham

60 Erophila DC.
   1 verna (L.) Chevall.   14-64   Da   Winge, 1940
     (a) verna
         simplex   14   Da   Winge, 1940
     (b) macrocarpa (Boiss. & Heldr.) Walters
     (c) praecox (Steven) Walters
     (d) spathulata (A.F. Lang) Walters
         duplex   30-40   Da   Winge, 1940
         quadruplex   52-64   Da   Winge, 1940
   2 minima C.A. Meyer

61 Petrocallis R. Br.
   1 pyrenaica (L.) R. Br.   14   He   Favarger, 1953

62 Andrzeiowskia Reichenb.
   1 cardamine Reichenb.

63 Cochlearia L.
   1 danica L.   42   Su   Lövkvist, 1963
   2 officinalis L.   24   Su   Saunte, 1955
   3 pyrenaica DC.   12   Ga   Rohner, 1954
                                            28
     alpina   12   Br   Gill, 1965
                                            28   ?Br   Crane &
                                                             Gairdner, 1923
   4 polonica Frohlich   36   Po   Bajer, 1950
   5 aestuaria (Lloyd) Heywood   12   Hs   Gill, 1965
   6 tatrae Borbis   42   Po   Bajer, 1950
   7 anglica L.   48   Da   Saunte, 1955
                                            54
   8 aragonensis Coste & Soulie
   9 fenestrata R. Br.
  10 groenlandica L.   14   Is   Löve & Löve, 1956
  11 scotica Druce   14   Br   Maude, 1940
  12 glastifolia L.

64 Kernera Medicus
   1 saxatilis (L.) Reichenb.   16   Au   Mattick, 1950
                                            32   Au   Mattick, 1950

65

65 Rhizobotrya Tausch
　1 alpina Tausch　　　　　　　　　　　　　14　　It　　Chiarugi, 1933

66 Camelina Crantz
　1 sativa (L.) Crantz　　　　　　　　　　40　　Is　　Löve & Löve, 1956
　2 microcarpa Andrz. ex DC.　　　　　　40　　Is　　Löve & Löve, 1956
　3 rumelica Velen.　　　　　　　　　　　12　　Hu　　Baksay, 1957b
　4 alyssum (Miller) Thell.　　　　　　　40　　Hu　　Baksay, 1961
　5 macrocarpa Wierzb. ex Reichenb.

67 Neslia Desv.
　1 paniculata (L.) Desv.
　　(a) paniculata　　　　　　　　　　　14　　Su　　Lövkvist, 1963
　　(b) thracica (Velen.) Bornm.

68 Capsella Medicus
　1 bursa-pastoris (L.) Medicus　　　　32　　Is　　Löve & Löve, 1956
　2 rubella Reuter
　3 thracica Velen.
　4 grandiflora (Fauche & Chaub.) Boiss.
　5 orientalis Klokov

69 Hutchinsia R. Br.
　1 alpina (L.) R. Br.
　　(a) alpina　　　　　　　　　　　　　12　　He　　Larsen, 1954a
　　(b) brevicaulis (Hoppe) Arcangeli　12　　Au　　Mattick, 1950
　　(c) auerswaldii (Willk.) Laínz

70 Hymenolobus Nutt. ex Torrey & A. Gray
　1 procumbens (L.) Nutt. ex Torrey &
　　Gray　　　　　　　　　　　　　　　12　　Si　　Manton, 1932
　　　　　　　　　　　　　　　　　　　24　　Si　　Manton, 1932
　2 pauciflorus (Koch) Schinz & Thell.

71 Hornungia Reichenb.
　1 petraea (L.) Reichenb.　　　　　　　12　　Su　　Lövkvist, 1963
　2 aragonensis (Loscos & Pardo) Heywood

72 Ionopsidium Reichenb.
　1 acaule (Desf.) Reichenb.　　　　　24　　Bot. Gard.　Chiarugi, 1928
　2 albiflorum Durieu
　3 prolongoi (Boiss.) Batt.
　4 abulense (Pau) Rothm.
　5 savianum (Caruel) Ball ex Arcangeli　32　　It　　Chiarugi, 1928

73 Teesdalia R. Br.
　1 nudicaulis (L.) R. Br.　　　　　　　36　　Su　　Lövkvist, 1963
　2 coronopifolia (J.P. Bergeret) Thell.

74 Thlaspi L.
　1 arvense L.　　　　　　　　　　　　14　　Su　　Lövkvist, 1963
　2 alliaceum L.
　3 perfoliatum L.　　　　　　　　　　70　　Su　　Lövkvist, 1963
　4 brachypetalum Jordan
　5 alpestre L.
　　(a) alpestre　　　　　　　　　　　14　　Su　　Lövkvist, 1963
　　(b) virens (Jordan) Hooker fil.
　6 stenopterum Boiss. & Reuter
　7 dacicum Heuffel
　　(a) dacicum
　　(b) banaticum (Uechtr.) Jáv.

```
  8 brevistylum (DC.) Jordan
  9 rivale J. & C. Presl
 10 graecum Jordan
 11 microphyllum Boiss. & Orph.
 12 stylosum (Ten.) Mutel
 13 bulbosum Spruner ex Boiss.
 14 praecox Wulfen
    (a) praecox
    (b) cuneifolium (Griseb.) Clapham
 15 jankae Kerner
 16 goesingense Halácsy
 17 epirotum Halácsy
 18 ochroleucum Boiss. & Heldr.
 19 montanum L.                          28      Hu       Baksay, 1961
 20 alpinum Crantz                     c.54      Au       Mattick, 1950
    (a) alpinum
    (b) sylvium (Gaudin) Clapham
 21 kerneri Huter
 22 avalanum Pančić                      14      Balkans  Manton, 1932
 23 macranthum N. Busch
 24 rotundiflolium (L.) Gaudin
    (a) rotundifolium                    14      Au       Mattick, 1950
    (b) cepaeifolium (Wulfen) Rouy & Fouc.
 25 bellidifolium Griseb.
 26 nevadense Boiss. & Reuter
```

```
75 Aethionema R. Br.
   1 orbiculatum (Boiss.) Hayek
   2 polygaloides DC.
   3 cordatum (Desf.) Boiss.
   4 iberideum (Boiss.) Boiss.
   5 arabicum (L.) Andrz. ex O.E. Schulz
   6 saxatile (L.) R. Br.                48      It       Larsen, 1955b
```

```
76 Teesdaliopsis (Willk.) Gand.
   1 conferta (Lag.) Rothm.
```

```
77 Iberis L.
   1 semperflorens L.                    22      Si       Manton, 1932
                                         44      Si       Manton, 1932
   2 sempervirens L.
   3 saxatilis L.
     (a) saxatilis
     (b) cinerea (Poiret) P.W. Ball &
         Heywood
   4 linifolia Loefl.
     (a) linifolia
     (b) welwitschii (Boiss.) Franco &
         P. Silva
   5 pruitii Tineo                       22  Bot. Gard.   Manton, 1932
   6 spathulata J.P. Bergeret
     (a) spathulata
     (b) nana (All.) Heywood
   7 aurosica Chaix
     (a) aurosica
     (b) cantabrica Franco & P. Silva
   8 procumbens Lange
     (a) procumbens
     (b) microcarpa Franco & P. Silva
   9 gibraltarica L.
  10 amara L.
```

67

      (a) amara
      (b) forestieri (Jordan) Heywood
  11 intermedia Guersent
      (a) intermedia
      (b) timeroyi (Jordan) Rouy & Fouc.
      (c) prostii (Soyer-Willemet ex Godron)
          Rouy & Fouc.
  12 stricta Jordan
      (a) stricta
      (b) leptophylla (Jordan) Franco &
          P. Silva
  13 umbellata L.
  14 ciliata All.
  15 fontqueri Pau
  16 pinnata L.
  17 crenata Lam..
  18 odorata L.
  19 sampaiana Franco & P. Silva

78 Biscutella L.
   1 laevigata L.
      (a) laevigata   36   Au   Manton, 1937
      (b) austriaca (Jordan) Mach.-Laur.   18   Au   Manton, 1937
      (c) kerneri Mach.-Laur.   18   Ce   Bresinskya &
      (d) gracilis Mach.-Laur.   Grau, 1970
      (e) tenuifolia (Bluff & Fingerh.) Mach.-
          Laur.   18   Ge   Manton, 1937
      (f) guestphalica Mach.-Laur.   18   Ge   Manton, 1937
      (g) varia (Dumort.) Rouy & Fouc.   18   ?error
      (h) subaphylla Mach.-Laur.
      (i) tirolensis (Mach.-Laur.) Heywood
      (j) angustifolia (Mach.-Laur.) Heywood
      (k) illyrica Mach.-Laur.
      (l) montenegrina Rohlena
      (m) lucida (DC.) Mach.-Laur.   18   It   Schönfelder, 1968
   2 scaposa Sennen ex Mach.-Laur.
   3 flexuosa Jordan
   4 variegata Boiss. & Reuter
   5 megacarpaea Boiss. & Reuter
   6 foliosa Mach.-Laur.
   7 gredensis Guinea
   8 neustriaca Bonnet
   9 frutescens Cosson
  10 vincentina (Samp.) Rothm. ex Guinea
  11 sempervirens L.   18   Hs   Schönfelder, 1968
  12 glacialis (Boiss. & Reuter) Jordan   18   Cult.   Manton, 1937
  13 brevifolia Rouy & Fouc.
  14 cuneata (Font Quer) Font Quer ex
      Mach.-Laur.
  15 rotgesii Fouc.   18   Co   Contandriopoulos, 1962
  16 lamottii Jordan   18   Ga   Manton, 1937
  17 sclerocarpa Revel
  18 arvernensis Jordan   18   Ga   Manton, 1937
  19 divionensis Jordan   18   Ga   Manton, 1937
  20 controversa Boreau   18   Ga   Manton, 1937
  21 brevicaulis Jordan
  22 coronopifolia L.   18   Ga   Manton, 1937
  23 intricata Jordan
  24 apricorum Jordan
  25 granitica Boreau ex Perard
  26 pinnatifida Jordan

```
27 polyclada Jordan
28 lusitanica Jordan                           54      Hs      Manton, 1937
29 mediterranea Jordan                         18      Ga      Manton, 1937
30 nicaeensis Jordan
31 intermedia Gouan
32 guillonii Jordan
33 valentina (L.) Heywood
34 microcarpa DC.
35 baetica Boiss. & Reuter
36 eriocarpa DC.
37 lyrata L.                                   16      Si      Manton, 1932
38 didyma L.                                   16      N. Afr. Reese, 1957
39 radicata Cosson
40 auriculata L.                               16      Hs      Schönfelder, 1968
41 cichoriifolia Loisel.

79 Megacarpaea DC.
   1 megalocarpa (Fischer ex DC.) Schischkin
     ex B. Fedstch.

80 Lepidium L.
   1 campestre (L.) R. Br.                      16      Is      Löve & Löve, 1956
   2 villarsii Gren. & Godron
     (a) villarsii
     (b) reverchonii (Debeaux) Breistr.
   3 heterophyllum Bentham                      16      Br      Manton, 1932
   4 hirtum (L.) Sm.
     (a) hirtum
     (b) nebrodense (Rafin.) Thell.
     (c) petrophilum (Cosson) Thell.
     (d) stylatum (Lag. & Rodr.) Thell.
     (e) oxyotum (DC.) Thell.
     (f) calycotrichum (Kunze) Thell.
   5 spinosum Ard.
   6 sativum L.
   7 subulatum L.
   8 cardamines L.
   9 virginicum L.
  10 bonariense L.
  11 densiflorum Schrader
  12 neglectum Thell.
  13 ruderale L.                                32      Su      Lövkvist, 1963
  14 divaricatum Solander
  15 pinnatifidum Ledeb.
  16 schinzii Thell.
  17 perfoliatum L.                             16      Is      Löve & Löve, 1956
  18 cartilagineum (J. Mayer) Thell.
     (a) cartilagineum
     (b) crassifolium (Waldst. & Kit.) Thell.16 Rm      Tarnavschi, 1938
  19 latifolium L.                              24      Is      Löve & Löve, 1956
  20 lyratum L.
     (a) lacerum (C.A.Meyer) Thell.
     (b) coronopifolium (Fischer) Thell.
  21 graminifolium L.
     (a) graminifolium
     (b) suffruticosum (L.) P. Monts.

81 Cardaria Desv.
   1 draba (L.) Desv.                           64      Su      Lövkvist, 1963

82 Coronopus Haller
```

```
 1 navasii Pau
 2 squamatus (Forskål) Ascherson
 3 didymus (L.) Sm.                        32        Br        Manton, 1932

83 Subularia L.
 1 aquatica L.                           c.36        Is        Löve & Löve , 1956

84 Conringia Adanson
 1 orientalis (L.) Dumort.                 14        Is        Löve & Löve, 1956
 2 austriaca (Jacq.) Sweet

85 Moricandia DC.
 1 arvensis (L.) DC.
 2 moricandioides (Boiss.) Heywood
 3 foetida Bourgeau ex Cosson

86 Euzomodendron Cosson
 1 bourgaeanum Cosson

87 Diplotaxis DC.
 1 catholica (L.) DC.
 2 siifolia G. Kunze
 3 crassifolia (Rafin.) DC.
 4 tenuifolia (L.) DC.                     22        Su        Lövkvist, 1963
 5 cretacea Kotov
 6 erucoides (L.) DC.                      14        Si        Manton, 1932
 7 virgata (Cav.) DC.
 8 viminea (L.) DC.
 9 muralis (L.) DC.                        44        Su        Lövkvist, 1963

88 Brassica L.
 1 elongata Ehrh.
   (a) elongata
   (b) integrifolia (Boiss.) Breistr.
 2 balearica Pers.
 3 macrocarpa Guss.
 4 oleracea L.
   (a) oleracea                           18        Ge        Netroufal, 1927
   (b) robertiana (Gay) Rouy & Fouc.       18        It        Netroufal, 1927
 5 rupestris Rafin.                        18        Si        Manton, 1932
 6 villosa Biv.
 7 incana Ten.
 8 insularis Moris                         18        Co        Contandriopoulos,
                                                               1962
 9 cretica Lam.
10 napus L.                                38        Po        Bijok, 1959
   subsp. rapifera                         38
11 rapa L.
   (a) sylvestris (L.) Janchen            20        Su        Lövkvist, 1963
   (b) rapa
12 juncea (L.) Czern.
13 fruticulosa Cyr.
   (a) fruticulosa
   (b) cossoniana (Boiss. & Reuter) Maire
14 barrelieri (L.) Janka
   (a) barrelieri
   (b) oxyrrhina (Cosson) P.W. Ball &
       Heywood
15 tournefortii Gouan
16 cadmea Heldr. ex O.E. Schulz
17 procumbens (Poiret) O.E. Schulz
```

```
18 souliei (Batt.) Batt.
19 nigra (L.) Koch                           16    Su    Löve & Löve, 1942
20 gravinae Ten.
21 repanda (Willd.) DC.
   (a) repanda
   (b) saxatilis (DC.) Heywood
   (c) galissieri (Giraud.) Heywood
   (d) maritima (Rouy) Heywood
   (e) confusa (Emberger & Maire) Heywood
   (f) cantabrica (Font Quer) Heywood
   (g) cadevallii (Font Quer) Heywood
   (h) blancoana (Boiss.) Heywood
   (i) latisiliqua (Boiss. & Reuter)
       Heywood
   (j) nudicaulis (Lag.) Heywood

89 Sinapis L.
   1 arvensis L.                             18    Su    Lövkvist, 1963
   2 pubescens L.
   3 alba L.
     (a) alba                                24    Is    Löve & Löve, 1956
     (b) dissecta (Lag.) Bonnier
   4 flexuosa Poiret

90 Eruca Miller
   1 vesicaria (L.) Cav.
     (a) vesicaria
     (b) sativa (Miller) Thell.              22    Hu    Baksay, 1961

91 Erucastrum C. Presl
   1 laevigatum (L.) O.E. Schulz
   2 virgatum (J. & C. Presl) C. Presl
   3 nasturtiifolium Poiret) O.E. Schulz
   4 palustre (Pirona) Vis.
   5 gallicum (Willd.) O.E. Schulz           30    Au    Manton, 1932

92 Rhynchosinapis Hayek
   1 richeri (Vill.) Heywood
   2 nivalis (Boiss. & Heldr.) Heywood
   3 cheiranthos (Vill.) Dandy               48    Br    Wright, 1936
     (a) cheiranthos
     (b) nevadensis (Willk.) Heywood
   4 wrightii (O.E. Schulz) Dandy ex
       Clapham                               24    Br    Wright, 1936
   5 monensis (L.) Dandy ex Clapham          24    Br    Sikka, 1940
   6 hispida (Cav.) Heywood
     (a) hispida
     (b) transtagana (Coutinho) Heywood
   7 granatensis (O.E. Schulz) Heywood
   8 longirostra (Boiss.) Heywood
   9 pseuderucastrum (Brot.) Franco
     (a) pseuderucastrum
     (b) setigera (Gay ex Lange) Heywood
     (c) cintrana (Coutinho) Franco & P.
         Silva
  10 johnstonii (Samp.) Heywood

93 Hirschfeldia Moench
   1 incana (L.) Lagreze-Fossat

94 Hutera Porta
```

```
       1 rupestris Porta
       2 leptocarpa González-Albo

  95 Carrichtera DC.
       1 annua (L.) DC.

  96 Vella L.
       1 pseudocytisus L.
       2 spinosa Boiss.

  97 Succowia Medicus
       1 balearica (L.) Medicus

  98 Boleum Desv.
       1 asperum (Pers.) Desv.

  99 Erucaria Gaertner
       1 hispanica (L.) Druce

 100 Cakile Miller
       1 maritima Scop.
         (a) maritima                          18      Lu      Rodrigues, 1953
         (b) baltica (Jordan ex Rouy & Fouc.)
             Hyl. ex P.W. Ball
         (c) aegyptiaca (Willd.) Nyman
         (d) euxina (Pobed.) E.I. Nyárády
       2 edentula (Bigelow) Hooker
         (a) edentula
         (b) islandica (Gand.) Á. & D. Löve     18      Is      Löve & Löve, 1956

 101 Rapistrum Crantz
       1 perenne (L.) All.
       2 rugosum (L.) All.
         (a) rugosum                            16     ?Su      Manton, 1932
         (b) orientale (L.) Arcangeli
         (c) linnaeanum Rouy & Fouc.

 102 Didesmus Desv.
       1 aegyptius (L.) Desv.

 103 Crambe L.
       1 maritima L.                            30      Rm      Tarnavschi, 1947
                                                60      Su      Lövkvist, 1963
       2 tataria Sebeók
       3 steveniana Rupr.
       4 grandiflora DC.
       5 aspera Bieb.
       6 koktebelica (Junge) N. Busch
       7 hispanica L.
       8 filiformis Jacq.

 104 Calepina Adanson
       1 irregularis (Asso) Thell.

 105 Morisia Gay
       1 monanthos (Viv.) Ascherson             14      Sa      Martinoli, 1955

 106 Guiraoa Cosson
       1 arvensis Cosson

 107 Enarthrocarpus Labill.
```

```
 1 arcuatus Labill.
 2 lyratus (Forskål) DC.

108 Raphanus L.
  1 raphanistrum L.
    (a) raphanistrum              18    Su    Lövkvist, 1963
    (b) microcarpus (Lange) Thell.
    (c) rostratus (DC.) Thell.
    (d) maritimus (Sm.) Thell.    18    Br    Manton, 1932
    (e) landra (Moretti ex DC.) Bonnier &
        Layens                    18    It    Manton, 1932


LXIX RESEDACEAE

 1 Reseda L.
    1 luteola L.                  28    Su    Lövkvist, 1963
    2 glauca L.
    3 gredensis (Cutanda & Willk.) Müller Arg.
    4 complicata Bory
    5 virgata Boiss. & Reuter
    6 alba L.
    7 undata L.
    8 decursiva Forskål
    9 suffruticosa Loefl.
   10 phyteuma L.
   11 media Lag.
   12 inodora Reichenb.
   13 tymphaea Hausskn.
   14 odorata L.
   15 orientalis (Müller Arg.) Boiss.
   16 stricta Pers.
   17 lutea L.
   18 lanceolata Lag.
   19 jacquinii Reichenb.
   20 arabica Boiss.

 2 Sesamoides Ortega
    1 pygmaea (Scheele) O. Kuntze
      (a) pygmaea
      (b) minor (Lange) Heywood
    2 canescens (L.) O. Kuntze
      (a) canescens
      (b) suffruticosa (Lange) Heywood


LXX SARRACENIACEAE

 1 Sarracenia L.
    1 purpurea L.


LXXI DROSERACEAE

 1 Aldrovanda L.
    1 vesiculosa L.

 2 Drosera L.
    1 rotundifolia L.             20    Lu    Fernandes, 1950
    2 anglica Hudson              40    Ge    Tischler, 1934
    3 intermedia Hayne            20    Ge    Behre, 1929
```

3 Drosophyllum Link
　1 lusitanicum (L.) Link

LXXII CRASSULACEAE

1 Crassula L.
　1 tillaea Lester-Garland
　2 aquatica (L.) Schönl.　　　　　42　　　Is　　Löve & Löve, 1956
　3 vaillantii (Willd.) Roth

2 Bryophyllum Salisb.
　1 pinnatum (Lam.) Oken

3 Umbilicus DC.
　1 parviflorus (Desf.) DC.
　2 chloranthus Heldr. & Sart. ex Boiss.
　3 erectus DC.
　4 rupestris (Salisb.) Dandy
　5 horizontalis (Guss.) DC.　　　24　　　It　　Uhl, 1948
　6 heylandianus Webb & Berth.

4 Pistorinia DC.
　1 hispanica (L.) DC.
　2 breviflora Boiss.

5 Mucizonia (DC.) A. Berger
　1 sedoides (DC.) D.A. Webb
　2 hispida (Lam.) A. Berger

6 Sempervivum L.
　1 wulfenii Hoppe ex Mert. & Koch　　36　　　He　　Zésiger, 1961
　2 grandiflorum Haw.　　　　　　　80　　Europe　Zésiger, 1961
　3 ciliosum Craib　　　　　　　　34　　　Bu　　Zésiger, 1961
　4 thompsonianum Wale
　5 octopodes Turrill　　　　　　34　　　Ju　　Zésiger, ined.
　6 zeleborii Schott　　　　　　　64　　Europe　Zésiger, 1961
　7 kindingeri Adamović
　8 leucanthum Pančić　　　　　　64　　Europe　Zésiger, 1961
　9 pittonii Schott　　　　　　　64　　Europe　Zésiger, 1961
10 arachnoideum L.
　(a) arachnoideum　　　　　　　32　　　Ga Hs　Zésiger, 1961
　(b) tomentosum (C.B. Lehm. &
　　　Schnittspahn) Schinz & Thell.　32　　　　　Zésiger, ined.
　　　　　　　　　　　　　　　　64　　　Hs　　Zésiger, 1961
11 montanum L.
　(a) montanum　　　　　　　　42　　　Ga　　Zésiger, 1961
　(b) burnatii Wettst. ex Hayek　42　　　Ga　　Zésiger, 1961
　(c) stiriacum Wettst. ex Hayek　64　　　Au　　Zésiger, 1961
12 dolomiticum Facch.　　　　　72　　　It　　Zésiger, ined.
13 cantabricum J.A. Huber
14 kosaninii Praeger
15 macedonicum Praeger　　　　　34　Bot. Gard.　Uhl, 1961
16 erythraeum Velen.　　　　　　34　　　Bu　　Zésiger, ined.
17 giuseppii Wale　　　　　　　72　　　Hs　　Zésiger, ined.
18 ballsii Wale　　　　　　　　34　　　Gr　　Zésiger, ined.
19 marmoreum Griseb.　　　　　　34　　　　　Zésiger, ined.
20 nevadense Wale　　　　　　　108　　　Hs　　Zésiger, ined.
21 andreanum Wale
22 tectorum L.　　　　　　　　36　　Alps　Zésiger, 1961
　　　　　　　　　　　　　　　　72　　Alps　Zésiger, 1961

```
23 calcareum Jordan                        38    Ga      Favarger &
                                                         Zésiger, ined.

 7 Jovibarba Opiz
   1 allionii (Jordan & Fourr.) D.A. Webb  38    Europe  Zésiger, 1961
   2 arenaria (Koch) Opiz                  38    Europe  Zésiger, 1961
   3 hirta (L.) Opiz
     (a) hirta                             38    Europe  Zésiger, 1961
     (b) glabrescens (Sabr.) Soó & Jáv.    38    Europe  Zésiger, 1961
   4 sobolifera (J. Sims) Opiz             38    Europe  Zésiger, 1961
   5 heuffelii (Schott) Á. & D. Löve       38    Europe  Zésiger, 1961

 8 Aeonium Webb & Berth.
   1 arboreum (L.) Webb & Berth.

 9 Aichryson Webb & Berth.
   1 villosum (Aiton) Webb & Berth.
   2 dichotomum (DC.) Webb & Berth.

10 Sedum L.
   1 praealtum A.DC.
   2 aizoon L.
   3 hybridum L.
   4 telephium L.                          24    Fe      Jalas &
                                                         Ronkko, 1960
                                           36    Ga      Delay, 1947
                                           48    Po      Banach-Pogan,
                                                         1958
     (a) telephium
     (b) fabaria (Koch) Kirschleger
     (c) maximum (L.) Krocker
     (d) ruprechtii Jalas
   5 ewersii Ledeb.
   6 anacampseros L.
   7 spurium Bieb.
   8 sediforme (Jacq.) Pau                 32    Lu      Rodrigues, 1953
   9 ochroleucum Chaix
     (a) ochroleucum
     (b) montanum (Song. & Perr.) D.A. Webb
  10 reflexum L.                          108    Su      Uhl, ined.
  11 forsteranum Sm.
  12 tenuifolium (Sibth. & Sm.) Strobl
  13 pruinatum Link ex Brot.
  14 acre L.                               40    Au      Uhl, 1961
                                           48    Su      Löve & Löve, 1944
                                           80    Cz      Uhl, ined.
  15 sartorianum Boiss.                    64    Gr      Uhl, ined.
     (a) sartorianum
     (b) stribrnyi (Velen.) D.A. Webb
     (c) hillebrandtii (Fenzl) D.A. Webb
     (d) ponticum (Velen.) D.A. Webb
  16 sexangulare L.                        74    Su      Uhl, 1961
  17 borissovae Balk.
  18 laconicum Boiss.
  19 alpestre Vill.                        16    He      Favarger, 1953
  20 flexuosum Wettst.
  21 kostovii Stefanov
  22 idaeum D.A. Webb
  23 tuberiferum Stoj. & Stefanov
  24 zollikoferi F. Hermann & Stefanov
  25 album L.                              68    He      Uhl, 1961
                                          136    Lu      Uhl, 1961

  26 serpentini Janchen
```

27 gypsicola Boiss. & Reuter
28 subulatum (C.A. Meyer) Boiss.
29 anglicum Hudson

(a) anglicum	120	Br	Uhl, 1961
	c.144	Da	Turesson, 1963
(b) pyrenaicum Lange	24-36	Lu	Uhl, 1961

(c) melanantherum (DC.) Maire
30 stefco Stefanov

31 dasyphyllum L.	28	N. Africa	Uhl, 1961
	42	Su	Uhl, 1961
	56	It	Larsen, 1955b

32 brevifolium DC.
33 monregalense Balbis
34 magellense Ten.
35 alsinefolium All.
36 hierapetrae Rech. fil.
37 tristriatum Boiss.
38 hirsutum All.
  (a) hirsutum
  (b) baeticum Rouy

39 villosum L.	30	No	Knaben, 1950

40 creticum Boiss.
41 cepaea L.
42 lagascae Pau
43 pedicellatum Boiss. & Reuter
44 arenarium Brot.

45 atratum L.	16	Ge	Mattick, 1950

  (a) atratum
  (b) carinthiacum (Hoppe ex Pacher)
      D.A. Webb
46 confertiflorum Boiss.
47 stellatum L.

48 annuum L.	22	Is	Löve & Löve, 1956

49 litoreum Guss.
50 aetnense Tineo
51 rubens L.
52 caespitosum (Cav.) DC.
53 andegavense (DC.) Desv.
54 nevadense Cosson

55 hispanicum L.	40	Tu	Uhl, pers. comm.

56 pallidum Bieb.
57 caeruleum L.

11 Rhodiola L.

1 rosea L.	22	Is	Löve & Löve, 1956

  2 quadrifida (Pallas) Fischer & C.A. Meyer

12 Orostachys (DC.) Fischer ex Sweet
  1 spinosa (L.) Sweet
  2 thyrsiflora (DC.) Fischer ex Sweet

13 Rosularia (DC.) Stapf
  1 serrata (L.) A. Berger

LXXIII SAXIFRAGACEAE

1 Saxifraga L.

1 hieracifolia Waldst. & Kit.	112	Sb	Flovik, 1940
	120	Ga	Hamel, 1953
2 nivalis L.	60	Is	Löve & Löve, 1951

```
 3 tenuis (Wahlenb.) H. Smith ex Lindman  20      Is      Löve & Löve, 1951
 4 clusii Gouan
   (a) clusii
   (b) lepismigena (Planellas) D.A. Webb
 5 stellaris L.                            28      Fa      Böcher, 1938b
   (a) stellaris
   (b) alpigena Temesy
 6 foliolosa R. Br.                        56      Is      Löve & Löve, 1951
 7 cuneifolia L.                           28      Ga      Hamel, 1953
 8 spathularis Brot.                       28      Hb      Packer, 1961
 9 umbrosa L.                              28      Ga      Hamel, 1953
10 hirsuta L.                              28      Hb      Packer, 1961
   (a) hirsuta
   (b) paucicrenata (Gillot) D.A. Webb
11 x polita (Haw.) Link                    28      Hb      Packer, 1961
12 x geum L.                               28
13 x urbium D.A. Webb                      28      Br      Packer, 1961
14 nelsoniana D. Don
15 rotundifolia L.
16 chrysosplenifolia Boiss.                22      Europe  K. Jones, ined.
17 taygetea Boiss. & Heldr.
18 aspera L.                             c.26      Ga      Hamel, 1953
19 bryoides L.                             26      He      Favarger, 1949a
20 bronchialis L.
21 flagellaris Sternb. & Willd.            32      Sb      Flovik, 1940
22 hirculus L.                             32      Su      Löve & Löve, 1951
23 hederacea L.
24 cymbalaria L.
   subsp. huetiana (Boiss.) Engler & Irmsch.18    Ga      Hamel, 1953
25 sibthorpii Boiss.
26 latepetiolata Willk.                  c.66      Hs      K. Jones, ined.
27 tridactylites L.                        22      Da      Skovsted, 1934
28 osloensis Knaben                        44      No      Knaben, 1954
29 adscendens L.
   (a) adscendens                         22      Ga      Hamel, 1953
   (b) parnassica (Boiss. & Heldr.) Hayek
   (c) blavii (Engler) Hayek
30 petraea L.
31 berica (Béguinot) D.A. Webb
32 irrigua Bieb.                           44      Rs( )   K. Jones, ined.
33 arachnoidea Sternb.                     56      It      Damboldt & Podlech,1963
34 paradoxa Sternb.
35 aizoides L.                             26      No      Skovsted, 1934
36 tenella Wulfen
37 glabella Bertol.
38 praetermissa D.A. Webb                  44      Ga      Hamel, 1953
39 aquatica Lapeyr.                        28      Ga      Hamel, 1953
                                           66      Ga      K. Jones, ined.
40 x capitata Lapeyr.
41 cuneata Willd.                          28      Hs      Hamel, 1953
42 pentadactylis Lapeyr.
43 corbariensis Timb.-Lagr.
   (a) corbariensis                     60-66     Ga      K. Jones, ined.
   (b) valentina (Willk.) D.A. Webb        64      Hs      K. Jones, ined.
44 camposii Boiss. & Reuter                64      Hs      K. Jones, ined.
   (a) camposii
   (b) leptophylla (Willk.) D.A. Webb
45 trifurcata Schrader                     28      Hs      Hamel, 1954
46 canaliculata Boiss. & Reuter ex Engler 52      Hs      K. Jones, ined.
47 geranioides L.                        c.52      Ga      K. Jones, ined.
48 moncayensis                             60      Hs      K. Jones, ined.
```

49 vayredana Luizet	c.64	Hs	K. Jones, ined.
50 nervosa Lapeyr.	34	Ga	K. Jones, ined.
51 pedemontana All			
(a) pedemontana			
(b) cymosa Engler			
(c) cervicornis (Viv.) Engler	26	Co	Hamel, 1954
	44	Co	K. Jones, ined.
(d) prostii (Sternb.) D.A. Webb	32	Ga	Hamel, 1954
52 wahlenbergii Ball			
53 androsacea L.	16	Ga	Hamel, 1954
	c.128	He	Favarger, 1957a
54 seguieri Sprengel			
55 italica D.A. Webb			
56 depressa Sternb.			
57 sedoides L.			
(a) sedoides			
(b) prenja			
58 aphylla Sternb.			
59 presolanensis Engler	16	It	Damboldt & Podlech, 1963
60 muscoides All.			
61 facchinii Koch			
62 moschata Wulfen	26	Au/He	Damboldt & Podlech, 1963
	52	Au	Damboldt, 1968b
63 hariotii Luizet & Soulie	34	Ga	K. Jones, ined.
64 exarata Vill.	20-22	Ga	K. Jones, ined.
65 cebennensis Rouy & Camus	26	Ga	Hamel, 1954
	32	Ga	K. Jones, ined.
66 pubescens Pourret			
(a) pubescens			
(b) iratiana (F.W. Schultz) Engler & Irmscher	26	Ga	Hamel, 1954
67 nevadensis Boiss.			
68 cespitosa L.	80	Sb	Flovik, 1940
69 hartii D.A. Webb	50	Hb	Packer, 1961
70 rosacea Moench			
(a) rosacea	56	Ga	K. Jones, ined.
	64	Is	Löve & Löve, 1951
(b) sponhemica (C.C. Gmelin) D.A. Webb	50	Ga	K. Jones, ined.
	52	Ga	Hamel, 1954
71 hypnoides L.	c.30	Hb	Packer, 1961
	48	Is	Löve & Löve, 1951
72 continentalis (Engler & Irmscher) D.A. Webb	52	Ga	Hamel, 1953
73 conifera Cosson & Durieu			
74 rigoi Porta			
75 globulifera Desf.			
76 reuterana Boiss.			
77 erioblasta Boiss. & Reuter	34	Hs	K. Jones, ined.
78 boissieri Engler	64	Hs	K. Jones, ined.
79 biternata Boiss.	66	Hs	K. Jones, ined.
80 gemmulosa Boiss.	c.64	Hs	K. Jones, ined.
81 haenseleri Boiss. & Reuter			
82 dichotoma Sternb.	32 + 2B	Hs	K. Jones, ined.
(a) dichotoma			
(b) albarracinensis (Pau) D.A. Webb			
83 carpetana Boiss. & Reuter	20	Hs	K. Jones, ined.
84 graeca Boiss.			
85 bulbifera L.	28	Cz	K. Jones, ined.
	32	Br	Whyte, 1930

```
86 granulata L.
   (a) granulata              46-60    Da      Skovsted, 1934
   (b) graniticola D.A. Webb
87 corsica (Duby) Gren. & Godron
   (a) corsica                52       Co      Contandriopoulos,
                                                 1962
                              62-66    Co      K. Jones, ined.
   (b) cossoniana (Boiss.) D.A. Webb  64-66  Hs  K. Jones, ined.
88 cintrana Kuzinsky ex Willk.
89 sibirica L.
90 carpathica Reichenb.
91 rivularis L.               52       Is      Löve & Löve, 1951
   hyperborea R. Br.          26       Sb      Flovik, 1940
92 cernua L.                  50       Alps    Chiarugi, 1935
                              64       Is      Löve & Löve, 1951
93 oppositifolia L.           26       Ga      Hamel, 1953
94 retusa Gouan               26       Ga      Hamel, 1953
   (a) retusa
   (b) augustana (Vacc.) D.A. Webb
95 biflora All.               26       Ga      Hamel, 1953
96 marginata Sternb.
97 scardica Griseb.
98 spruneri Boiss.
99 diapensioides Bellardi
100 tombeanensis Boiss. ex Engler
101 vandellii Sternb.         26       It      Damboldt & Podlech,1963
102 burserana L.              26       It      Damboldt & Podlech,1963
103 juniperifolia Adams
   (a) juniperifolia
   (b) sancta (Griseb.) D.A. Webb
104 ferdinandi-coburgi J. Kellerer & Sund.
105 aretioides Lapeyr.        26       Ga      Hamel, 1953
106 caesia L.                 26       Ga      Hamel, 1953
107 squarrosa Sieber
108 luteoviridis Schott & Kotschy
109 sempervivum C. Koch
110 grisebachii Degen & Dörfler
111 porophylla Bertol.
112 stribrnyi (Velen.) Podp.
113 media Gouan
114 valdensis DC.             28       Ga      Hamel, 1953
115 callosa Sm.
   (a) callosa                28       Ga      Hamel, 1953
   (b) catalaunica (Boiss.) D.A. Webb
116 cochlearis Reichenb.      28       Ga      Hamel, 1953
117 crustata Vest
118 cotyledon L.              28       Ga      Hamel, 1953
119 paniculata Miller         28       Ga      Hamel, 1953
120 hostii Tausch
   (a) hostii
   (b) rhaetica (Kerner) Br.-Bl.
121 mutata L.
   (a) mutata                 28       Ga      Hamel, 1953
   (b) demissa (Schott & Kotschy) D.A. Webb
122 longifolia Lapeyr.        28       Ga      Hamel, 1953
123 florulenta Moretti        28       Ga      Hamel, 1953

2 Bergenia Moench
  1 crassifolia (L.) Fritsch

3 Chrysosplenium L.
```

```
1 alternifolium L.                          48     Ga       Hamel, 1953
2 tetrandrum (N. Lund) Th. Fries            24     Europe   Løve, 1954a
3 oppositifolium L.                         42     Ga       Hamel, 1953
4 alpinum Schur
5 dubium Gay ex Ser.
```

LXXIV PARNASSIACEAE

```
1 Parnassia L.
  1 palustris L.
    (a) palustris                           18     Ga       Hamel, 1953
    (b) obtusiflora (Rupr.) D.A. Webb       36     Rs(N)    Sokolovskaya, 1958
```

LXXV HYDRANGEACEAE

```
1 Philadelphus L.
  1 coronarius L.

2 Deutzia Thunb.
  1 scabra Thunb.
```

LXXVI ESCALLONIACEAE

```
1 Escallonia Mutis ex L. fil.
  1 macrantha Hooker & Arnott
  2 rubra (Ruiz & Pavón) Pers.
```

LXXVII GROSSULARIACEAE

```
1 Ribes L.
  1 multiflorum Kit. ex Roemer & Schultes
  2 rubrum L.                               16     Is       Løve & Løve, 1956
  3 spicatum Robson                         16     Su       Løve & Løve, 1956
  4 petraeum Wulfen
  5 nigrum L.                               16     Su       Lövkvist, 1963
  6 uva-crispa L.                           16     Su       Lövkvist, 1963
  7 sardoum U. Martelli
  8 alpinum L.                              16     Su       Lövkvist, 1963
  9 orientale Desf.
```

LXXVIII PITTOSPORACEAE

```
1 Pittosporum Gaertner
  1 crassifolium Putterl.
  2 undulatum Vent.
```

LXXIX. PLATANACEAE

```
1 Platanus L.
  1 orientalis L.
  2 hybrida Brot.                           42     Europe   Winge, 1917
```

LXXX ROSACEAE

1 Sorbaria (Ser.) A. Braun
  1 sorbifolia (L.) A. Braun
  2 tomentosa (Lindley) Rehder

2 Physocarpus (Camb.) Maxim.
  1 opulifolius (L.) Maxim.

3 Spiraea L.
  1 salicifolia L.
  2 alba Duroi
  3 douglasii Hooker
  4 tomentosa L.
  5 decumbens Koch
    (a) decumbens
    (b) tomentosa (Poech) Dostál
  6 japonica L. fil.
  7 corymbosa Rafin.
  8 chamaedryfolia L.
  9 media Franz Schmidt   10   Hu   Baksay, 1957b
    (a) media
    (b) polonica (Blocki) Pawl.
 10 cana Waldst. & Kit.
 11 crenata L.
 12 x vanhouttei (Briot) Zabel
 13 hypericifolia L.
    (a) hypericifolia
    (b) obovata (Waldst. & Kit. ex Willd.)
       Dostál

4 Sibiraea Maxim.
  1 altaiensis (Laxm.) C.K. Schneider

5 Aruncus L.
  1 dioicus (Walter) Fernald

6 Filipendula Miller
  1 vulgaris Moench   14   Hu   Pólya, 1950
                            16   Su   Lövkvist, 1963
  2 ulmaria (L.) Maxim.   14   Ho   Gadella &
                                            Kliphuis, 1963
                            16   Ho   Gadella &
                                            Kliphuis, 1963
                            24   Su   Lovkvist, 1963

    (a) ulmaria
    (b) picbaueri (Podp.) Smejkal
    (c) denudata (J. & C. Presl) Hayek

7 Rhodotypos Siebold & Zucc.
  1 scandens (Thunb.) Makino

8 Kerria DC.
  1 japonica (L.) DC.

9 Rubus L.
  1 chamaemorus L.   56   Br   Heslop-Harrison,
                                        1953
  2 humulifolius C.A. Meyer
  3 arcticus L.   14   Rs(N)   Sokolovskaya &
                                        Strelkova, 1960
  4 saxatilis L.   28   Fe   Sorsa, 1963
  5 odoratus L.

6 spectabilis Pursh			
7 idaeus L.	14	Br	Heslop-Harrison, 1953
loganobaccus L.H. Bailey	42	Br	Thomas, 1940
8 sachalinensis Léveillé			
9 phoenicolasius Maxim.			
10 nessensis W. Hall	28	Su	Heslop-Harrison, 1953
11 scissus W.C.R. Watson	28	Br	Heslop-Harrison, 1953
12 plicatus Weihe & Nees	28	Br	Heslop-Harrison, 1953
bertramii G. Braun ex Focke	28	Br	Maude, 1939
opacus Focke	28	Br	Datta, 1932
13 sulcatus Vest ex Tratt.	28	Scand.	Gustafsson, 1943
14 divaricatus P.J. Mueller	21	Scand.	Gustafsson, 1943
15 affinis Weihe & Nees	28	Br	Heslop-Harrison, 1953
16 lentiginosus Lees	28	Europe	Maude, 1939
17 vulgaris Weihe & Nees	21	Br	Heslop-Harrison, 1953
selmeri Lindeb. ex F. Aresch.	28	Scand.	Gustafsson, 1939
incurvatus Bab.	28	Br	Heslop-Harrison, 1953
18 pedatifolius Genev.	28	Scand.	Gustafsson, 1939
19 gratus Focke	28	Br	Heslop-Harrison, 1953
sciocharis (Sudre) W.C.R. Watson	28	Br	Heslop-Harrison, 1953
20 chaerophyllus Sagorski & W. Schultze			
21 hypomalacus Focke			
22 arrhenii (Lange) Lange	28	Br	Heslop-Harrison, 1953
23 sprengelii Weihe	28	Br	Heslop-Harrison, 1953
24 myricae Focke			
25 chlorothyrsos Focke	28	Br	Heslop-Harrison, 1953
axillaris Lej.	28	Europe	Maude, 1939
26 questieri P.J. Mueller & Lefèvre	28	Br	Maude, 1939
mercicus Bagnall	28	Br	Heslop-Harrison, 1953
scheutzii Lindeb. ex F. Aresch.	28	Scand.	Gustafsson, 1943
27 laciniatus Willd.	28	Br	Heslop-Harrison, 1953
28 rhombifolius Weihe ex Boenn.			
alterniflorus P.J. Mueller & Lefèvre	28	Br	Heslop-Harrison, 1953
silesiacus Weihe	28	Br	Heslop-Harrison, 1953
29 pyramidalis Kaltenb.	28	Br	Heslop-Harrison, 1953
dumnoniensis Bab.	28	Br	Heslop-Harrison, 1953
30 macrophyllus Weihe & Nees	28	Br	Heslop-Harrison, 1953
schlechtendalii Weihe ex Link	28	Europe	Maude, 1939
31 silvaticus Weihe & Nees			
32 egregius Focke	28	Br	Heslop-Harrison, 1953
lindleianus Lees	28	Br	Datta, 1932

33 villicaulis Koehler ex Weihe & Nees	28	Br	Datta, 1932
34 polyanthemus Lindeb.	28	Br	Heslop-Harrison, 1953
35 rhamnifolius Weihe & Nees			
cardiophyllus P.J. Mueller & Lefèvre	28	Br	Heslop-Harrison, 1953
36 hochstetterorum Seub.			
37 lindebergii P.J. Mueller	28	Br	Heslop-Harrison, 1953
38 ulmifolius Schott	14	Br	Heslop-Harrison, 1953
39 godronii Lecoq & Lamotte			
winteri P.J. Mueller ex Focke	28	Europe	Maude, 1939
40 bifrons Vest ex Tratt.			
cuspidifer P.J. Mueller & Lefèvre	28	Br	Heslop-Harrison, 1953
41 discolor Weihe & Nees			
42 chloocladus W.C.R. Watson			
43 candicans Weihe ex Reichenb.	21	Scand.	Gustafsson, 1943
44 canescens DC.			
45 vestitus Weihe & Nees			
podophyllos P.J. Mueller	28	Br	Heslop-Harrison, 1953
46 boraeanus Genev.	28	Br	Heslop-Harrison, 1953
47 adscitus Genev.			
leucostachys Schleicher ex Sm.	28	Br	Heslop-Harrison, 1953
macrothyrsus Lange	28	Br	Heslop-Harrison, 1953
48 mucronulatus Boreau			
49 gremlii Focke			
schmidelyanus Sudre	35	Br	Heslop-Harrison, 1953
50 radula Weihe ex Boenn.	28	Br	Heslop-Harrison, 1953
51 genevieri Boreau			
echinatus Lindley	28	Europe	Maude, 1939
52 apiculatus Weihe & Nees	28	Br	Heslop-Harrison, 1953
53 fuscus Weihe & Nees			
granulatus P.J. Mueller & Lefèvre	28	Br	Heslop-Harrison, 1953
54 foliosus Weihe & Nees			
insericatus P.J. Mueller ex Wirtgen	28	Br	Heslop-Harrison, 1953
55 infestus Weihe ex Boenn.	28	Br	Heslop-Harrison, 1953
56 thyrsiflorus Weihe & Nees	28	Br	Heslop-Harrison, 1953
57 pallidus Weihe & Nees	28	Br	Heslop-Harrison, 1953
bloxamii Lees	28	Br	Datta, 1932
58 obscurus Kaltenb.	28	Br	Heslop-Harrison, 1953
59 menkei Weihe & Nees			
60 rudis Weihe & Nees			
61 melanoxylon P.J. Mueller & Wirtgen ex Genev.			
62 vallisparsus Sudre	28	Br	Heslop-Harrison, 1953

63	fuscater Weihe & Nees	28	Europe	Maude, 1939
	adornatus P.J. Mueller ex Wirtgen	28	Br	Heslop-Harrison, 1953
	hartmanii Gand. ex Sudre	35	Br	Heslop-Harrison, 1953
64	pilocarpus Gremli			
65	lejeunei Weihe & Nees	35	Br	Heslop-Harrison, 1953
66	rosaceus Weihe & Nees	28	Br	Heslop-Harrison, 1953
67	histrix Weihe & Nees	28	Br	Heslop-Harrison, 1953
	rufescens P.J. Mueller & Lefevre	28	Br	Heslop-Harrison, 1953
68	koehleri Weihe & Nees	28	Br	Maude, 1939
	apricus Wimmer	28	Br	Heslop-Harrison, 1953
	dasyphyllus (Rogers) Rogers	28	Scand.	Gustafsson, 1943
69	incanescens Bertol.			
70	scaber Weihe & Nees	28	Br	Heslop-Harrison, 1953
71	schleicheri Weihe ex Tratt.	28	Br	Maude, 1939
72	glandulosus Bellardi	28	Br	Heslop-Harrison, 1953
73	serpens Weihe ex Lej. & Court.	28	Scand.	Gustafsson, 1943
	angustifrons Sudre	28	Br	Heslop-Harrison, 1953
74	hirtus Waldst. & Kit.	28	Br	Heslop-Harrison, 1953
75	caesius L.	28	Br	Heslop-Harrison, 1953
10	Rosa L.			
1	sempervirens L.			
2	arvensis Hudson			
3	phoenicia Boiss.			
4	moschata J. Herrmann			
5	pimpinellifolia L.	28	Br	Blackburn & H.-Harrison, 1921
6	foetida J. Herrmann			
7	acicularis Lindley			
8	majalis J. Herrmann	14	Su	Täckholm, 1922
9	glauca Pourret			
10	rugosa Thunb.	14	Br	Blackburn & H.-Harrison, 1921
11	blanda Aiton			
12	pendulina L.			
13	virginiana J. Herrmann			
14	gallica L.			
15	stylosa Desv.			
16	jundzillii Besser			
17	montana Chaix			
18	canina L.	35	Br	Blackburn & H.-Harrison, 1921
19	squarrosa (Rau) Boreau			
20	andegavensis Bast.			
21	nitidula Besser			
22	pouzinii Tratt.			
23	vosagiaca Desportes			
24	subcanina (Christ) Dalla Torre & Sarnth.			
25	caesia Sm.	35		

```
26 subcolina (Christ) Dalla Torre & Sarnth.
27 rhaetica Gremli
28 obtusifolia Desv.
29 corymbifera Borkh.                          35    Br    Blackburn & H.-
30 deseglisei Boreau                                        Harrison, 1921
31 abietina Gren. ex Christ
32 tomentosa Sm.
33 scabriuscula Sm.
34 sherardii Davies                            28    Br    Blackburn & H.-
                                                            Harrison, 1921
35 villosa L.
36 mollis Sm.                                  28    Br    Blackburn & H.-
                                                            Harrison, 1921
37 heckeliana Tratt.
38 orientalis Dupont ex Ser.
39 rubiginosa L.                               35    Br    Blackburn & H.-
                                                            Harrison, 1921
40 elliptica Tausch
41 agrestis Savi
42 micrantha Borrer ex Sm.
43 caryophyllaceae Besser
44 sicula Tratt.
45 glutinosa Sibth. & Sm.
46 turcica Rouy
47 serafinii Viv.


11 Agrimonia L.
   1 pilosa Ledeb.
   2 eupatoria L.
     (a) eupatoria                             28    Hu    Pólya, 1950
     (b) grandis (Andrz. ex Ascherson &
         Graebner) Bornm.
     (c) asiatica (Juz.) Skalický
   3 procera Wallr.
   4 repens L.


12 Aremonia Nestler
   1 agrimonoides (L.) DC.
     (a) agrimonoides
     (b) pouzarii Skalický


13 Sanguisorba L.
   1 officinalis L.                            28    Su    Nordborg, 1958
                                               56    Su    Nordborg, 1958
   2 albanica Andrasovszky & Jáv.
   3 dodecandra Moretti                        56    It    Nordborg, 1966
   4 hybrida (L.) Nordborg                     56    Hs Lu Nordborg, 1966
   5 ancistroides (Desf.) Cesati               28    Hs    Nordborg, 1967
   6 minor Scop.
     (a) minor                                 28    Su    Nordborg, 1958
     (b) lasiocarpa (Boiss. & Hausskn.)
         Nordborg                              56    Tu    Nordborg, 1967
     (c) lateriflora (Cosson) M.C.F. Proctor
     (d) magnolii (Spach) Briq.                28  Ga Hs It Lu Nordborg, 1967
     (e) muricata Briq.                        28    Su    Erdtman &
                                                            Nordborg, 1961
                                               56    Su    Erdtman &
                                                            Nordborg, 1961
     (f) rupicola (Boiss. & Reuter)
         Nordborg                              28    Hs    Nordborg, 1967
                                               56    Hs    Nordborg, 1967
```

7 cretica Hayek	28	Cr	Nordborg, 1967

14 Sarcopoterium Spach

1 spinosum (L.) Spach	28	Si	Larsen & Laeggard, 1971

15 Acaena Mutis ex L.
  1 anserinifolia (J.R. & G. Forster) Druce

16 Dryas L.

1 octopetala L.	18	Po	Skalińska et al., 1959
	36	He	Böcher & Larsen, 1955

17 Geum L.

1 reptans L.	42	Po	Gajewski, 1957
2 montanum L.	28	Po	Gajewski, 1957
3 bulgaricum Pančić	56	Bu	Gajewski, 1957
4 heterocarpum Boiss.			
5 rivale L.	42	Is Po	Gajewski, 1957
6 sylvaticum Pourret	42	Lu	Gajewski, 1957
7 coccineum Sibth. & Sm.	42	Bu	Gajewski, 1957
8 pyrenaicum Miller			
9 urbanum L.			
10 aleppicum Jacq.	42	Europe	Gajewski, 1957
11 molle Vis. & Pančić	42	Bu	Gajewski, 1957
12 macrophyllum Willd.			
13 hispidum Fries	42	Hs Su	Gajewski, 1957

18 Waldsteinia Willd.

1 geoides Willd.	14	Hu	Pólya, 1949
2 ternata (Stephan) Fritsch			

19 Potentilla L.

1 fruticosa L.	14	Ga	Elkington & Woodell, 1963
	28	Br	Elkington & Woodell, 1963
2 bifurca L.			
3 palustris (L.) Scop.	28	Su	Lövkvist, 1963
	35	Su	Lövkvist, 1963
	42	Su	Lövkvist, 1963
	62-64	Rs(N)	Sokolovskaya & Strelkova, 1960
4 anserina L.			
(a) anserina	28	Su	Lövkvist, 1963
	35	Su	Lövkvist, 1963
	42	Su	Lövkvist, 1963
(b) egedii (Wormsk.) Hiitonen	28	No	Rousi, 1965a
5 rupestris L.	14	Ga	Gagnieu et al., 1959
6 geoides Bieb.			
7 pulchella R. Br.	28	Sb	Flovik, 1940
8 rubricaulis Lehm.			
9 multifida L.			
10 eversmanniana Fischer ex Ledeb.			
11 pensylvanica L.			
12 longifolia Willd.			
13 pimpinelloides L.			
14 visianii Pančić			
15 nivea L.	56	No	Knaben & Engelskjön, 1967

16 chamissonis Hultén	77	Su	Hultén, 1945
17 hookerana Lehm.			
18 argentea L.	14	Su	Muntzing, 1931b
19 neglecta Baumg.	42	Su	A. & G. Muntzing,
20 calabra Ten.			1941
21 inclinata Vill.	42	Ga	Gagnieu et al.,
			1959
22 collina Wibel	42	Su	Muntzing, 1931b
23 supina L.	28	Hu	Pólya, 1949
24 norvegica L.			
25 intermedia L.			
26 astracanica Jacq.			
27 detommasii Ten			
28 recta L.	42	Ga	Gagnieu et al.,
			1959
29 hirta L.			
30 nevadensis Boiss.			
31 grandiflora L.			
32 montenegrina Pant.			
33 umbrosa Steven ex Bieb.			
34 delphinensis Gren. & Godron			
35 pyrenaica Ramond ex DC.			
36 stipularis L.			
37 longipes Ledeb.			
38 chrysantha Trev.			
39 thuringiaca Bernh. ex Link			
40 brauniana Hoppe	14	Au	Mattick, 1950
41 frigida Vill.	28	He	Favarger, 1965b
42 hyparctica Malte			
43 crantzii (Crantz) G. Beck ex Fritsch	28	Cz	Czapik, 1961
	42	Fe	Sorsa, 1963
	48	Su	Muntzing, 1958
	64	Br	Smith, 1963
44 aurea L.			
(a) aurea	14	Au Cz	Walters, 1967
(b) chrysocraspeda (Lehm.) Nyman			
45 heptaphylla L.	14	Po	Skalińska, 1950
	28	Cz	Walters, 1967
	30	Cz	Walters, 1967
	35	Cz	Walters, 1967
46 humifusa Willd.			
47 patula Waldst. & Kit.	42	Ju	Walters, 1967
48 australis Krašan	42	Cz	Walters, 1967
49 tabernaemontani Ascherson	42	Ho	Gadella &
			Kliphuis, 1968
	49	Su	Muntzing, 1958
	50	Su	Muntzing, 1958
	56	Su	Muntzing, 1958
	c.60	Su	Muntzing, 1958
	63	Su	Muntzing, 1958
	70	Br	Smith, 1963
	c.84	Su	Muntzing, 1931b
50 pusilla Host			
51 cinerea Chaix ex Vill.	14	Ga	Guinochet, 1964
	28	Hs	Küpfer, 1968
52 erecta (L.) Rauschel	28	Po	Czapik, 1961
53 anglica Laicharding	56	Su	Lövkvist, 1963
54 reptans L.	28	Ho	Gadella &
			Kliphuis, 1963
55 caulescens L.			

56 petrophila Boiss.
57 clusiana Jacq.   42   It   Contandriopoulos, 1962
58 crassinervia Viv.   14   Co   Contandriopoulos, 1962
59 nitida L.   42   Ga   Favarger, 1962
60 alchemilloides Lapeyr.
61 valderia L.   14   Ga   Contandriopoulos, 1962
62 haynaldiana Janka   14   Bu   Tombal, 1969
63 doerfleri Wettst.
64 nivalis Lapeyr.
65 grammopetala Moretti
66 speciosa Willd.
67 apennina Ten.
   (a) apennina
   (b) stoianovii Urum. & Jáv.
68 kionaea Halácsy
69 deorum Boiss. & Heldr.
70 alba L.   28   Po   Czapik, 1961
71 montana Brot.
72 sterilis (L.) Garcke   28   Ge   Wulff, 1938
73 micrantha Ramond ex DC.   12   Ga   Gagnieu et al.,
                                                                   1959
                                               14   Ga   Gagnieu et al.,
                                                                   1959
74 carniolica A. Kerner
75 saxifraga Ardoino ex De Not.   14   Ga   Contandriopoulos, 1962

20 Sibbaldia L.
   1 procumbens L.   14   Fe   Sorsa, 1963
   2 parviflora Willd.

21 Fragaria L.
   1 vesca L.   14   Ho   Gadella &
                                                                    Kliphuis, 1967a
   2 moschata Duchesne
   3 viridis Duchesne
     (a) viridis   14   Po   Skalińska et al.,
                                                                    1959
     (b) campestris (Steven) Pawl.
   4 virginiana Duchesne
   5 x ananassa Duchesne   56   cult.   Staudt, 1962

22 Duchesnea Sm.
   1 indica (Andrews) Focke

23 Alchemilla L.
   1 pentaphyllea L.   64   Ga   Strasburger, 1904
   2 saxatilis Buser
   3 transiens (Buser) Buser
   4 alpina L.   c.120   Au   Turesson, 1958
                                               c.128   Is   Löve & Löve, 1956
                                               c.140   Su   Lövkvist, 1963
                                               c.152   Fa   Bozman & Walters
   5 basaltica Buser
   6 subsericea Reuter
   7 plicatula Gand.
   8 pallens Buser
   9 anisiaca Wettst.
  10 hoppeana (Reichenb.) Dalla Torre
  11 conjuncta Bab.
  12 grossidens Buser   64   Ga   Strasburger, 1904

13 faeroensis (Lange) Buser	c.220	Is	Walters & Bozman, 1967
	c.224	Is	Löve & Löve, 1956
14 paicheana (Buser) Rothm.			
15 splendens Christ ex Favrat	c.163	He	Walters & Bozman, 1967
16 infravalesiaca (Buser) Rothm.			
17 schmidelyana Buser			
18 jaquetiana Buser			
19 fulgens Buser	c.140	Ga Hs	Walters & Bozman, 1967
20 kerneri Rothm.			
21 glaucescens Wallr.	103-110	Su	Turesson, 1957
22 hirsuticaulis H. Lindb.			
23 flabellata Buser			
24 cinerea Buser			
25 erythropoda Juz.			
26 lithophila Juz.			
27 bulgarica Rothm.			
28 lapeyrousii Buser			
29 plicata Buser			
30 colorata Buser			
31 illyrica Rothm.			
32 exigua Buser ex Paulin			
33 helvetica Brugger			
34 vetteri Buser			
35 hebescens Juz.			
36 gibberulosa H. Lindb.			
37 bungei Juz.			
38 hirsutissima Juz.			
39 propinqua H. Lindb.			
40 conglobata H. Lindb.			
41 monticola Opiz	c.101	Su	Ehrenberg, 1945
	103-109	He No Su	Turesson, 1957
42 schistophylla Juz.			
43 crinita Buser			
44 strigosula Buser			
45 subglobosa C.G. Westerlund	c.102	Su	Turesson, 1957
	108	Su	Turesson, 1957
46 sarmatica Juz.	105-106	Su	Turesson, 1957
47 tytthantha Juz.			
48 subcrenata Buser	96	Is	Löve & Löve, 1956
	104-110	He Su	Turesson, 1957
49 cymatophylla Juz.	106-107	Su	Turesson, 1957
50 heptagona Juz.			
51 acutiloba Opiz	c.100	Su	Ehrenberg, 1945
	105-109	Su	Lövkvist, 1963
52 nemoralis Alechin			
53 gracilis Opiz	c.93	Su	Ehrenberg, 1945
	104-110	Su	Turesson, 1957
54 xanthochlora Rothm.	c.105	Su	Turesson, 1957
55 heterophylla Rothm.			
56 curtiloba Buser			
57 filicaulis Buser			
(a) filicaulis Buser	c.96	Is	Löve & Löve, 1956
	103-110	Su	Lövkvist, 1963
	c.150	Br	Bradshaw, 1963
(b) vestita (Buser) M.E. Bradshaw	c.96	Is	Löve & Löve, 1956
	104-110	Br	Bradshaw, 1963
58 minima Walters	103-108	Br	Bradshaw, 1964
59 compta Buser			

60 rhododendrophila Buser
61 polonica Pawl.
62 decumbens Buser
63 undulata Buser
64 rubristipula Buser
65 tirolensis Buser
66 tatricola Pawl.
67 heteropoda Buser
68 tenuis Buser

69 glomerulans Buser	101-109	No Su	Turesson, 1957
	c.144	Is	Löve & Löve, ined.
borealis Sam. ex Juz.	130-152	No Su	Turesson, 1957

70 camptopoda Juz.
71 crebridens Juz.
72 buschii Juz.
73 controversa Buser
74 wallischii Pawl.
75 connivens Buser
76 baltica Sam. ex Juz.

77 wichurae (Buser) Stéfansson	103-106	No Su	Turesson, 1957
78 murbeckiana Buser	102-109	Su	Turesson, 1957

79 acutidens Buser
80 reniformis Buser
81 lineata Buser

82 obtusa Buser	c.103	Su	Turesson, 1957
83 glabra Neygenf.	96	Is	Löve & Löve, 1956
	c.100	Su	Ehrenberg, 1945
	102-110 Fa	No Su	Turesson, 1957

84 glabricaulis H. Lindb.
85 versipila Buser
86 coriacea Buser
87 inconcinna Buser
88 stanislaae Pawl.
89 trunciloba Buser
90 sinuata Buser
91 straminea Buser
92 aequidens Pawl.
93 demissa Buser
94 pseudincisa Pawl.
95 longana Buser
96 achtarowii Pawl.
97 junrukczalica Pawl.
98 mollis (Buser) Rothm.
99 catachnoa Rothm.
100 viridiflora Rothm.
101 albanica Rothm.
102 phegophila Juz.
103 indivisa (Buser) Rothm.
104 heterotricha Rothm.
105 peristerica Pawl.
106 aroanica (Buser) Rothm.
107 zapalowiczii Pawl.
108 haraldii Juz.
109 fallax Buser
110 sericoneura Buser
111 flexicaulis Buser
112 gorcensis Pawl.
113 cuspidens Buser
114 othmarii Buser
115 oculimarina Pawl.
116 pyrenaica Dufour

117 incisa Buser
118 fissa Günther & Schummel

24 Aphanes L.
  1 arvensis L.   48   Su   Hjelmquist, 1959
  2 microcarpa (Boiss. & Reuter) Rothm. 16   Lu Su   Hjelmquist, 1959
  3 cornucopioides Lag.
  4 floribunda (Murb.) Rothm.

25 Cydonia Miller
  1 oblonga Miller   34   It   Bagnoli et al.,
                                                       1963

26 Pyrus L.
  1 cordata Desv.
  2 magyarica Terpó
  3 rossica Danilov
  4 pyraster Burgsd.
  5 caucasica Fedorov
  6 bourgaeana Decne
  7 syriaca Boiss.
  8 amygdaliformis Vill.
  9 salvifolia DC.
  10 elaeagrifolia Pallas
  11 nivalis Jacq.
  12 austriaca A. Kerner
  13 communis L.

27 Malus Miller
  1 trilobata (Labill.) C.K. Schneider
  2 florentina (Zuccagni) C.K. Schneider
  3 sylvestris Miller
  4 dasyphylla Borkh.
  5 praecox (Pallas) Borkh.
  6 domestica Borkh.

28 Sorbus L.
  1 domestica L.
  2 aucuparia L.   34   Fe   V. Sorsa, 1962
    (a) aucuparia
    (b) glabrata (Wimmer & Grab.) Cajander
    (c) sibirica (Hedl.) Krylov
    (d) praemorsa (Guss.) Nyman
    (e) fenenkiana Georgiev & Stoj.
  3 torminalis (L.) Crantz   34   Hu   Baksay, 1956
  4 chamaemespilus (L.) Crantz   34   Au   Liljefors, 1953
  5 aria (L.) Crantz
    leptophylla E.F. Warburg   68   Br   Warburg, 1952
  6 graeca (Spach) Kotschy   34   Hu   Baksay, 1956
    eminens E.F. Warburg   68   Br   Warburg, 1952
    porrigentiformis E.F. Warburg   51   Br   Warburg, 1952
                                         68   Br   Warburg, 1952
  7 rupicola (Syme) Hedl.   68   Europe Liljefors, 1953
  8 umbellata (Desf.) Fritsch
    (a) umbellata
    (b) banatica (Jáv.) Kárpáti
    (c) flabellifolia (Spach) Kárpáti
    (d) koevessii (Penzes) Kárpáti
  9 minima (A. Ley) Hedl.
  10 mougeotii Soyer-Willemet & Godron
  11 austriaca (G. Beck) Hedl.

    (a) austriaca
    (b) hazslinszkyana (Soo) Karpati
    (c) croatica Karpati
    (d) serpentini Karpati
        anglica Hedl.   68   Br   Warburg, 1952
 12 dacica Borbas
 13 hybrida L.
 14 meinichii (Lindeb.) Hedl.
        teodori Liljefors   51   Su   Liljefors, 1953
 15 intermedia (Ehrh.) Pers.   68   Su   Liljefors, 1955
 16 latifolia (Lam.) Pers.
        bristoliensis Wilmott   51
        semiincisa Borbas   34   Hu   Baksay, 1956
 17 sudetica (Tausch) Fritsch
 18 margittaiana (Jav.) Karpati

29 Eriobotrya Lindley
    1 japonica (Thunb.) Lindley

30 Amelanchier Medicus
    1 ovalis Medicus
    2 spicata (Lam.) C. Koch   68   Su   Lövkvist, 1963
    3 grandiflora Rehder

31 Cotoneaster Medicus
    1 horizontalis Decne
    2 simonsii Baker
    3 integerrimus Medicus
    4 cinnabarinus Juz.
    5 nebrodensis (Guss.) C. Koch
    6 acuminatus Lindley
    7 niger (Thunb.) Fries
    8 granatensis Boiss.
    9 nummularia Fischer & C.A. Meyer
   10 tauricus Pojark.
   11 microphyllus Wallich ex Lindley

32 Pyracantha M.J. Roemer
    1 coccinea M.J. Roemer

33 Mespilus L.
    1 germanica L.

34 Crataegus L.
    1 intricata Lange
    2 crus-galli L.
    3 sanguinea Pallas
    4 altaica (Loudon) Lange
    5 laevigata (Poiret) DC.
     (a) laevigata
     (b) palmstruchii (Lindman) Franco
    6 macrocarpa Hegetschw.
    7 taurica Pojark.
    8 ucrainica Pojark.
    9 calycina Peterm.
     (a) calycina
     (b) curvisepala (Lindman) Franco
   10 microphylla C. Koch
   11 pallasii Griseb.
   12 karadaghensis Pojark.
   13 ambigua C.A. Meyer ex A. Becker

14 monogyna Jacq.
   (a) nordica Franco   34   Su   Lövkvist, 1963
   (b) monogyna
   (c) leiomonogyna (Klokov) Franco
   (d) brevispina (G. Kunze) Franco
   (e) aegeica (Pojark.) Franco
   (f) azarella (Griseb.) Franco
15 sphaenophylla Pojark.
16 pentagyna Waldst. & Kit.
17 nigra Waldst. & Kit.
18 schraderana Ledeb.
19 heldreichii Boiss.
20 laciniata Ucria
   (a) laciniata
   (b) pojarkovae (Kossych) Franco
21 azarolus L.
22 pycnoloba Boiss. & Heldr.

35 Prunus L.
   1 persica (L.) Batsch
   2 dulcis (Miller) D.A. Webb
   3 webbii (Spach) Vierh.
   4 tenella Batsch
   5 armeniaca L.
   6 brigantina Vill.
   7 cerasifera Ehrh.
   8 spinosa L.   32   Su   Lövkvist, 1963
   9 ramburii Boiss.
  10 domestica L.
     (a) domestica
     (b) insititia (L.) C.K. Schneider
  11 cocomilia Ten.
  12 prostrata Labill.
  13 fruticosa Pallas
  14 avium L.   16   Su   Lövkvist, 1963
  15 cerasus L.
  16 mahaleb L.
  17 padus L.   32   Su   Lövkvist, 1963
     (a) padus
     (b) borealis Cajander   32   No   Knaben &
                                                                 Engelskjon, 1967

  18 serotina Ehrh.
  19 virginiana L.
  20 lusitanica L.
     (a) lusitanica
     (b) azorica (Mouillefert) Franco
  21 laurocerasus L.

LXXXI LEGUMINOSAE

1 Cercis L.
  1 siliquastrum L.

2 Ceratonia L.
  1 siliqua L.

3 Gleditsia L.
  1 triacanthos L.

4 Acacia Miller

```
  1 farnesiana (L.) Willd.
  2 karoo Hayne
  3 dealbata Link
  4 mearnsii De Wild.
  5 longifolia (Andrews) Willd.
  6 cyclops A. Cunn. ex G. Don fil.
  7 melanoxylon R. Br.
  8 pycnantha Bentham
  9 cyanophylla Lindley
 10 retinodes Schlecht.

 5 Sophora L.
  1 japonica L.
  2 alopecuroides L.
  3 jaubertii Spach

 6 Thermopsis R. Br.
  1 lanceolata R. Br.

 7 Anagyris L.
  1 foetida L.

 8 Laburnum Fabr.
  1 anagyroides Medicus
  2 alpinum (Miller) Berchtold & J. Presl

 9 Podocytisus Boiss. & Heldr.
  1 caramanicus Boiss. & Heldr.

10 Calicotome Link
  1 spinosa (L.) Link
  2 villosa (Poiret) Link                    48      It      Larsen, 1956a

11 Lembotropis Griseb.
  1 nigricans (L.) Griseb.
   (a) nigricans
   (b) australis (Freyn ex Wohlf.) J. Holub

12 Cytisus L.
  1 sessilifolius L.
  2 aeolicus Guss. ex Lindley
  3 villosus Pourret
  4 emeriflorus Reichenb.
  5 purgans (L.) Boiss.
  6 multiflorus (L'Hér.) Sweet
  7 ardoini E. Fourn.
  8 sauzeanus Burnat & Briq.
  9 procumbens (Waldst. & Kit. ex Willd.)
     Sprengel
 10 decumbens (Durande) Spach
 11 agnipilus Velen.
 12 pseudoprocumbens Markgraf
 13 commutatus (Willk.) Briq.
 14 ingramii Blakelock
 15 malacitanus Boiss.
   (a) malacitanus
   (b) catalaunicus (Webb) Heywood
 16 baeticus (Webb) Steudel
 17 striatus (Hill) Rothm.
 18 grandiflorus DC.
 19 cantabricus (Willk.) Reichenb. fil.
```

94

20 scoparius (L.) Link
   (a) scoparius   46   Po   Skalińska et al.,
                                                   1964
   (b) maritimus (Rouy) Heywood   46   Ga   Böcher & Larsen,
21 reverchonii (Degen & Hervier) Bean   1958b
22 patens L.
23 tribracteolatus Webb

13 Chamaecytisus Link
  1 creticus (Boiss. & Heldr.) Rothm.
  2 subidaeus (Gand.) Rothm.
  3 spinescens (C. Presl) Rothm.
  4 purpureus (Scop.) Link
  5 hirsutus (L.) Link
  6 polytrichus (Bieb.) Rothm.
  7 wulfii (V. Krecz.) A. Klásková
  8 ciliatus (Wahlenb.) Rothm.
  9 glaber (L. fil.) Rothm.
 10 leiocarpus (A. Kerner) Rothm.
 11 borysthenicus (Gruner) A. Klásková
 12 lindemannii (V. Krecz.) A. Klásková
 13 ratisbonensis (Schaeffer) Rothm.
 14 zingeri (Nenukow ex Litv.) A.
    Klaskova
 15 paczoskii (V. Krecz.) A. Klásková
 16 ruthenicus (Fischer ex Woloszczak)
    A. Klásková
 17 graniticus (Rehmann) Rothm.
 18 skrobiszewskii (Pacz.) A. Klásková
 19 supinus (L.) Link
 20 eriocarpus (Boiss.) Rothm.
 21 austriacus (L.) Link
 22 tommasinii (Vis.) Rothm.
 23 heuffelii (Wierzb.) Rothm.
 24 pygmaeus (Willd.) Rothm.
 25 jankae (Velen.) Rothm.
 26 danubialis (Velen.) Rothm.
 27 dorycnioides (Davidov) Frodin &
    Heywood
 28 litwinowii (V. Krecz.) A. Klásková
 29 blockianus (Pawl.) A. Klásková
 30 albus (Hacq.) Rothm.
 31 banaticus (Griseb. & Schenk) Rothm.
 32 rochelii (Wierzb.) Rothm.
 33 podolicus (Blocki) A. Klásková
 34 kovacevii (Velen.) Rothm.
 35 nejceffii (Urum.) Rothm.

14 Chronanthus (DC.) C. Koch
  1 biflorus (Desf.) Frodin & Heywood

15 Teline Medicus
  1 monspessulana (L.) C. Koch
  2 linifolia (L.) Webb & Berth.

16 Genista L.
  1 tinctoria L.   48   Ho   Gadella &
                                                 Kliphuis, 1966
  2 januensis Viv.
  3 lydia Boiss.

```
 4 ramosissima (Desf.) Poiret
 5 cinerea (Vill.) DC.
   (a) cinerea
   (b) leptoclada (Willk.) O. Bolós &
       Molinier
 6 florida L.
 7 valentina (Willd. ex Sprengel) Steudel
 8 obtusiramea Gay ex Spach
 9 pseudopilosa Cosson
10 teretifolia Willk.
11 sakellariadis Boiss. & Orph.
12 sericea Wulfen
13 subcapitata Pančić
14 pulchella Vis.
15 albida Willd.
16 pilosa L.
17 lobelii DC.
18 pumila (Debeaux & Reverchon ex Hervier)
   Vierh.
19 baetica Spach
20 salzmannii DC.
21 hystrix Lange
   (a) hystrix
   (b) legionensis (Pau) P. Gibbs
22 polyanthos R. de Roemer ex Willk.
23 aspalathoides Lam.
24 scorpius (L.) DC.              40    Hs    Lorenzo-Andreu &
                                              Garcia 1950
                                48    Ga    Favarger &
25 corsica (Loisel.) DC.                       Contandriopoulos, 1961
26 carpetana Leresche ex Lange
27 morisii Colla
28 anglica L.                     42    Ho    Gadella &
                                              Kliphuis, 1966
29 falcata Brot.
30 berberidea Lange
31 micrantha Ortega
32 carinalis Griseb.
33 sylvestris Scop.
34 aristata C. Presl
35 hispanica L.
   (a) hispanica
   (b) occidentalis Rouy
36 germanica L.                 c.46    Su    Lövkvist, 1963
37 hirsuta Vahl
38 tournefortii Spach
39 lucida Camb.
40 anatolica Boiss.
41 triacanthos Brot.
42 tridens (Cav.) DC.
43 cupanii Guss.
44 radiata (L.) Scop.
45 holopetala (Fleischm. ex Koch) Bald.
46 hassertiana (Bald.) Bald. ex
   Buchegger
47 dorycnifolia Font Quer           48    Hs    Santos, 1945
48 sessilifolia DC.
49 nissaba Petrović
50 ephedroides DC.
51 aetnensis (Biv.) DC.
52 spartioides Spach
```

53 haenseleri Boiss.
54 acanthoclada DC.
55 fasselata Decne
56 umbellata (L'Hér.) Poiret

17 Chamaespartium Adanson
  1 sagittale (L.) P. Gibbs
  2 tridentatum (L.) P. Gibbs

18 Echinospartum (Spach) Rothm.
  1 horridum (Vahl) Rothm.   44   Hs   Castro, 1945
  2 boissieri (Spach) Rothm.   44   Hs   Castro, 1945
  3 lusitanicum (L.) Rothm.
    (a) lusitanicum   c.52   Lu   Castro, 1945
    (b) barnadesii (Graells) C. Vicioso   c.52   Hs   Castro, 1945

19 Gonocytisus Spach
  1 angulatus (L.) Spach

20 Lygos Adanson
  1 sphaerocarpa (L.) Heywood
  2 monosperma (L.) Heywood
  3 raetam (Forskål) Heywood

21 Spartium L.
  1 junceum L.

22 Petteria C. Presl
  1 ramentacea (Sieber) C. Presl

23 Erinacea Adanson
  1 anthyllis Link

24 Ulex L.
  1 europaeus L.
    (a) europaeus   96   Su   Lövkvist, 1963
    (b) latebracteatus (Mariz) Rothm.   64   Lu   Castro, 1943
  2 minor Roth   32   Lu   Castro, 1943
  3 gallii   80   Hs   Castro, 1943
  4 densus Welw. ex Webb   64   Lu   Castro, 1945
  5 parviflorus Pourret   32   Lu   Castro, 1943
                                                    64   Lu   Castro, 1945
                                                    96   Lu   Castro, 1945

    (a) parviflorus
    (b) jussiaei (Webb) D.A. Webb
    (c) funkii (Webb) Guinea
    (d) eriocladus (C. Vicioso) D.A. Webb
  6 micranthus Lange   32   Hs   Castro, 1945
  7 argenteus Welw. ex Webb   64   Lu   Castro, 1941
                                                    96   Lu   Castro, 1943

    (a) argenteus
    (b) subsericeus (Coutinho) Rothm.
    (c) erinaceus (Welw. ex Webb) D.A. Webb

25 Stauracanthus Link
  1 boivinii (Webb) Samp.   c.128   Lu   Castro, 1941
  2 genistoides (Brot.) Samp.   48   Lu   Castro, 1943
    (a) genistoides
    (b) spectabilis (Webb) Rothm.

26 Adenocarpus DC.

1 complicatus (L.) Gay
  (a) complicatus
  (b) aureus (Cav.) C. Vicioso
  (c) commutatus (Guss.) Coutinho
2 telonensis (Loisel.) DC.
3 hispanicus (Lam.) DC.
  (a) hispanicus
  (b) argyrophyllus Rivas Goday
4 decorticans Boiss.

27 Lotononis (DC.) Ecklon & Zeyher
  1 lupinifolia (Boiss.) Bentham
  2 genistoides (Fenzl) Bentham

28 Lupinus L.
  1 luteus L.
  2 hispanicus Boiss. & Reuter   52   Lu   Malheiros, 1942
  3 angustifolius L.
    (a) angustifolius
    (b) reticulatus (Desv.) Coutinho
  4 micranthus Guss.
  5 albus L.
    (a) albus
    (b) graecus (Boiss. & Spruner) Franco
       & P. Silva
  6 varius L.
    (a) varius   32   Lu   Malheiros, 1942
    (b) orientalis Franco & P. Silva
  7 perennis L.
  8 nootkatensis Donn ex Sims   48   Br   Maude, 1940
  9 polyphyllus Lindley
10 arboreus Sims

29 Argyrolobium Ecklon & Zeyher
  1 zanonii (Turra) P.W. Ball   48   Ju   Larsen, 1956a
  2 biebersteinii P.W. Ball

30 Robinia L.
  1 pseudacacia L.
  2 hispida L.

31 Wisteria Nutt.
  1 sinensis (Sims) Sweet

32 Galega L.
  1 officinalis L.
  2 orientalis Lam.

33 Colutea L.
  1 arborescens L.
    (a) arborescens
    (b) gallica Browicz
  2 atlantica Browicz
  3 cilicica Boiss. & Balansa
  4 orientalis Miller

34 Eremosparton Fischer & C.A. Meyer
  1 aphyllum (Pallas) Fischer & C.A. Meyer

35 Halimodendron Fischer ex DC.
  1 halodendron (Pallas) Voss

36 Caragana Fabr.
 1 arborescens Lam.
 2 frutex (L.) C. Koch
 3 grandiflora (Bieb.) DC.

37 Calophaca Fischer
 1 wolgarica (L. fil.) Fischer

38 Astragalus L.
 1 contortuplicatus L.
 2 boeticus L.
 3 striatellus Pallas ex Bieb.
 4 arpilobus Kar. & Kir.
 5 reticulatus Bieb.
 6 ankylotus Fischer & C.A. Meyer
 7 stalinskyi Sirj.
 8 scorpioides Pourret ex Willd.
 9 longidentatus Chater
 10 peregrinus Vahl
 11 pamphylicus Boiss.
 12 haarbachii Spruner ex Boiss.
 13 suberosus Banks & Solander
 14 sinaicus Boiss.
 15 tribuloides Delile
 16 polyactinus Boiss.
 17 stella Gouan
 18 sesameus L.
 19 oxyglottis Steven ex Bieb.
 20 echinatus Murray
 21 epiglottis L.
   (a) epiglottis
   (b) asperulus (Dufour) Nyman
 22 edulis Durieu ex Bunge
 23 cymbicarpos Brot.
 24 hamosus L.
 25 cicer L.
 26 glaux L.

27 danicus Retz.	16	Ga	Guinochet & Logeois, 1962

 28 purpureus Lam.
 29 bourgaeanus Cosson
 30 pseudopurpureus Gusuleac
 31 setosulus Gontsch.
 32 austraegaeus Rech. fil.
 33 turolensis Pau
 34 galegiformis L.
 35 frigidus (L.) A. Gray

(a) frigidus	16	No	Knaben, 1950

   (b) grigorjewii (B. Fedtsch.) Chater
 36 penduliflorus Lam.
 37 umbellatus Bunge
 38 alpinus L.

(a) alpinus	16	He	Favarger, 1965
(b) arcticus Lindman	16	Rs(N)	Sokolovskaja & Strelkova, 1960
39 depressus L.	16	Ga	Favarger, 1965

 40 norvegicus Weber

41 australis (L.) Lam.	32 48	It	Favarger, 1965

 42 lusitanicus Lam.
   (a) lusitanicus
   (b) orientalis Chater & Meikle

99

```
43 glycyphyllos L.                              16      It      Larsen, 1955
44 glycyphylloides DC.
45 dasyanthus Pallas
46 pubiflorus DC.
47 tanaiticus C. Koch
48 longipetalus Chater
49 exscapus L.
50 huetii Bunge
51 utriger Pallas
52 henningii (Steven) Boriss.
53 wolgensis Bunge
54 nummularius Lam.
55 hellenicus Boiss.
56 anatolicus Boiss.
57 tremolsianus Pau
58 graecus Boiss. & Spruner
59 drupaceus Orph. ex Boiss.
60 creticus Lam.
   (a) creticus
   (b) rumelicus (Bunge) Maire & Petitmengin
61 granatensis Lam.
   (a) granatensis
   (b) siculus (Biv.) Franco & P. Silva
62 arnacantha Bieb.
63 parnassi Boiss.
   (a) parnassi
   (b) calabrus (Fiori) Chater
   (c) cylleneus (Boiss. & Heldr. ex
       Fischer) Hayek
64 thracicus Griseb.
65 trojanus Steven ex Fischer
66 physocalyx Fischer                           16      Bu      Kozuharov &
                                                                Kuzmanov, 1965b

67 ponticus Pallas
68 vulpinus Willd.
69 alopecurus Pallas
70 alopecuroides L.
71 grossii Pau
72 centralpinus Br.-Bl.
73 ajubensis Bunge
74 clusii Boiss.
75 sempervirens Lam.
   (a) sempervirens
   (b) cephalonicus (C. Presl) Ascherson
       & Graebner
   (c) muticus (Pau) Rivas Goday & Borja
   (d) nevadensis (Boiss.) P. Monts.
76 giennensis Heywood
77 angustifolius Lam.
   (a) angustifolius
   (b) pungens (Willd.) Hayek
78 balearicus Chater
79 massiliensis (Miller) Lam.
80 sirinicus Ten.
   (a) sirinicus
   (b) genargenteus (Moris) Arcangeli
81 odoratus Lam.
82 algarbiensis Cosson ex Bunge
83 falcatus Lam.
84 asper Jacq.
85 roemeri Simonkai
```

```
 86 austriacus Jacq.
 87 sulcatus L.
 88 clerceanus Iljin & Krasch.
 89 tenuifolius L.
 90 arenarius L.
 91 baionensis Loisel.
 92 tenuifoliosus Maire
 93 mesopterus Griseb.
 94 onobrychis L.
 95 leontinus Wulfen                          32      He      Favarger, 1959a
 96 amarus Pallas
 97 helmii Fischer ex DC.
 98 idaeus Bunge
 99 autranii Bald.
100 agraniotii Orph. ex Boiss.
101 arcuatus Kar. & Kir.
102 reduncus Pallas
103 dolichophyllus Pallas
104 lacteus Heldr. & Sart. ex Boiss.
105 baldaccii Degen
106 wilmottianus Stoj.
107 testiculatus Pallas
108 rupifragus Pallas
109 physodes L.
110 monspessulanus L.
    (a) monspessulanus                        16      Ga      Guinochet &
                                                              Logeois, 1962

    (b) illyricus (Bernh.) Chater
111 spruneri Boiss.
112 incanus L.                                16      Hs      Lorenzo-Andreu &
                                                              Garcia, 1950

    (a) incanus
    (b) incurvus (Desf.) Chater
    (c) nummularioides (Desf. ex DC.) Maire
    (d) macrorhizus (Cav.) Chater
113 sericophyllus Griseb.
114 apollineus Boiss. & Heldr.
115 karelinianus M. Popov
116 subuliformis DC.
117 corniculatus Bieb.
118 muelleri Steudel & Hochst.
119 pugionifer Fischer ex Bunge
120 cornutus Pallas
121 brachylobus DC.
122 varius S.G. Gmelin
123 macropus Bunge
124 pallescens Bieb.
125 glaucus Bieb.
126 zingeri Korsh.
127 vesicarius L.
    (a) vesicarius                            16      Ga      Guinochet &
                                                              Logeois, 1962

    (b) carniolicus (A. Kerner) Chater        32
    (c) pastellianus (Pollini) Arcangeli
128 peterfii Jáv.
129 hispanicus Cosson ex Bunge
130 hegelmaieri Willk.
131 albicaulis DC.
132 medius Schrenk
133 fialae Degen
```

39 Oxytropis DC.
1 lapponica (Wahlenb.) Gay	16	No	Knaben & Engelskjön, 1967
2 deflexa (Pallas) DC.	16	No	Knaben & Engelskjön, 1967
3 jacquinii Bunge			
4 carpatica Uechtr.			
5 gaudinii Bunge			
6 amethystea Arvet-Touvet	16	Ga	Damboldt, 1964
7 pyrenaica Godron & Gren.	16	It	Favarger, 1959a
8 triflora Hoppe			
9 mertensiana Turcz.			
10 hippolyti Boriss.			
11 campestris (L.) DC.	48	Fe	Sorsa, 1963
(a) campestris	48	Europe	Jalas, 1950
(b) tiroliensis (Sieber ex Fritsch) Leins & Merxm.			
(c) sordida (Willd.) Hartman fil.	48	No	Laane, 1966
12 prenja (G. Beck) G. Beck			
13 urumovii Jáv.			
14 foucaudii Gillot	32	Ga	Favarger & Huynh, 1964
15 spicata (Pallas) O. & B. Fedtsch.			
16 songorica (Pallas) DC.			
17 ambigua (Pallas) DC.			
18 halleri Bunge ex Koch			
(a) halleri	32	He	Favarger, 1965
(b) velutina (Sieber) O. Schwarz	16	Ga	Favarger & Huynh, 1964
19 uralensis (L.) DC.			
20 pilosa (L.) DC.	16	Hu	Baksay, 1958
21 pallasii Pers.			
22 floribunda (Pallas) DC.			
23 purpurea (Bald.) Markgraf			
24 fetida (Vill.) DC.	16	He	Favarger, 1959a

40 Biserrula L.
  1 pelecinus L.

41 Glycyrrhiza L.
  1 foetida Desf.
  2 glabra L.
  3 korshinskyi Grigoriev
  4 aspera Pallas
  5 echinata L.

42 Amorpha L.
  1 fruticosa L.

43 Psoralea L.
  1 bituminosa L.
  2 americana L.

44 Apios Fabr.
  1 americana Medicus

45 Phaseolus L.
  1 vulgaris L.
  2 coccineus L.

46 Vigna Savi

1 unguiculata (L.) Walpers

47 Glycine Willd.
   1 max (L.) Merr.

48 Cicer L.
   1 arietinum L.
   2 incisum (Willd.) K. Malý
   3 montbretii Jaub. & Spach
   4 graecum Orph. ex Boiss.

49 Vicia L.
   1 orobus DC.
   2 montenegrina Rohlena
   3 sparsiflora Ten.   12   Hu   Baksay, 1956
   4 pinetorum Boiss. & Spruner
   5 ochroleuca Ten.
     (a) ochroleuca   12   It   Larsen, 1955b
     (b) dinara (Borbás) K. Malý ex Rohlena
   6 pisiformis L.
   7 argentea Lapeyr.
   8 serinica Uechtr. & Huter
   9 sicula (Rafin.) Guss.
   10 cracca L.   14   Su   Lövkvist, 1963
                                                         27   Ge   Rousi, 1961
                                                         28   Fe   V. Sorsa, 1962
                                                         30   Da   Rousi, 1961
      oreophila Zertova   28   Cz   Žertová, 1962
   11 incana Gouan   12   Cz   Žertová, 1963
   12 tenuifolia Roth   24   Hu   Baksay, 1954
   13 dalmatica A. Kerner   12   Hu   Baksay, 1954
   14 sibthorpii Boiss.
   15 cassubica L.   12   Su   Lövkvist, 1963
   16 sylvatica L.   14   Su   Lövkvist, 1963
   17 multicaulis Ledeb.
   18 onobrychioides L.
   19 altissima Desf.
   20 dumetorum L.
   21 villosa Roth
      (a) villosa
      (b) varia (Host) Corb.
      (c) eriocarpa (Hausskn.) P.W. Ball
      (d) microphylla (D'Urv.) P.W. Ball
      (e) pseudocracca (Bertol.) P.W. Ball   14   It   Larsen, 1955b
   22 benghalensis L.   14   Lu   Fernandes &
   23 glauca C. Presl   Santos, 1971
   24 biennis L.
   25 cretica Boiss. & Heldr.
      (a) cretica
      (b) aegaea (Halácsy) P.W. Ball
   26 monantha Retz.
      (a) monantha
      (b) triflora (Ten.) B.L. Burtt &
          P. Lewis
   27 articulata Hornem.
   28 ervilia (L.) Willd.
   29 leucantha Biv.
   30 vicioides (Desf.) Coutinho
   31 hirsuta (L.) S.F. Gray
   32 meyeri Boiss.
   33 durandii Boiss.

34 disperma DC.
35 tenuissima (Bieb.) Schinz & Thell.
36 tetrasperma (L.) Schreber   14   Ho   Gadella &
                                                          Kliphuis, 1966
37 pubescens (DC.) Link
38 bifoliolata Rodr.
39 oroboides Wulfen   14   Hu   Baksay, 1956
40 truncatula Fischer ex Bieb.
41 pyrenaica Pourret
42 sepium L.   14   Ho   Gadella &
                                                          Kliphuis, 1966
43 grandiflora Scop.
44 barbazitae Ten. & Guss.
45 pannonica Crantz
   (a) pannonica
   (b) striata (Bieb.) Nyman
46 sativa L.
   (a) nigra (L.) Ehrh.   12   It   Larsen, 1956a
                                             14   Su   Lövkvist, 1963
   (b) amphicarpa (Dorthes) Ascherson &
       Graebner
   (c) cordata (Wulfen ex Hoppe)   10   Al Ga Hanelt & Mettin,
       Ascherson & Graebner   Gr It 1966
   (d) incisa (Bieb.) Arcangeli
   (e) sativa   12   It   Larsen, 1956a
   (f) macrocarpa (Moris) Arcangeli
47 lathyroides L.   12   Su   Lövkvist, 1963
48 cuspidata Boiss.
49 peregrina L.
50 melanops Sibth. & Sm.
51 lutea L.
   (a) lutea   14   Su   Lövkvist, 1963
   (b) vestita (Boiss.) Rouy
52 hybrida L
53 bithynica (L.) L.   14   It   Larsen, 1955b
54 narbonensis L.
55 faba L.

50 Lens Miller
   1 nigricans (Bieb.) Godron
   2 culinaris Medicus
   3 orientalis (Boiss.) M. Popov
   4 ervoides (Brign.) Grande

51 Lathyrus L.
   1 vernus (L.) Bernh.   14   Fe   Sorsa, 1962
   2 venetus (Miller) Wohlf.   14   He Ju Rm Brunsberg, 1965
   3 niger (L.) Bernh.
     (a) niger   14   Cz Hu Rm Brunsberg, 1965
     (b) jordanii (Ten.) Arcangeli
   4 japonicus Willd.
     (a) japonicus   14   Is   Löve & Löve, 1956
     (b) maritimus (L.) P.W. Ball   14   Fe Po Su Brunsberg, 1967
   5 laevigatus (Waldst. & Kit.) Gren.
     (a) laevigatus   14   Rm   Brunsberg, 1965
     (b) occidentalis (Fischer & C.A. Meyer)
         Breistr.   14   Ga   Favarger &
                                                             Huynh, 1964
   6 transsilvanicus (Sprengel) Fritsch   14   Rm   Brunsberg, 1965
   7 gmelinii Fritsch
   8 aureus (Steven) Brandza

9 pisiformis L.	14	Hu	Brunsberg, 1967
10 humilis (Ser.) Sprengel			
11 incurvus (Roth) Willd.	14	Ju Su	Brunsberg, 1967
12 filiformis (Lam.) Gay			
13 bauhinii Genty			
14 digitatus (Bieb.) Fiori	14	Rs(K)	Trankovsky, 1940
15 pannonicus (Jacq.) Garcke	14	Hu	Baksay, 1961
(a) pannonicus			
(b) collinus (Ortmann) Soó			
(c) asphodeloides (Gouan) Bassler			
(d) hispanicus (Lacaita) Bassler			
(e) varius (C. Koch) P.W. Ball			
16 pallescens (Bieb.) C. Koch			
17 pancicii (Juriš.) Adamović			
18 alpestris (Waldst. & Kit.) Kit. ex Čelak.			
19 montanus Bernh.	14	Ga Ge	Brunsberg, 1965
20 pratensis L.	7	Fe	Simola, 1964
	8	Fe	Simola, 1964
	9	Fe	Simola, 1964
	12	Fe	Simola, 1964
	14	Au Cz Da	Larsen, 1957a
	16	Su	Simola, 1964
	21	Su	Brunsberg, 1965
	28	Be Ga He Ho	Brunsberg, 1965
	42	Ga	Brunsberg, 1965
21 binatus Pančić	14	Su	Brunsberg, 1965
22 hallersteinii Baumg.	14	Rm Su	Brunsberg, 1965
23 palustris L.			
(a) palustris	42	Br Fe Ge Ho	Brunsberg, 1965
(b) nudicaulis (Willk.) P.W. Ball			
24 laxiflorus (Desf.) O. Kuntze	14	Rs(K)	Trankovsky, 1940
25 neurolobus Boiss. & Heldr.	14	Gr	Brunsberg, 1965
26 cirrhosus Ser.			
27 tuberosus L.	14	Cz Ga	Brunsberg, 1965
28 roseus Steven			
29 grandiflorus Sibth. & Sm.			
30 rotundifolius Willd.	14	Rs(K)	Trankovsky, 1940
31 undulatus Boiss.			
32 sylvestris L.	14	Br Cz Su	Brunsberg, 1965
33 latifolius L.	14	Ga Ju	Brunsberg, 1965
34 heterophyllus L.	14	He	Brunsberg, 1965
35 tremolsianus Pau			
36 tingitanus L.			
37 odoratus L.	14	It Tu	Brunsberg, 1965
38 saxatilis (Vent.) Vis.			
39 sphaericus Retz.	14	Da	Larsen, 1954
40 angulatus L.			
41 inconspicuus L.			
42 setifolius L.			
43 cicera L.	14	Ga Gr	Brunsberg, 1965
44 sativus L.	14	Ga Ge Ju	Brunsberg, 1965
45 amphicarpos L.			
46 annuus L.			
47 hierosolymitanus Boiss.			
48 gorgoni Parl.			
49 hirsutus L.	14	Ga Lu	Brunsberg, 1965
50 clymenum L.			
51 articulatus L.			
52 ochrus (L.) DC.	14	Ga	Brunsberg, 1965
53 nissolia L.			

54 aphaca L.   14   Ga Lu   Brunsberg, 1965

52 Pisum L.
  1 sativum L.
    (a) sativum
    (b) elatius (Bieb.) Ascherson & Graebner

53 Ononis L.
  1 rotundifolia L.
  2 tridentata L.   30   Hs   Lorenzo-Andreu, 1951
  3 fruticosa L.
  4 cristata Miller
  5 adenotricha Boiss.
  6 natrix L.
    (a) natrix
    (b) ramosissima (Desf.) Batt.
    (c) hispanica (L. fil.) Coutinho
  7 crispa L.
  8 ornithopodiodes L.
  9 biflora Desf.
 10 maweana Ball
 11 sicula Guss.
 12 reclinata L.
 13 dentata Solander ex Lowe
 14 pendula Desf.
 15 laxiflora Desf.
 16 verae Širj.
 17 pubescens L.
 18 viscosa L.   32   It   Larsen, 1956a
    (a) viscosa
    (b) brachycarpa (DC.) Batt.
    (c) breviflora (DC.) Nyman
    (d) sieberi (Besser ex DC.) Širj.
    (e) subcordata (Cav.) Širj.
    (f) foetida (Schousboe ex DC.) Širj.
 19 crotalarioides Cosson
 20 cintrana Brot.
 21 speciosa Lag.
 22 aragonensis Asso
 23 reuteri Boiss.
 24 pinnata Brot.
 25 leucotricha Cosson
 26 pusilla L.
 27 saxicola Boiss. & Reuter
 28 minutissima L.   30   Ga   Morisset, 1964
 29 striata Gouan
 30 cephalotes Boiss.
 31 hispida Desf.
 32 spinosa L.   30   It   Larsen, 1956a
    (a) spinosa
    (b) antiquorum (L.) Arcangeli
    (c) leiosperma (Boiss.) Širj.
    (d) austriaca (G. Beck) Gams
 33 repens L.   30   Ga   Larsen, 1956a
                                                 60   Ho   Gadella &
                                                                 Kliphuis, 1967a
 34 arvensis L.   30   Po   Morisset, 1964
 35 masquillierii Bertol.   30   It   Morisset, 1964
 36 gilivsulid Salzm. ex Boiss.
 37 alba Poiret
 38 oligophylla Ten.

39 variegata L.
40 euphrasiifolia Desf.
41 hirta Poiret
42 cossoniana Boiss.
43 diffusa Ten.
44 serrata Forskål
45 tournefortii
46 subspicata Lag.
47 mitissima L.
48 alopecuroides L.
49 baetica Clemente

54 Melilotus Miller
   1 dentata (Waldst. & Kit.) Pers.   16   Su   Lövkvist, 1963
   2 altissima Thuill.   16   Ge   Scheerer, 1939
   3 alba Medicus   16   Su   Lövkvist, 1963
   4 wolgica Poiret
   5 officinalis (L.) Pallas   16   Hu   Pólya, 1949
   6 polonica (L.) Pallas
   7 taurica (Bieb.) Ser.
   8 italica (L.) Lam.
   9 neapolitana Ten.   16   It   Larsen, 1955b
  10 indica (L.) All.
  11 elegans Salzm. ex Ser.
  12 infesta Guss.
  13 sulcata Desf.
  14 segetalis (Brot.) Ser.
  15 messanensis (L.) All.
  16 physocarpa Stefanov

55 Trigonella L.
   1 graeca (Boiss. & Spruner) Boiss.
   2 cretica (L.) Boiss.
   3 maritima Delile ex Poiret
   4 corniculata (L.) L.   16   It   Larsen, 1956a
   5 balansae Boiss. & Reuter
   6 rechingeri Širj.
   7 grandiflora Bunge
   8 sprunerana Boiss.
   9 smyrnaea Boiss.
  10 spinosa L.
  11 spicata Sibth. & Sm.
  12 striata L. fil.
  13 arcuata C.A. Meyer
  14 fischerana Ser.
  15 polyceratia L.
  16 orthoceras Kar. & Kir.
  17 monspeliaca L.
  18 caerulea (L.) Ser.
  19 procumbens (Besser) Reichenb.
  20 coerulescens (Bieb.) Halácsy
  21 gladiata Steven ex Bieb.
  22 cariensis Boiss.
  23 foenum-graecum L.   16   Lu   Azevedo Coutinho
                                                      & Santos, 1943

56 Medicago L.
   1 lupulina L.   16   Hu   Pólya, 1948
   2 secundiflora Durieu
   3 hybrida (Pourret) Trautv.
   4 cretacea Bieb.

5 sativa L.
  (a) sativa   32   Su   Lövkvist, 1963
  (b) ambigua (Trautv.) Tutin
  (c) caerulea (Less. ex Ledeb.) Schmalh.   16   Rs(W)   Sinskaya &
                                                                Maleyeva, 1959
  (d) falcata (L.) Arcangeli   16   Bu   Kozuharov &
                                                                Kuzmanov, 1965a
                                                32   Su   Lövkvist, 1963
  (e) glomerata (Balbis) Tutin
6 prostrata Jacq.   32   Hu   Baksay, 1958
7 rupestris Bieb.
8 saxatilis Bieb.
9 cancellata Bieb.
10 suffruticosa Ramond ex DC.
  (a) suffruticosa
  (b) leiocarpa (Bentham) P. Fourn.
11 arborea L.
12 orbicularis (L.) Bartal.   16   Bu   Kozuharov &
                                                                Kuzmanov, 1965a
13 carstiensis Jacq.
14 intertexta (L.) Miller
15 ciliaris (L.) All.
16 scutellata (L.) Miller
17 blancheana Boiss.
18 rugosa Desr.
19 soleirolii Duby
20 tornata (L.) Miller
21 marina L.
22 truncatula Gaertner
23 rigidula (L.) All.   14   Bu   Kozuharov &
                                                                Kuzmanov, 1965a
                                                16   Bu   Kozuharov &
                                                                Kuzmanov, 1965a
24 pironae Vis.   16   It   Lesins & Lesins,
                                                                1961
25 littoralis Rohde ex Loisel.
26 aculeata Gaertner
27 globosa C. Presl
28 turbinata (L.) All.
29 murex Willd.
30 arabica (L.) Hudson
31 polymorpha L.   14   It   Larsen, 1956a
32 praecox DC.
33 laciniata (L.) Miller
  (a) laciniata
  (b) schimperana P. Fourn.
34 coronata (L.) Bartal.   16   Bu   Kozuharov &
                                                                Kuzmanov, 1965a
35 disciformis DC.
36 tenoreana Ser.
37 minima (L.) Bartal.   16   It   Larsen, 1956a

57 Trifolium L.
  1 ornithopodioides L.   16   Ho   Kliphuis, 1962
  2 lupinaster L.   32   Rs(C)   Iljin &
    ciswolgense Sprygin ex Iljin & Truch.   16   Rs(C)   Trukhaleva,
    litwinowii Iljin   32   Rs(C)   1960
  3 alpinum L.   16   Au   Reese, 1953
  4 pilczii Adamović
  5 strictum L.
  6 nervulosum Boiss. & Heldr.

108

```
 7 montanum L.                              16    Hu    Pólya, 1950
   balbisianum Ser.                         16    Ga    Guinochet &
                                                        Logeois, 1962
 8 ambiguum Bieb.
 9 parnassi Boiss. & Spruner
10 repens L.
   (a) repens                               32    Br    Coombe, 1961
   (b) nevadense (Boiss.) D.E. Coombe
   (c) orbelicum (Velen.) Pawl.
   (d) ochranthum E.I. Nyárády
   (e) orphanideum (Boiss.) D.E. Coombe
   (f) prostratum Nyman
11 occidentale D.E. Coombe                  16    Br    Coombe, 1961
12 pallescens Schreber                      16    Ga    Favarger, 1965b
13 thalii Vill.
14 hybridum L.
   (a) hybridum                             16    Ho    Gadella &
                                                        Kliphuis, 1966
   (b) elegans (Savi) Ascherson & Graebner
   (c) anatolicum (Boiss.) Hossain
15 bivonae Guss.
16 isthmocarpum Brot.
   (a) isthmocarpum
   (b) jaminianum (Boiss.) Murb.
17 nigrescens Viv.                          16    It    Larsen, 1956a
   (a) nigrescens
   (b) petrisavii (G.C. Clementi)
       Holmboe
18 michelianum Savi
19 angulatum Waldst. & Kit.
20 retusum L.
21 cernuum Brot.
22 glomeratum L.
23 suffocatum L.
24 uniflorum L.
25 spumosum L.
26 vesiculosum Savi
27 mutabile Portenschl.
28 physodes Steven ex Bieb.
29 fragiferum L.
   (a) fragiferum                           16    Hu    Pólya, 1948
   (b) bonannii (C. Presl) Soják
30 resupinatum L.
31 tomentosum L.
32 badium Schreber
33 spadiceum L.
34 speciosum Willd.
35 boissieri Guss. ex Boiss.
36 lagrangei Boiss.
37 brutium Ten.
38 mesogitanum Boiss.
39 aurantiacum Boiss. & Spruner
40 dolopium Heldr. & Hausskn. ex Gibelli
   & Belli
41 patens Schreber
42 aureum Pollich                           16    Su    Lövkvist, 1962
43 velenovskyi Vandas
44 campestre Schreber                       14    Hs    Kliphuis, 1962
45 sebastianii Savi
46 dubium Sibth.                            28    Ho    Kliphuis, 1962
                                            32    Ho    Gadella &
                                                        Kliphuis, 1966
```

47 micranthum Viv.   16   Ho   Kliphuis, 1962
48 striatum L.   14   Ho   Kliphuis, 1962
49 arvense L.   14 Au Da Hu Ju Bücher & Larsen,
                                                                           1958
50 affine C. Presl
51 saxatile All.
52 bocconei Savi
53 tenuifolium Ten.
54 trichopterum Pančić
55 phleoides Pourret ex Willd.
56 gemellum Pourret ex Willd.
57 ligusticum Balbis ex Loisel.
58 scabrum L.   10   Ga   Gadella &
                                                                           Kliphuis, 1968
59 dalmaticum Vis.
60 stellatum L.   14   It   Larsen, 1956a
61 dasyurum C. Presl
62 incarnatum L.
   (a) incarnatum
   (b) molinerii (Balbis ex Hornem.) Syme
63 pratense L.   14   Ce   Reese, 1952
   var. frigidum Gaudin   14   Au   Reese, 1952
64 pallidum Waldst. & Kit.
65 diffusum Ehrh.
66 noricum Wulfen
67 wettsteinii Dörfler & Hayek
68 ottonis Spruner ex Boiss.
69 lappaceum L.
70 congestum Guss.
71 barbeyi Gibelli & Belli
72 hirtum All.
73 cherleri L.
74 medium L.   c.70   Su   Lövkvist, 1963
                                                    78-80   Su   Lövkvist, 1963
                                                       84   Su   Lövkvist, 1963

   (a) medium
   (b) banaticum (Heuffel) Hendrych
   (c) sarosiense (Hazsl.) Simonkai
   (d) balcanicum Velen.
75 heldreichianum Hausskn.
76 patulum Tausch
77 velebiticum Degen
78 pignantii Fauché & Chaub.
79 alpestre L.
80 rubens L.
81 angustifolium L.
82 purpureum Loisel.
83 desvauxii Boiss. & Blanche
84 smyrnaeum Boiss.
85 ochroleucon Hudson
86 pannonicum Jacq.
87 canescens Willd.
88 alexandrinum L.
89 apertum Bobrov
90 echinatum Bieb.
91 latinum Sebastiani
92 leucanthum Bieb.
93 squamosum L.
94 squarrosum L.
95 obscurum Savi
   (a) obscurum

```
        (b) aequidentatum (Pérez Lara) C.
            Vicioso
    96 clypeatum L.
    97 subterraneum L.                    16      Ho      Kliphuis, 1962
    98 globosum L.
    99 pauciflorum D'Urv.


    58 Dorycnium Miller
        1 hirsutum (L.) Ser.              14      Ju      Larsen, 1956a
        2 rectum (L.) Ser.
        3 graecum (L.) Ser.
        4 pentaphyllum Scop.
          (a) pentaphyllum
          (b) germanicum (Gremli) Gams
          (c) gracile (Jordan) Rouy
          (d) herbaceum (Vill.) Rouy


    59 Lotus L.
        1 tenuis Waldst. & Kit. ex Willd.  12      Hu      Pólya, 1949
        2 krylovii Schischkin & Serg.
        3 borbasii Ujhelyi                 12      Hu      Ujhelyi, 1960
        4 delortii Timb.-Lagr. ex F.W. Schultz
        5 stenodon (Boiss. & Heldr.) Heldr.
        6 glareosus Boiss. & Reuter
        7 corniculatus L.                  24      Au      Titz, 1964
        8 alpinus (DC.) Schleicher ex Ramond  12   Ga      Favarger &
                                                           Huynh, 1964
        9 uliginosus Schkuhr               12   Ga Ge Lu   Larsen, 1955a
       10 pedunculatus Cav.
       11 preslii Ten.
       12 palustris Willd.
       13 aegaeus (Griseb.) Boiss.
       14 parviflorus Desf.                12      Lu      Larsen, 1956b
       15 subbiflorus Lag.
          (a) subbiflorus                  24      Br      Rutland, 1941
          (b) castellanus (Boiss. & Reuter)
              P.W. Ball
       16 angustissimus L.                 12      Hu      Larsen, 1955a
       17 edulis L.                        14      It      Larsen, 1955a
       18 conimbricensis Brot.             12      Lu      Larsen, 1956b
       19 aduncus (Griseb.) Nyman
       20 cytisoides L.                    14     Hs Ju    Larsen, 1958a
       21 drepanocarpus Durieu
       22 creticus L.                      28      Lu      Larsen, 1958a
       23 collinus (Boiss.) Heldr.
       24 strictus Fischer & C.A. Meyer
       25 halophilus Boiss. & Spruner
       26 ornithopodioides L.              14      It      Larsen, 1955a
       27 peregrinus L.
       28 azoricus P.W. Ball
       29 arenarius Brot.
       30 tetraphyllus L.


    60 Tetragonolobus Scop.
        1 maritimus (L.) Roth              14      Su      Lövkvist, 1963
        2 biflorus (Desr.) Ser.
        3 purpureus Moench
        4 conjugatus (L.) Link
        5 requienii (Mauri ex Sanguinetti)
          Sanguinetti
```

61 Hymenocarpos Savi
  1 circinnatus (L.) Savi

62 Securigera DC.
  1 securidaca (L.) Degen & Dörfler   12   It   Larsen, 1955b

63 Anthyllis L.
  1 cytisoides L.
  2 terniflora (Lag.) Pau
  3 hermanniae L.
  4 barba-jovis L.
  5 aegaea Turrill
  6 henoniana Cosson ex Batt.
  7 montana L.   12   Ga   Guinochet &
                                                             Logeois, 1962
    (a) montana
    (b) jacquinii (A. Kerner) Hayek
    (c) hispanica (Degen & Hervier) Cullen
  8 aurea Welden
  9 polycephala Desf.
  10 tejedensis Boiss.
  11 rupestris Cosson
  12 ramburii Boiss.
  13 onobrychoides Cav.
  14 gerardi L.
  15 vulneraria L.
    (a) vulneraria
    (b) maritima (Schweigger) Corb.
    (c) polyphylla (DC.) Nyman   12   Hu   Baksay, 1956
    (d) bulgarica (Sagorski) Cullen
    (e) boissieri (Sagorski) Bornm.
    (f) pulchella (Vis.) Bornm.
    (g) arundana (Boiss. & Reuter) Vasc.
    (h) argyrophylla (Rothm.) Cullen
    (i) reuteri Cullen
    (j) atlantis Emberger & Maire
    (k) vulnerarioides (All.) Arcangeli
    (l) forondae (Sennen) Cullen
    (m) pindicola Cullen
    (n) praepropera (A. Kerner) Bornm.
    (o) weldeniana (Reichenb.) Cullen
    (p) maura (G. Beck) Lindb.
    (q) hispidissima (Sagorski) Cullen
    (r) corbierei (Salmon & Travis) Cullen
    (s) carpatica (Pant.) Nyman
    (t) alpestris Ascherson & Graebner   12   Hu   Baksay, 1956
    (u) pyrenaica (G. Beck) Cullen
    (v) iberica (W. Becker) Jalas
    (w) lapponica (Hyl.) Jalas   12   Su   Jalas, 1950
    (x) borealis (Rouy) Jalas
  16 tetraphylla L.
  17 lotoides L.
  18 cornicina L.
  19 hamosa Desf.

64 Ornithopus L.
  1 compressus L.
  2 sativus Brot.
    (a) sativus
    (b) isthmocarpus (Cosson) Dostál
  3 perpusillus L.   14   Ge   Scheerer, 1939

```
  4 pinnatus (Miller) Druce                  14      Ga     Ball, 1961

65 Coronilla L.
   1 emerus L.
     (a) emerus                              14      It     Larsen, 1955b
     (b) emeroides (Boiss. & Spruner) Hayek
   2 valentina L.
     (a) valentina
     (b) glauca (L.) Batt.
   3 vaginalis Lam.                          12      Hu     Baksay, 1956
   4 minima L.                               24      Ga     Guinochet &
                                                            Logeois, 1962
   5 juncea L.
   6 coronata L.                             10      Hu     Baksay, 1956
   7 varia L.                                24      Hu     Baksay, 1956
   8 elegans Pančić                          12      Hu     Baksay, 1956
   9 globosa Lam.
  10 cretica L.
  11 rostrata Boiss. & Spruner
  12 scorpioides (L.) Koch                   12      It     Larsen, 1955b
  13 repanda (Poiret) Guss.
     (a) repanda
     (b) dura (Cav.) Coutinho

66 Hippocrepis L.
   1 glauca Ten.
   2 scabra DC.
   3 comosa L.                               28      Br     Maude, 1940
   4 squamata (Cav.) Cosson
     (a) squamata
     (b) eriocarpa (Boiss.) Nyman
   5 valentina Boiss.
   6 balearica Jacq.
   7 ciliata Willd.
   8 multisiliquosa L.                       14      Hs     Lorenzo-Andreu &
                                                            Garcia, 1950
   9 salzmanii Boiss. & Reuter
  10 unisiliquosa L.

67 Hammatolobium Fenzl
   1 lotoides Fenzl

68 Scorpiurus L.
   1 muricatus L.                            28      Hs     Dominguez &
   2 vermiculatus L.                                        Galiano, 1974

69 Eversmannia Bunge
   1 subspinosa (Fischer ex DC.) B.
     Fedtsch.

70 Hedysarum L.
   1 coronarium L.                           16      It     Larsen, 1955b
   2 flexuosum L.
   3 spinosissimum L.
   4 glomeratum F.G. Dietrich
   5 hedysaroides (L.) Schinz & Thell.
     (a) hedysaroides                        14      Au     Reese, 1953
     (b) exaltatum (A. Kerner) Žertová
     (c) arcticum (B. Fedtsch.) P.W. Ball
   6 alpinum L.
   7 boutignyanum Alleiz.
```

```
 8 gmelinii Ledeb.
 9 razoumowianum Helm & Fischer ex DC.
10 cretaceum Fischer ex DC.
11 ucrainicum B. Kaschm.
12 tauricum Pallas ex Willd.
13 humile L.
14 varium Willd.
15 candidum Bieb.
16 grandiflorum Pallas
17 biebersteinii Žertová
18 macedonicum Bornm.              16      Ju      Šopova, 1966

71 Onobrychis Miller
 1 hypargyrea Boiss.
 2 pallasii (Willd.) Bieb.
 3 radiata (Desf.) Bieb.
 4 saxatilis (L.) Lam.
 5 petraea (Bieb. ex Willd.) Fischer
 6 ebenoides Boiss. & Spruner
 7 supina (Chaix) DC.              14      Ga      Guinochet &
                                                   Logeois, 1962
 8 gracilis Besser
 9 pindicola Hausskn.
10 stenorhiza DC.
11 alba (Waldst. & Kit.) Desv.
   (a) alba
   (b) laconica (Orph. ex Boiss.) Hayek
   (c) echinata (G. Don fil.) P.W. Ball
   (d) calcarea (Vandas) P.W. Ball
12 degenii Dörfler
13 sphaciotica W. Greuter
14 peduncularis (Cav.) DC.
   (a) peduncularis
   (b) matritensis (Boiss. & Reuter) Maire
15 argentea Boiss.
   (a) argentea
   (b) hispanica (Širj.) P.W. Ball
16 reuteri Leresche
17 montana DC.
   (a) montana                     28      It      Larsen, 1960
   (b) scardica (Griseb.) P.W. Ball
   (c) cadmea (Boiss.) P.W. Ball
18 arenaria (Kit.) DC.
   (a) arenaria                     14      He      Favarger, 1953
                                    28      Hu      Baksay, 1956
   (b) taurerica Hand.-Mazz.
   (c) sibirica (Turcz. ex Besser)
       P.W. Ball
   (d) miniata (Steven) P.W. Ball
   (e) tommasinii (Jordan) Ascherson &
       Graebner                     14      It      Larsen, 1955b
   (f) lasiostachya (Boiss.) Hayek
   (g) cana (Boiss.) Hayek
19 oxyodonta Boiss.
20 inermis Steven
21 viciifolia Scop.                 28      Hu      Pólya, 1950
22 caput-galli (L.) Lam.            14      It      Larsen, 1956a
23 aequidentata (Sibth. & Sm.) D'Urv.

72 Ebenus L.
 1 cretica L.
```

114

2 sibthorpii DC.

73 Alhagi Gagnebin
   1 pseudalhagi (Bieb.) Desv.
   2 graecorum Boiss.

74 Arachis L.
   1 hypogaea L.

LXXXII OXALIDACEAE

1 Oxalis L.
   1 corniculata L.   24   Br   Rutland, 1941
   2 exilis A. Cunn.
   3 stricta L.   c.24   Po   Skalińska et al.,
                                                                 1959
   4 europaea Jordan   24   Ho   Gadella &
                                                                 Kliphuis, 1966
   5 articulata Savigny
   6 acetosella L.   22   Br   Marks, 1956
   7 corymbosa DC.
   8 latifolia Kunth
   9 tetraphylla Cav.
  10 pes-caprae L.
  11 purpurea L.
  12 incarnata L.

LXXXIII GERANIACEAE

1 Geranium L.
   1 macrorrhizum L.   46   Al   Warburg, 1938
   2 dalmaticum (G. Beck) Rech. fil.
   3 cinereum Cav.
     (a) cinereum
     (b) subcaulescens (L'Hér. ex DC.) Hayek
   4 humbertii Beauverd
   5 argenteum L.
   6 sanguineum L.   84   Su   Lövkvist, 1963
                                             (52-56)   Fe   Sorsa, 1963
   7 pratense L.   28   Br   Warburg, 1938
   8 sylvaticum L.
     (a) sylvaticum   28   Po   Skalińska et al.,
                                                                 1959
     (b) rivulare (Vill.) Rouy   28   He   Favarger, 1965
     (c) caeruleatum (Schur) D.A. Webb
     (d) pseudosibiricum (J. Mayer) D.A. Webb
           & I.K. Ferguson
   9 endressii Gay
  10 versicolor L.
  11 nodosum L.
  12 albiflorum Ledeb.
  13 peloponesiacum Boiss.
  14 ibericum Cav.
  15 phaeum L.   28   Po   Skalińska et al.,
                                                                 1959
  16 reflexum L.
  17 aristatum Freyn & Sint.
  18 tuberosum L.
  19 macrostylum Boiss.

```
20 linearilobum DC.
21 malviflorum Boiss. & Reuter
22 asphodeloides Burm. fil.
23 palustre L.                          28     Hu      Pólya, 1950
24 collinum Stephan ex Willd.
25 divaricatum Ehrh.
26 bohemicum L.                         28     No Su   Dahlgren, 1952
27 lanuginosum Lam.                     48     Su      Dahlgren, 1952
28 sibiricum L.
29 pyrenaicum Burm. fil.                26     Ge      Gauger, 1937
30 rotundifolium L.                     26     It      Larsen, 1956a
31 molle L.                             26     Ho      Gadella & Kliphuis,
                                                       1966
32 brutium Gasparr.
33 pusillum L.                          26     Hu      Pólya, 1950
                                        36             error
34 columbinum L.                        18     Br      Warburg, 1938
35 dissectum L.                         22     Br      Warburg, 1938
36 lucidum L.                           20     Br Lu   Warburg, 1938
37 robertianum L.                       32     Fe      Sorsa, 1962
                                        64     Hu      Pólya, 1950
38 purpureum Vill.                      32     Ga Lu   Bocher & Larsen,
                                                       1955
39 cataractarum Cosson

2 Erodium L'Hér.
  1 gussonii Ten.                       20     It      Guittonneau, 1967
  2 guttatum (Desf.) Willd.
  3 boissieri Cosson                    20     Hs      Guittonneau, 1965
  4 chium (L.) Willd.                   20     Ga      Guittonneau, 1964
    (a) chium
    (b) littoreum (Leman) Ball
  5 laciniatum (Cav.) Willd.            20     Hs      Guittonneau, 1965
  6 malacoides (L.) L'Hér.              40     Lu      Fernandes &
                                                       Queirós, 1971
  7 alnifolium Guss.                    20     Si      Guittonneau, 1967
  8 maritimum (L.) L'Hér.               20     Br      Larsen, 1958
  9 sanguis-christi Sennen              20     Hs      Guittonneau, 1964
 10 corsicum Leman                      20     Co      Contandriopoulos,
                                                       1957
                                        18     Co      Negodi, 1937
 11 reichardii (Murray) DC.             20     Bl      Guittonneau, 1967
 12 ruthenicum Bieb.
 13 hoefftianum C.A. Meyer
 14 botrys (Cav.) Bertol.               40     Hs      Guittonneau, 1965
 15 gruinum (L.) L'Hér.
 16 ciconium (L.) L'Hér.                18     Ga      Guittonneau, 1964
 17 guicciardii Heldr. ex Boiss.
 18 chrysanthum L'Hér. ex DC.
 19 absinthoides Willd.
 20 beketowii Schmalh.
 21 alpinum L'Hér.                       18     It      Guittonneau, 1966
 22 rupestre (Pourret ex Cav.)
    Guittonneau                         20     Hs      Guittonneau, 1965
 23 petraeum (Gouan) Willd.             20     Ga      Guittonneau, 1964
    (a) petraeum
    (b) lucidum (Lapeyr.) D.A. Webb & Chater
    (c) glandulosum (Cav.) Bonnier
    (d) crispum (Lapeyr.) Rouy
    (e) valentinum (Lange) D.A. Webb &
        Chater
```

24 rodiei (Br.-Bl.) Poirion
25 cicutarium (L.) L'Hér.

(a) cicutarium	20	Ge	Rottgardt, 1956
	36	Ge	Rottgardt, 1956
	40	Ge	Rottgardt, 1956
	48	Ge	Rottgardt, 1956
	54	Ge	Rottgardt, 1956
(b) bipinnatum Tourlet	20		
	40	Lu	Guittonneau, 1966

(c) jacquinianum (Fischer, C.A. Meyer
& Ave-Lall.) Briq.

subsp. dunense	40	Br Da	Larsen, 1958
danicum K. Larsen	60	Da	Larsen, 1958
26 moschatum (L.) L'Hér.	20	?Europe	Warburg, 1938
27 acaule (L.) Becherer & Thell.	40	It	Guittonneau, 1966
28 carvifolium Boiss. & Reuter	20		
	40	Hs	Guittonneau, 1967
29 paui Sennen			
30 rupicola Boiss.	20		
31 astragaloides Boiss. & Reuter			
32 daucoides Boiss.	40	Hs	Guittonneau, 1967
	60	Hs	Guittonneau, 1967
	80	Hs	Guittonneau, 1967
33 manescavi Cosson	40	Ga	Guittonneau, 1966
34 hirtum (Forskål) Willd.			

3 Biebersteinia Stephan
  1 orphanidis Boiss.

LXXXIV TROPAEOLACEAE

1 Tropaeolum L.
  1 majus L.

LXXXV ZYGOPHYLLACEAE

1 Peganum L. 1 harmala L.	24	Hs	Lorenzo-Andreu & Garcia, 1950

2 Fagonia L.
  1 cretica L.

3 Zygophyllum L.
  1 macropterum C.A. Meyer
  2 fabago L.
  3 ovigerum Fischer & C.A. Meyer ex Bunge
  4 album L. fil.

4 Tribulus L.
  1 terrestris L.

5 Nitraria L.
  1 schoberi L.

6 Tetradiclis Steven ex Bieb.
  1 tenella (Ehrenb.) Litv.

LXXXVI LINACEAE

1 Linum L.
  1 arboreum L.
  2 capitatum Kit. ex Schultes
  3 flavum L.
  4 thracicum Degen
  5 tauricum Willd.
  6 uninerve (Rochel) Jáv.
  7 ucranicum Czern.
  8 campanulatum L.
  9 elegans Spruner ex Boiss.

dolomiticum Borbás	28	Hu	Baksay, 1956
10 leucanthum Boiss. & Spruner			
11 pallasianum Schultes	28	Rm	Ockendon, ined.
12 nodiflorum L.			
13 narbonense L.	30	Ga	Ockendon, ined.
14 nervosum Waldst. & Kit.	18		Ockendon, ined.
15 aroanium Boiss. & Orph.			
16 hologynum Reichenb.			
17 virgultorum Boiss. & Heldr. ex Planchon			
18 perenne L.			
(a) perenne	18	Ge	Ockendon, 1971
(b) anglicum (Miller) Ockendon	36	Br	Ockendon, 1971
(c) alpinum (Jacq.) Ockendon	18	He	Ockendon, 1971
(d) montanum (DC.) Ockendon	36	Ga	Ockendon, 1971
(e) extraaxillare (Kit.) Nyman	18	Cz	Ockendon, 1971
19 austriacum L.			
(a) austriacum	18	Au	Ockendon, 1971
(b) collinum Nyman	18	Ga	Ockendon, 1971
(c) euxinum (Juz.) Ockendon			
20 punctatum C. Presl			
21 leonii F.W. Schultz	18	Ga	Ockendon, 1971
22 decumbens Desf.			
23 bienne Miller	30	Ga	Ockendon, ined.
24 usitatissimum L.			
25 viscosum L.			
26 hirsutum L.	16	Cz	Majovsky et al., 1967
27 spathulatum (Halácsy & Bald.) Halácsy			
28 pubescens Banks & Solander			
29 maritumum L.	20	Ga	Ockendon, ined.
30 trigynum L.			
31 tenue Desf.			
32 tenuifolium L.	16	Hu	Baksay, 1956
	18	Cz	Májovský et al., 1967
33 suffruticosum L.	72	Hs	Lorenzo-Andreu & Garcia, 1950
(a) suffruticosum			
(b) salsoloides (Lam.) Rouy			
34 strictum L.	18	Hs	Lorenzo-Andreu & Garcia, 1950
(a) strictum			
(b) corymbulosum (Reichenb.) Rouy			
35 setaceum Brot.	18	Lu	Ockendon, 1964
36 catharticum L.	16	Is	Löve & Löve, 1956
2 Radiola Hill			
1 linoides Roth	18	Da	Hagerup, 1941

LXXXVII EUPHORBIACEAE

1 Andrachne L.
  1 telephioides L.

2 Securinega Commerson
  1 tinctoria (L.) Rothm.

3 Chrozophora A. Juss.
  1 tinctoria (L.) A. Juss.
  2 obliqua (Vahl) A. Juss. ex Sprengel

4 Mercurialis L.
  1 annua L.   16   Ho   Gadella & Kliphuis,
                                                                  1963
                                                48   Bl   Dahlgren et al.,
                                                                  1971
                                                64   Co   Contandriopoulos,
                                                                  1962
                                         80 96 112   Co   Durand, 1962
    ambigua L. fil.   48   Hs Lu   Durand, 1962
  2 elliptica Lam.
  3 corsica Cosson
  4 reverchonii Rouy
  5 tomentosa L.
  6 perennis L.   42   Hu   Baksay, 1957b
                                                64   Ho   Gadella & Kliphuis,
                                                                  1963
                                                66   Ho   Gadella & Kliphuis,
                                                                  1963
                                             c.80   Su   Löyvkvist, 1963
                                                84   Ho   Gadella & Kliphuis,
                                                                  1967a
  7 ovata Sternb. & Hoppe   32   Hu   Baksay, 1957b

5 Acalypha L.
  1 virginica L.

6 Ricinus L.
  1 communis L.

7 Euphorbia L.
  1 nutans Lag.
  2 peplis L.
  3 polygonifolia L.
  4 humifusa Willd.
  5 chamaesyce L.
    (a) chamaesyce
    (b) massiliensis (DC.) Thell.
  6 maculata L.
  7 prostrata Aiton
  8 stygiana H.C. Watson
  9 dendroides L.   18   It   D'Amato, 1946
 10 serrata L.
 11 isatidifolia Lam.
 12 villosa Waldst. & Kit. ex Willd.
 13 corallioides L.
 14 lagascae Sprengel
 15 arguta Banks & Solander
 16 microsphaera Boiss.
 17 akenocarpa Guss.
 18 palustris L.   20   Hu   Pólya, 1950
 19 velenovskyi Bornm.

119

```
20 soongarica Boiss.
21 ceratocarpa Ten.
22 hyberna L.                          36      It      Cesca, 1963
   (a) hyberna
   (b) insularis (Boiss.) Briq.
   (c) canuti (Parl.) Tutin
23 gregersenii K. Malý ex G. Beck
24 squamosa Willd.
25 epithymoides L.                     16      Hu      Pólya, 1950
26 lingulata Heuffel
27 gasparrinii Boiss.
28 fragifera Jan
29 montenegrina (Bald.) K. Malý ex Rohlena
30 oblongata Griseb.
31 apios L.                            12      Bu      Kozuharov &
                                                       Kuzmanov, 1962
32 polygalifolia Boiss. & Reuter
33 uliginosa Welw. ex Boiss.
34 dulcis L.                           12      It      Cesca, 1961
                                       24      It      Cesca, 1961
35 angulata Jacq.
36 carniolica Jacq.
37 duvalii Lecoq & Lamotte
38 brittingeri Opiz ex Samp.
39 ruscinonensis Boiss.
40 welwitschii Boiss. & Reuter
41 monchiquensis Franco & P. Silva
42 clementei Boiss.
43 bivonae Steudel
44 squamigera Loisel.
45 spinosa L.                          14      It      Cesca, 1963
46 glabriflora Vis.
47 acanthothamnos Heldr. & Sart. ex Boiss.
48 chamaebuxus Bernard ex Gren. & Godron
49 capitulata Reichenb.
50 platyphyllos L.
51 serrulata Thuill.
52 pubescens Vahl                      14      It      D'Amato, 1945
53 cuneifolia Guss.
54 pterococca Brot.
55 helioscopia L.                      42      Su      Lövkvist, 1963
56 phymatosperma Boiss. & Gaill.
57 myrsinites L.
58 rigida Bieb.                        20      It      D'Amato, 1946
59 broteri Daveau
60 lathyris L.
61 aleppica L.
62 medicaginea Boiss.
63 dracunculoides Lam.
64 exigua L.                           24      Br      Rutland, 1941
65 falcata L.                          16      Hu      Pólya, 1950
                                       36      It      D'Amato, 1939
66 sulcata De Lens ex Loisel.
67 peplus L.                           16      Ho      Gadella
                                                       & Kliphuis, 1963
68 ledebourii Boiss.
69 taurinensis All.
70 segetalis L.                        16      Ga      D'Amato, 1945
71 pinea L.
72 portlandica L.
73 deflexa Sibth. & Sm.
```

74 petrophila C.A. Meyer			
75 transtagana Boiss.			
76 matritensis Boiss.			
77 biumbellata Poiret			
78 boetica Boiss.			
79 bupleuroides Desf.			
80 nicaeensis All.			
(a) nicaeensis			
(b) glareosa (Pallas ex Bieb.) A.R. Sm.			
81 bessarabica Klokov			
82 orphanidis Boiss.			
83 barrelieri Savi			
84 triflora Schott			
85 saxatilis Jacq.	18	Au	Polatschek, 1966
86 valliniana Belli			
87 minuta Loscos & Pardo			
88 variabilis Cesati			
89 gayi Salis			
90 maresii Knoche			
91 herniariifolia Willd			
92 seguierana Necker			
(a) seguierana			
(b) niciciana (Borbás ex Novák) Rech. fil.			
93 pithyusa L.	36	It	D'Amato, 1939
(a) pithyusa			
(b) cupanii (Guss. ex Bertol.) A.R. Sm.			
94 paralias L.	16	It	D'Amato, 1939
95 nevadensis Boiss. & Reuter			
96 agraria Bieb.			
97 lucida Waldst. & Kit.	36	Hu	Pólya, 1950
98 salicifolia Host	36	Hu	Pólya, 1950
99 esula L.			
(a) esula	60	Ho	Gadella & Kliphuis, 1966
	64	Ge	Reese, 1952
(b) tommasiniana (Bertol.) Nyman	20	Hu	Baksay, 1958
100 undulata Bieb.			
101 cyparissias L.	20	Br	Pritchard, 1957
	40	Br	Pritchard, 1957
102 terracina L.			
103 amygdaloides L.	18	It	D'Amato, 1939
(a) amygdaloides			
(b) semiperfoliata (Viv.) A.R. Sm.			
104 heldreichii Orph. ex Boiss.			
105 characias L.			
(a) characias	20	It	Larsen, 1956a
(b) wulfenii (Hoppe ex Koch) A.R. Sm.			

LXXXVIII RUTACEAE

1 Ruta L.
 1 montana (L.) L.
 2 angustifolia Pers.
 3 chalepensis L.
 4 graveolens L.
 5 corsica DC.

2 Haplophyllum A. Juss
  1 linifolium (L.) G. Don fil.
  2 suaveolens (DC.) G. Don fil.
  3 thesioides (Fischer ex DC.) G. Don fil.
  4 coronatum Griseb.
  5 balcanicum Vandas
  6 patavinum (L.) G. Don fil.
  7 boissieranum Vis. & Pančić
  8 buxbaumii (Poiret) G. Don fil.

3 Dictamnus L.
  1 albus L.

4 Citrus L.
  1 medica L.
  2 limon (L.) Burm. fil.
  3 limetta Risso
  4 deliciosa Ten.
  5 paradisi Macfadyen
  6 grandis (L.) Osbeck
  7 aurantium L.
  8 sinensis (L.) Osbeck

5 Phellodendron Rupr.
  1 amurense Rupr.

6 Ptelea L.
  1 trifoliata L.

LXXXIX CNEORACEAE

1 Cneorum L.
  1 tricoccon L.

XC SIMAROUBACEAE

1 Ailanthus Desf.
  1 altissima (Miller) Swingle

XCI MELIACEAE

1 Melia L.
  1 azedarach L.

XCII POLYGALACEAE

1 Polygala L.
  1 microphylla L.
  2 chamaebuxus L.

38	Au	Mattick, 1950
c.46	He	Glendinning, 1960

  3 vayredae Costa
  4 myrtifolia L.
  5 rupestris Pourret
  6 exilis DC.
  7 sibirica L.
  8 subuniflora Boiss. & Heldr.
  9 supina Schreber

(a) supina
(b) hospita (Heuffel) McNeill
(c) rhodopea (Velen.) McNeill

10 monspeliaca L.	c.38	Lu	Glendinning, 1960

11 major Jacq.
12 anatolica Boiss. & Heldr.
13 boissieri Cosson
14 venulosa Sibth. & Sm.
15 preslii Sprengel
16 sardoa Chodat
17 lusitanica Welw. ex Chodat
18 baetica Willk.
19 doerfleri Hayek
20 flavescens DC.
21 nicaeensis Risso ex Koch
   (a) nicaeensis
   (b) mediterranea Chodat
   (c) tomentella (Boiss.) Chodat

(d) corsica (Boreau) Graebner	34	Co	Contandriopoulos, 1962

   (e) caesalpinii (Bubani) McNeill
   (f) gariodiana (Jordan & Fourr.) Chodat
   (g) carniolica (A. Kerner) Graebner
   (h) forojulensis (A. Kerner) Graebner
22 apiculata Porta

23 comosa Schkuhr	28-32	Au	Mattick, 1950
	34	Su	Lövkvist, 1963
24 vulgaris L.	28	Au	Mattick, 1950
	32	Au	Mattick, 1950
	48	Ge	Wulff, 1938
	c.56	Au	Mattick, 1950
	68	Br He Lu	Glendinning, 1960
	c.70	Scand.	Löve & Löve, 1944

25 alpestris Reichenb.

(a) alpestris	34	He	Glendinning, 1960

   (b) croatica (Chodat) Hayek

26 serpyllifolia J.A.C. Hose	32	Da No	Björse, 1961
	34	Br He	Glendinning, 1960
	c.68	Co	Glendinning, 1960

27 carueliana (A.W. Benn) Burnat ex
   Caruel
28 cristagalli Chodat
29 edmundii Chodat

30 calcarea F.W. Schultz	34	Br He	Glendinning, 1960

31 amara L.

(a) amara	28	Hu	Baksay, 1956
(b) brachyptera (Chodat) Hayek	28	Po	Skalińska et al., 1959
32 amarella Crantz	34	Fe	Sorsa, 1963
33 alpina (Poiret) Steudel	c.34	He	Glendinning, 1960

XCIII CORIARIACEAE

1 Coriaria L.
  1 myrtifolia L.

XCIV ANACARDIACEAE

1 Rhus L.
  1 coriaria L.

2 typhina L.
3 pentaphylla (Jacq.) Desf.
4 tripartita (Ucria) Grande

2 Cotinus Miller
1 coggygria Scop.

3 Pistacia L.
1 terebinthus L.
2 atlantica Desf.
3 vera L.
4 lentiscus L.

4 Schinus L.
1 molle L.
2 terebinthifolia Raddi

XCV ACERACEAE

1 Acer L.
1 platanoides L.	26	Su	Lövkvist, 1963

2 lobelii Ten.
3 campestre L.
4 tataricum L.
5 pseudoplatanus L.	52	Su	Lövkvist, 1963

6 heldreichii Orph. ex Boiss.
  (a) heldreichii
  (b) visianii K. Malý
7 trautvetteri Medv.
8 opalus Miller
9 obtusatum Waldst. & Kit. ex Willd.
10 granatense Boiss.
11 hyrcanum Fischer & C.A. Meyer
12 stevenii Pojark.
13 monspessulanum L.
14 sempervirens L.
15 negundo L.

XCVI SAPINDACEAE

1 Cardiospermum L.
1 halicacabum L.

XCVII HIPPOCASTANACEAE

1 Aesculus L.
1 hippocastanum L.

XCVIII BALSAMINACEAE

1 Impatiens L.

1 noli-tangere L.	20	Ho	Gadella & Kliphuis, 1966
	40	Po	Skalińska et al., 1959
2 capensis Meerb.			
3 parviflora DC.	24	Su	Löve & Löve, 1942
	26	Po	Skalińska et al., 1959

```
4 glandulifera Royle                 18    Br    Jones & Smith,
                                      20    Br      1966
5 balfourii Hooker fil.
6 balsamina L.
```

XCIX AQUIFOLIACEAE

```
1 Ilex L.
  1 aquifolium L.                     40    Br    Maude, 1940
  2 colchica Pojark.
  3 perado Aiton
```

C CELASTRACEAE

```
1 Euonymus L.
  1 europaeus L.                      64    Hu    Pólya, 1949
  2 latifolius (L.) Miller
  3 verrucosus Scop.
  4 nanus Bieb.
  5 japonicus L. fil.

2 Maytenus Molina
  1 senegalensis (Lam.) Exell
```

CI STAPHYLEACEAE

```
1 Staphylea L.
  1 pinnata L.
```

CII BUXACEAE

```
1 Buxus L.
  1 sempervirens L.
  2 balearica Lam.
```

CIII RHAMNACEAE

```
1 Paliurus Miller
  1 spina-christi Miller

2 Ziziphus Miller
  1 jujuba Miller
  2 lotus (L.) Lam.

3 Rhamnus L.
  1 alaternus L.
  2 ludovici-salvatoris Chodat
  3 lycioides L.
   (a) lycioides
   (b) velutinus (Boiss.) Tutin
   (c) oleoides (L.) Jahandiez & Maire
   (d) graecus (Boiss. & Reuter)
  4 saxatilis Jacq.
   (a) saxatilis
   (b) tinctorius (Waldst. & Kit.) Nyman
  5 rhodopeus Velen.
```

6 intermedius Steudel & Hochst.
7 prunifolius Sibth. & Sm.
8 catharticus L.   24   Ge   Wulff, 1939
9 orbiculatus Bornm.
10 persicifolius Moris
11 alpinus L.
   (a) alpinus
   (b) fallax (Boiss.) Maire & Petitmengin
   (c) glaucophyllus (Sommier) Tutin
12 pumilus Turra
13 sibthorpianus Roemer & Schultes

4 Frangula Miller
1 alnus Miller   20   Br   Rutland, 1941
                                          26   Ge   Wulff, 1937b
2 azorica Tutin
3 rupestris (Scop.) Schur

CIV VITACEAE

1 Vitis L.
   1 vinifera L.
   (a) sylvestris (C.C. Gmelin) Hegi
   (b) vinifera   38   Rm   Constantinescu
                                                             et al., 1964

2 Parthenocissus Planchon
   1 quinquefolia (L.) Planchon
   2 inserta (A. Kerner) Fritsch
   3 tricuspidata (Siebold & Zucc.) Planchon

CV TILIACEAE

1 Tilia L.
   1 tomentosa Moench
   2 dasystyla Steven
   3 platyphyllos Scop.
   (a) platyphyllos
   (b) cordifolia (Besser) C.K. Schneider
   (c) pseudorubra C.K. Schneider
   4 rubra DC.
   (a) rubra
   (b) caucasica (Rupr.) V. Engler
   5 cordata Miller
   6 x vulgaris Hayne

CVI MALVACEAE

1 Malope L.
   1 malacoides L.
   2 trifida Cav.

2 Kitaibela Willd.
   1 vitifolia Willd.

3 Sida L.
   1 rhombifolia L.

4 Malvella Jaub. & Spach
  1 sherardiana (L.) Jaub. & Spach

5 Malva L.
  1 hispanica L.
  2 aegyptia L.
  3 stipulacea Cav.
  4 cretica Cav.
    (a) cretica
    (b) althaeoides (Cav.) Dalby
  5 alcea L.   84   Su   Lövkvist, 1963
  6 moschata L.   42   Su   Lövkvist, 1963
  7 tournefortiana L.
  8 sylvestris L.   42   Su   Lövkvist, 1963
  9 nicaeensis All.
 10 parviflora L.
 11 pusilla Sm.
 12 neglecta Wallr.   42   Su   Lövkvist, 1963
 13 verticillata L.

6 Lavatera L.
  1 cretica L.
  2 mauritanica Durieu
  3 arborea L.
  4 maritima Gouan
  5 oblongifolia Boiss.
  6 olbia L.
  7 bryoniifolia Miller
  8 thuringiaca L.
    (a) thuringiaca
    (b) ambigua (DC.) Nyman
  9 punctata All.
 10 trimestris L.
 11 triloba L.
    (a) triloba
    (b) pallescens (Moris) Nyman
    (c) agrigentina (Tineo) R. Fernandes

7 Althaea L.
  1 hirsuta L.
  2 longiflora Boiss. & Reuter
  3 cannabina L.
  4 officinalis L.   42   Hu   Pólya, 1950
  5 armeniaca Ten.

8 Alcea L.
  1 setosa (Boiss.) Alef.
  2 rosea L.
  3 pallida (Willd.) Waldst. & Kit.
    (a) pallida
    (b) cretica (Weinm.) D.A. Webb
  4 heldreichii (Boiss.) Boiss.
  5 lavateriflora (DC.) Boiss.
  6 rugosa Alef.

9 Abutilon Miller
  1 theophrasti Medicus

10 Modiola Moench
   1 caroliniana (L.) G. Don fil.

127

11 Gossypium L.
  1 herbaceum L.
  2 hirsutum L.

12 Hibiscus L.
  1 syriacus L.
  2 palustris L.
  3 trionum L.
  4 cannabinus L.

13 Abelmoschus Medicus
  1 esculentus (L.) Moench

14 Kosteletzkya C. Presl
  1 pentacarpos (L.) Ledeb.

CVII THYMELAEACEAE

1 Daphne L.
  1 mezereum L.   18   Fe   V. Sorsa, 1962
  2 gnidium L.
  3 sophia Kalenicz
  4 laureola L.
    (a) laureola   18   Au   Fuchs, 1938
    (b) philippi (Gren.) Rouy
  5 pontica L.
  6 alpina L.
  7 oleoides Schreber
  8 blagayana Freyer   18   Ju   Nevling, 1962
  9 sericea Vahl
 10 cneorum L.   18   Hu   Baksay, 1956
 11 striata Tratt.
 12 arbuscula Čelak
 13 petraea Leybold
 14 rodriguezii Texidor
 15 jasminea Sibth. & Sm.
 16 malyana Blečić
 17 gnidioides Jaub. & Spach

2 Thymelaea Miller
  1 sanamunda All.
  2 pubescens (L.) Meissner
  3 hirsuta (L.) Endl.
  4 tartonraira (L.) All.
    (a) tartonraira
    (b) argentea (Sibth. & Sm.) Holmboe
    (c) thomasii (Duby) Briq.   18   Co   Contandriopoulos,
                                                         1964a
  5 nitida (Vahl) Endl.
  6 myrtifolia (Poiret) D.A. Webb
  7 lanuginosa (Lam.) Ceballos & C.
      Vicioso
  8 coridifolia (Lam.) Endl.
  9 procumbens A. & R. Fernandes
 10 broterana Coutinho
 11 calycina (Lapeyr.) Meissner
 12 ruizii Loscos ex Casav.
 13 subrepens Lange
 14 tinctoria (Pourret) Endl.
 15 dioica (Gouan) All.

16 villosa (L.) Endl.
17 passerina (L.) Cosson & Germ.

3 Diarthron Turcz.
  1 vesiculosum (Fischer & C.A. Meyer)
    C.A. Meyer

CVIII ELAEAGNACEAE

1 Hippophae L.
  1 rhamnoides L.   24   Fe Ga Ge   Rousi, 1965a
                                                  He It Po

2 Elaeagnus L.
  1 angustifolia L.

CIX GUTTIFERAE

1 Hypericum L.
   1 calycinum L.
   2 foliosum Aiton
   3 hircinum L.
   4 inodorum Miller
   5 androsaemum L.
   6 balearicum L.
   7 aciferum (W. Greuter) N.K.B. Robson
   8 aegypticum L.
   9 empetrifolium Willd.
  10 amblycalyx Coust. & Gand.
  11 coris L.
  12 ericoides L.
  13 haplophylloides Halácsy & Bald.
  14 hirsutum L.   18   Da   Nielson, 1924
  15 linarioides Bosse
  16 pulchrum L.   18   Da Fa   Bücher, 1940a
  17 nummularium L.
  18 fragile Heldr. & Sart. ex Boiss.
  19 taygeteum Quezel & Contandr.
  20 hyssopifolium Chaix
     (a) hyssopifolium
     (b) elongatum (Ledeb.) Woronow
  21 olympicum L.
  22 cerastoides (Spach) N.K.B. Robson
  23 montanum L.   16   Da   Nielson, 1964
  24 annulatum Moris
  25 delphicum Boiss. & Heldr.
  26 athoum Boiss. & Orph.
  27 atomarium Boiss.
  28 cuisinii W. Barbey
  29 tomentosum L.
  30 pubescens Boiss.
  31 caprifolium Boiss.
  32 elodes L.   32   Br   Robson, unpub.
  33 vesiculosum Griseb.
  34 perfoliatum L.
  35 montbretii Spach
  36 umbellatum A. Kerner
  37 bithynicum Boiss.
  38 richeri Vill.

(a) richeri	14	He	Favarger, 1959a
(b) burseri (DC.) Nyman			
(c) grisebachii (Boiss.) Nyman			
39 spruneri Boiss.			
40 rochelii Griseb. & Schenk			
41 barbatum Jacq.			
42 rumeliacum Boiss.			
43 trichocaulon Boiss. & Heldr.			
44 kelleri Bald.			
45 aviculariifolium Jaub. & Spach.			
46 thasium Griseb.			
47 aucheri Jaub. & Spach			
48 linarifolium Vahl	16	Br	Robson, unpub.
49 australe Ten.			
50 humifusum L.			
51 tetrapterum Fries	16	Hu	Pólya, 1950
52 undulatum Schousboe ex Willd.			
53 maculatum Crantz			
(a) maculatum	16	Fe	V. Sorsa, 1962
(b) obtusiusculum (Tourlet) Hayek	32	Br	Robson
54 perforatum L.			
55 triquetrifolium Turra			
56 elegans Stephan ex Willd.			
57 gymnanthum Engelm. & A. Gray			
58 mutilum L.			
59 majus (A. Gray) Britton			
60 canadense L.	16	Hb	Moore, 1973
61 gentianoides (L.) Britton			

CX VIOLACEAE

1 Viola L.			
1 odorata L.	20	Ge	Schöfer, 1954
2 suavis Bieb.	40	Au He It	Schmidt, 1961
3 alba Besser			
(a) alba	20	Ge	Schöfer, 1954
(b) scotophylla (Jordan) Nyman	20	Au	Schmidt, 1961
(c) dehnhardtii (Ten.) W. Becker	20	Bl	Schmidt, 1961
4 cretica Boiss. & Heldr.	20	Cr	Schmidt, 1961
5 jaubertiana Marès & Vigineix	20	Bl	Schmidt, 1961
6 hirta L.	20	Ge	Schöfer, 1954
7 collina Besser	20	Ge	Schöfer, 1954
8 ambigua Waldst. & Kit.			
9 thomasiana Song. & Perr.	40	Au	Schmidt, 1961
10 myrenaica Ramond ex DC.	20	Au	Schöfer, 1954
11 chelmea Boiss. & Heldr.			
(a) chelmea	20	Gr	Schmidt, 1964b
(b) vratnikensis Gáyer & Degen			
12 mirabilis L.	20	Hu	Pólya, 1950
13 willkommii R. de Roemer			
14 rupestris F.W. Schmidt			
(a) rupestris	20	Ge	Schöfer, 1954
(b) relicta Jalas	20	No	Knaben & Engelskjön, 1967
15 reichenbachiana Jordan ex Boreau	20	Ga	Larsen, 1954
16 tanaitica Grosset			
17 mauritii Tepl.			
18 riviniana Reichenb.	35 40 45	Ho	Gadella & Kliphuis, 1963

	46	Ho	Gadella & Kliphuis, 1963
	47	Ho	Gadella & Kliphuis, 1963

19 sieheana W. Becker
20 canina L.

(a) canina	40	Ge	Schöfer, 1954
(b) montana (L.) Hartman	40	Ge	Schöfer, 1954
(c) schultzii (Billot) Kirschleger	40	Ge	Schöfer, 1954
21 lactea Sm.	58	Br Lu	Moore, 1958
22 persicifolia Schreber	20	Da	Clausen, 1931
23 pumila Chaix	40	Ge	Schöfer, 1954
24 elatior Fries	40	Ge	Schmidt, 1961
25 jordanii Hanry	40	Ga	Schmidt, 1961
26 uliginosa Besser	20	Ju	Schmidt, 1961

27 palustris L.

(a) palustris	48	Be Ho	Gadella & Kliphuis, 1963

(b) juressi (Link ex K. Wein) Coutinho

28 epipsila Ledeb.	24	Fe	Sorsa, 1963
29 jooi Janka	24	Rm	Merxmüller, 1974

30 selkirkii Pursh ex Goldie
31 pinnata L.
32 obliqua Hill

33 biflora L.	12	Fe	Sorsa, 1963

34 crassiuscula Bory
35 diversifolia (DC.) W. Becker

36 cenisia L.	20	It	Schmidt, 1961

37 comollia Massara

38 valderia All.	20	It	Schmidt, 1961
39 magellensis Porta & Rigo ex Strobl	22	It	Schmidt, 1964a

40 brachyphylla W. Becker
41 perinensis W. Becker
42 stojanowii W. Becker
43 fragrans Sieber
44 poetica Boiss. & Spruner

45 grisebachiana Vis.	22	Ju	Schmidt, 1964a
46 alpina Jacq.	22	Po	Skalińska et al., 1959

47 nummulariifolia Vill.
48 calcarata L.

(a) calcarata	40	He It	Schmidt, 1961
(b) zoysii (Wulfen) Merxm.	40	Au	Schmidt, 1961

(c) villarsiana (Roemer & Schultes) Merxm.
49 bertolonii Pio

(a) bertolonii	40	It	Schmidt, 1965
(b) messanensis (W. Becker) A. Schmidt	40	It	Schmidt, 1961
heterophylla Bertol.	20	It	Schmidt, 1961
50 splendida W. Becker	40	It	Schmidt, 1965
51 aethnensis Parl.	40	Si	Schmidt, 1964a
52 nebrodensis C. Presl	20	Si	Merxmüller, 1974

53 corsica Nyman

(a) corsica	52	Sa	Schmidt, 1964
	c.120	Co	Contandriopoulos, 1962
(b) ilvensis (W. Becker) Merxm.	52	It	Merxmüller, 1974
54 munbyana Boiss. & Reuter	52	Si	Schmidt, 1965

55 pseudogracilis Strobl

(a) pseudogracilis	34	It	Schmidt, 1961
(b) cassinensis (Strobl) Merxm. & A. Schmidt	34	It	Schmidt, 1964a

56 eugeniae Parl.
(a) eugeniae	34	It	Schmidt, 1965
(b) levieri (Parl.) A. Schmidt	34	It	Schmidt, 1965

57 oreades Bieb.
58 eximia Form.
59 doerfleri Degen
60 allchariensis G. Beck
  (a) allchariensis
  (b) gostivarensis W. Becker & Bornm.
61 frondosa (Velen.) Hayek
62 arsenica G. Beck

63 cornuta L.	22	Hs	Merxmüller, 1974

64 orphanidis Boiss.
  (a) orphanidis
  (b) nicolai (Pant.) Valentine
65 dacica Borbás
66 elegantula Schott

67 athois W. Becker	20	Gr	Schmidt, 1963
68 gracilis Sibth. & Sm.	20	Ju	Merxmüller, 1974
orbelica Pancic	20	Bu	Griesinger, 1937

69 speciosa Pant.
70 beckiana Fiala
71 rhodopeia W. Becker
72 dubyana Burnat ex Gremli
73 declinata Waldst. & Kit.
74 lutea Hudson

(a) lutea	48	Br	Pettet, 1964

  (b) sudetica (Willd.) W. Becker

75 bubanii Timb.-Lagr.	c.128	Ga Hs	Schmidt, 1964a

76 hispida Lam.
77 langeana Valentine
78 tricolor L.

(a) tricolor	26	Su	Lövkvist, 1963
(b) curtisii (E. Forster) Syme	26	Ho	Gadella & Kliphuis, 1963
(c) macedonica (Boiss. & Heldr.) A. Schmidt	26	Gr	Schmidt, 1963
(d) subalpina Gaudin	26	Po	Skalińska et al., 1966

  (e) matutina (Klokov) Valentine

79 aetolica Boiss. & Heldr.	16	Gr	Schmidt, 1963
80 arvensis Murray	34	Ho	Gadella, 1963
81 kitaibeliana Schultes	16	Sb	Schmidt, 1965
	48	Br	Fothergill, 1944
82 hymettia Boiss. & Heldr.	16	Gr	Schmidt, 1964a
83 parvula Tineo	10	Si	Schmidt, 1962

84 heldreichiana Boiss.
85 mercurii Orph. ex Halácsy
86 demetria Prolongo ex Boiss.
87 occulta Lehm.

88 arborescens L.	c.140		

89 scorpiuroides Cosson

90 delphinantha Boiss.	20	Gr	Schmidt, 1964a

91 kosaninii (Degen) Hayek
92 cazorlensis Gand.

CXI PASSIFLORACEAE

1 Passiflora L.

1 caerulea L.

CXII CISTACEAE

1 Cistus L.
  1 albidus L.   18   Hs   Löve & Kjellquist, 1964

1 Cistus L.			
1 albidus L.	18	Hs	Löve & Kjellquist, 1964
2 crispus L.	18	Lu	Rodrigues, 1954
3 incanus L.			
4 heterophyllus Desf.			
5 parviflorus Lam.			
6 varius Pourret			
7 monspeliensis L.	18	Hs	Löve & Kjellquist, 1964
8 psilosepalus Sweet			
9 albanicus E.F. Warburg ex Heywood			
10 salvifolius L.	18	Lu	Löve & Kjellquist, 1964
11 populifolius L.	18	Hs	Löve & Kjellquist, 1964
(a) populifolius			
(b) major (Pourret ex Dunal) Heywood			
12 laurifolius L.	18	Hs	Löve & Kjellquist, 1964
13 ladanifer L.			
14 palhinhae Ingram			
15 clusii Dunal			
16 libanotis L.			
2 Halimium (Dunal) Spach			
1 ocymoides (Lam.) Willk.	18	Lu	Proctor, 1955
2 alyssoides (Lam.) C. Koch			
3 lasianthum (Lam.) Spach			
(a) lasianthum			
(b) formosum (Curtis) Heywood			
4 atriplicifolium (Lam.) Spach			
5 halimifolium (L.) Willk.	18	Lu	Proctor, 1955
(a) halimifolium			
(b) multiflorum (Salzm. ex Dunal) Maire			
6 umbellatum (L.) Spach	18	Lu	Proctor, 1955
7 viscosum (Willk.) P. Silva			
8 verticillatum (Brot.) Sennen			
9 commutatum Pau			
3 Tuberaria (Dunal) Spach			
1 lignosa (Sweet) Samp.			
2 globularifolia (Lam.) Willk.			
3 major (Willk.) P. Silva & Rozeira			
4 guttata (L.) Fourr.	36	Br Ga Hb Lu	Proctor, 1955
5 bupleurifolia (Lam.) Willk.			
6 villosissima (Pomel) Grosser			
7 praecox Grosser			
8 acuminata (Viv.) Grosser			
9 macrosepala (Cosson) Willk.			
10 echioides (Lam.) Willk.			
4 Helianthemum Miller			
1 lavandulifolium Miller	20	Hs	Coutinho & Lorenzo-Andreu, 1948
2 squamatum (L.) Pers.	10	Hs	Coutinho & Lorenzo-Andreu, 1948

3 caput-felis Boiss.			
4 piliferum Boiss.			
5 viscarium Boiss. & Reuter			
6 asperum Lag. ex Dunal			
7 hirtum (L.) Miller			
8 croceum (Desf.) Pers.	20	Hs	Löve & Kjellquist, 1964
9 nummularium (L.) Miller	20	Br Ge Hb	Proctor, 1955
(a) nummularium	20	He	Rohner, 1954
(b) tomentosum (Scop.) Schinz. & Thell.			
(c) pyrenaicum (Janchen) Schinz & Thell.			
(d) obscurum (Celak.) J. Holub	20	He	Rohner, 1954
(e) glabrum (Koch) Wilczek			
(f) semigalbrum (Badaro) M.C.F. Proctor			
(g) berterianum (Bertol.) Breistr.			
(h) grandiflorum (Scop.) Schinz & Thell.	20	He	Rohner, 1954
10 leptophyllum Dunal			
11 appeninum (L.) Miller	20	Br Ga	Proctor, 1955
12 pilosum (1.) Pers.			
13 virgatum (Desf.) Pers.			
14 sessiliflorum (Desf.) Pers.			
15 stipulatum (Forskål) C. Chr.			
16 villosum Thib.			
17 papillare Boiss.			
18 ledifolium (L.) Miller	20	Lu	Proctor, 1955
19 salicifolium (L.) Miller	20	Hs	Löve & Kjellquist, 1964
20 sanguineum (Lag.) Lag. ex Dunal			
21 aegyptiacum (L.) Miller			
22 oelandicum (L.) DC.			
(a) oelandicum	22	Su	Proctor, 1955
(b) alpestre (Jacq.) Breistr.	22	He	Proctor, 1955
(c) rupifragum (A. Kerner) Breistr.			
(d) italicum (L.) Font Quer & Rothm.	20	Ga	Guinochet & Logeois, 1962
(e) orientale (Grosser) M.C.F. Proctor			
23 canum (L.) Baumg.			
(a) canum	22	Br Ga Hb	Proctor, 1955
(b) nebrodense (Heldr. ex Guss.) Arcangeli			
(c) canescens (Hartman) M.C.F. Proctor			
(d) levigatum M.C.F. Proctor			
(e) piloselloides (Lapeyr.) M.C.F. Proctor			
(f) pourretii (Timb.Lagr.) M.C.F. Proctor			
(g) stevenii (Rupr. ex Juz. & Pozd.) M.C.F. Proctor			
24 origanifolium (Lam.) Pers.			
(a) origanifolium			
(b) glabratum (Willk.) Guinea & Heywood			
(c) serrae (Camb.) Guinea & Heywood			
(d) molle (Cav.) Font Quer & Rothm.			
25 marifolium (L.) Miller			
26 pannosum Boiss.			
(a) pannosum			
(b) frigidulum (Cuatrec.) Font Quer & Rothm.			
27 cinereum (Cav.) Pers.	20	Hs	Lorenzo-Andreu, 1951
	22	Hs	Coutinho & Lorenzo-Andreu, 1948

28 hymettium Boiss. & Heldr.
29 rossmaessleri Willk.
30 viscidulum Boiss.
  (a) viscidulum
  (b) guadiccianum Font Quer & Rothm.
  (c) viscarioides (Debeaux & Reverchon)
     Guinea & Heywood
31 lunulatum (All.) DC.

5 Fumana (Dunal) Spach
 1 arabica (L.) Spach
 2 procumbens (Dunal) Gren. & Godron
 3 ericoides (Cav.) Gand.
 4 scoparia Pomel
 5 bonapartei Maire & Petitmengin
 6 thymifolia (L.) Spach ex Webb   32   Hs   Löve & Kjellquist,
                                                    1964
 7 paradoxa Heywood
 8 laevipes (L.) Spach   32   Lu   Proctor, 1955
 9 aciphylla Boiss.

CXIII TAMARICACEAE

1 Reaumuria Hasselq. ex L.
 1 vermiculata L.

2 Tamarix L.
 1 africana Poiret
 2 canariensis Willd.
 3 gallica L.
 4 tetrandra Pallas ex Bieb.
 5 parviflora DC.
 6 hampeana Boiss. & Heldr.
 7 dalmatica Baum.
 8 ramosissima Ledeb.
 9 smyrnensis Bunge
10 boveana Bunge
11 meyeri Boiss.
12 laxa Willd.
13 gracilis Willd.
14 hispida Willd.

3 Myricaria Desv.
 1 germanica (L.) Desv.   24   No   Knaben, 1950

CXIV FRANKENIACEAE

1 Frankenia L.
 1 pulverulenta L.
 ·2 boissieri Reuter ex Boiss.
 3 corymbosa Desf.
 4 thymifolia Desf.
 5 laevis L.
 6 hirsuta L.

CXV ELATINACEAE

1 Bergia L.

```
   1 capensis L.

 2 Elatine L.
   1 alsinastrum L.
   2 hydropiper L.                      c.40      Su      Frisendahl, 1927
   3 macropoda Guss.
   4 hungarica Moesz
   5 triandra Schkuhr                   c.40      No Su   Frisendahl, 1927
   6 ambigua Wight
   7 hexandra (Lapierre) DC.            72        Da      Frisendahl, 1927
   8 brochonii Clavaud


CXVI DATISCACEAE

 1 Datisca L.
   1 cannabina L.


CXVII CUCURBITACEAE

 1 Thladiantha Bunge
   1 dubia Bunge

 2 Ecballium A. Richard
   1 elaterium (L.) A. Richard

 3 Bryonia L.
   1 alba L.
   2 cretica L.
     (a) cretica
     (b) dioica (Jacq.) Tutin          20        Ho      Gadella & Kliphuis,
                                                            1966
                                       40        Co      Contandriopoulos,
                                                            1964b

     (c) acuta (Desf.) Tutin

 4 Citrullus Schrader
   1 lanatus (Thunb.) Mansfeld
   2 colocynthis (L.) Schrader

 5 Cucumis L.
   1 melo L.
   2 myriocarpus Naudin
   3 sativus L.

 6 Cucurbita L.
   1 pepo L.
   2 moschata Duchesne ex Poiret
   3 maxima Duchesne

 7 Echinocystis Torrey & A. Gray
   1 lobata (Michx) Torrey & A. Gray

 8 Sicyos L.
   1 angulatus L.


CXVIII CACTACEAE

 1 Opuntia Miller
```

1 tuna (L.) Miller
2 vulgaris Miller
3 monacantha Haw.
4 stricta (Haw.) Haw.
5 ficus-indica (1.) Miller
6 maxima Miller

2 Cereus Miller
peruvianus (L.) Miller

CXIX LYTHRACEAE

1 Lythrum L.
1 salicaria L.   60   Ho   Gadella & Kliphuis,
                                                                 1966

2 virgatum L.
3 junceum Banks & Solander
4 acutangulum Lag.
5 castellanum González-Albo ex Borja
6 flexuosum Lag.
7 hyssopifolia L.
8 thymifolia L.
9 tribracteatum Salzm. ex Sprengel
10 thesioides Bieb.
11 borysthenicum (Schrank) Litv.   30   Lu   Cook, 1968
12 portula (L.) D.A. Webb   10   Hu   Pólya, 1949
13 volgense D.A. Webb

2 Ammannia L.
1 auriculata Willd.
2 coccinea Rottb.
3 baccifera L.
4 verticillata (Ard.) Lam.

3 Rotala L.
1 indica (Willd.) Koehne
2 filiformis (Bellardi) Hiern

CXX TRAPACEAE

1 Trapa L.
1 natans L.   c.48   Po   Skalińska et al.,
                                                                 1959

CXXI MYRTACEAE

1 Myrtus L.
1 communis L.

2 Eucalyptus L'Hér.
1 botryoides Sm.   22   It   Ruggeri, 1961
2 robustus Sm.
3 resinifer Sm.
4 x trabutii Vilmorin ex Trabut   22   It   Ruggeri, 1961
5 gomphocephalus DC.
6 tereticornis Sm.
7 camaldulensis Dehnh.   22   It   Ruggeri, 1961
8 rudis Endl.

```
 9 globulus Labill.
10 maidenii F. Mueller
11 viminalis Labill.


CXXII PUNICACEAE

1 Punica L.
  1 granatum L.


CXXIII ONAGRACEAE

1 Fuchsia L.
  1 magellanica Lam.                          44      Hb      Raven, ined.

2 Circaea L.
  1 lutetiana L.                              22      Su      Lövkvist, 1963
  2 x intermedia Ehrh.                        22      Su      Lövkvist, 1963
  3 alpina L.                                 22      Su      Lövkvist, 1963

3 Oenothera L.
  1 biennis L.                                14      Hu      Pólya, 1949
  2 rubricaulis Klebahn                       14      Rs(C)   Renner, 1951
  3 chicagoensis Renner ex Cleland &
    Blakeslee                                 14      Ga      Renner, 1951
  4 strigosa (Rydb.) Mackenzie & Bush         14      Ge      Oehkers, 1926
    renneri H. Scholz                         14      He      Steiner, 1955
  5 erythrosepala Borbás                      14      Ho      Gadella & Kliphuis,
                                                              1966
    coronifera Renner                         14      Ge      Rossmann, 1963
  6 suaveolens Pers.                          14      Ge      Renner, 1951
  7 parviflora L.                             14      Ge      Renner, 1951
    issleri Renner ex Rostanski               14      Ga      Renner, 1951
  8 silesiaca Renner                          14      Ge      Renner, 1951
    atrovirens Shull & Bartlett               14      Ge      Renner, 1937b
  9 ammophila Focke                           14      Ge      Renner, 1951
    syrticola Bartlett                        14      Au      Renner, 1951
    rubricuspis Renner ex Rostanski           14      Ge      Renner, 1951
 10 longiflora L.
 11 affinis Camb. in St.-Hil.                 14
 12 stricta Ledeb. ex Link                    14      Lu      Rodrigues, 1953
 13 rosea L'Hér. ex Aiton

4 Ludwigia L.
  1 uruguayensis (Camb.) Hara                 80      N. Amer.  Raven, 1963
  2 peploides (Kunth) P.H. Raven              16      Ga      Raven, ined.
  3 palustris (L.) Elliott                    16

5 Epilobium L.
  1 angustifolium L.                          36      Br      Mosquin, 1967
  2 latifólium L.                             72      Is      Löve & Löve, 1956
  3 dodonaei Vill.                            36      Cz      Májovský et al.,
                                                              1970
  4 fleischeri Hochst.                        36      Ca      Zickler, 1967
  5 hirsutum L.                               36      Br      Raven & Moore,
                                                              1964
  6 parviflorum Schreber                      36      Br      Raven & Moore,
                                                              1964
  7 duriaei Gay ex Godron
  8 montanum L.                               36      Br      Raven & Moore,
                                                              1964
```

9 collinum C.C. Gmelin	36	Au	Griesinger, 1937
10 lanceolatum Sebastiani & Mauri	36	Br	Raven & Moore, 1964
11 alpestre (Jacq.) Krocker	36	Ge	Michaelis, 1925
12 tetragonum L.			
(a) tetragonum	36	Br	Raven & Moore, 1964
(b) lamyi (F.W. Schultz) Nyman	36	Br	Raven & Moore, 1964
(c) tournefortii (Michalet) Léveillé			
13 obscurum Schreber	36	Br	Raven & Moore, 1964
14 roseum Schreber			
(a) roseum	36	Br	Raven & Moore, 1964
(b) subsessile (Boiss.) P.H. Raven			
15 palustre L.	36	Br	Raven & Moore, 1964
16 davuricum Fischer ex Hornem.			
(a) davuricum	36	No	Knaben, 1950
(b) arcticum (Sam.) P.H. Raven			
17 nutans F.W. Schmidt			
18 tundrarum Sam.			
19 atlanticum Litard. & Maire			
20 anagallidifolium Lam.	36	Br	Raven & Moore, 1964
21 hornemanni Reichenb.	36	No	Knaben & Engelskjon, 1967
22 lactiflorum Hausskn.	36	Is	Löve & Löve, 1956
23 alsinifolium Vill.	36	No	Knaben, 1950
24 gemmascens C.A. Meyer			
25 glandulosum Lehm.	36	Non-Europe	Sokolovskaya, 1963
26 adenocaulon Hausskn.	36	Br	Raven & Moore, 1964
27 nerterioides A. Cunn.	36	Br	Raven & Moore, 1964

CXXIV HALORAGACEAE

1 Gunnera L.
  1 tinctoria (Molina) Mirbel

2 Myriophyllum L.			
1 verticillatum L.	28	Ge	Scheerer, 1940
2 spicatum L.			
3 alterniflorum DC.	14	Da	Larsen, 1965
4 brasiliense Camb. in St.-Hil.			
5 heterophyllum Michx			

CXXV THELIGONACEAE

1 Theligonum L.			
1 cynocrambe L.	20	It	Schneider, 1913

CXXVI HIPPURIDACEAE

1 Hippuris L.			
1 vulgaris L.	32	Ho	Gadella & Kliphuis, 1963

CXXVII CORNACEAE

1 Cornus L.
  1 sanguinea L.   22   Hu   Pólya, 1949
    (a) sanguinea
    (b) australis (C.A. Meyer) Jav.
  2 alba L.
  3 sericea L.
  4 mas L.
  5 suecica L.   22   Fe   Sorsa, 1963

CXXVIII ARALIACEAE

1 Hedera L.
  1 helix
    (a) helix   48   Da   Jacobsen, 1954
    (b) poetarum Nyman
    (c) canariensis (Willd.) Coutinho
       hibernica hort.   96   Bot. Gdn. Jacobsen, 1954
  2 colchica (C. Koch) C. Koch

CXXIX UMBELLIFERAE

1 Hydrocotyle L.
  1 ranunculoides L. fil.
  2 bonariensis Commerson ex Lam.
  3 vulgaris L.   96   Ho   Gadella &
                                        Kliphuis, 1966
  4 sibthorpioides Lam.
  5 moschata G. Forster

2 Bowlesia Ruiz & Pavón
  1 incana Ruiz & Pavón

3 Naufraga Constance & Cannon
  1 balearica Constance & Cannon

4 Sanicula L.
  1 europaea L.   16   Hu   Pólya, 1949
  2 azorica Guthnick ex Seub.

5 Hacquetia DC.
  1 epipactis (Scop.) DC.   16   Po   Skalińska et al.,
                                        1959

6 Astrantia L.
  1 major L.
    (a) major   28   Hu   Baksay, 1957b
    (b) carinthiaca Arcangeli
  2 minor L.   16   It   Favarger &
                                        Huynh, 1964
  3 pauciflora Bertol.
  4 bavarica F.W. Schultz   14   Au   Favarger &
                                        Huynh, 1964
  5 carniolica Jacq.   14   Au   Favarger &
                                        Huynh, 1964

7 Eryngium L.
  1 tenue Lam.

```
 2 barrelieri Boiss.
 3 galioides Lam.
 4 viviparum Gay
 5 duriaei Gay ex Boiss.
 6 ilicifolium Lam.
 7 aquifolium Cav.
 8 maritimum L.                            16      Ge      Wulff, 1937a
 9 alpinum L.
10 planum L.                               16      Po      Skalińska et al.,
                                                           1964
11 dichotomum Desf.
12 creticum Lam.
13 tricuspidatum L.
14 triquetrum Vahl
15 ternatum Poiret
16 serbicum Pančić
17 palmatum Pančić & Vis.
18 glaciale Boiss.
19 dilatatum Lam.
20 spinalba Vill.
21 bourgatii Gouan                         16      He      Burdet & Miége, 1968
22 amorginum Rech. fil.
23 amethystinum L.
24 campestre L.                            14      Po      Skalińska et al.,
                                                           1964
                                           28      Hu      Pólya, 1950
25 corniculatum Lam.
26 pandanifolium Cham. & Schlecht.

 8 Lagoecia L.
   1 cuminoides L.

 9 Petagnia Guss.
   1 saniculifolia Guss.                   42      Si      Moore, ined.

10 Echinophora L.
   1 spinosa L.
   2 tenuifolia L.
    (a) tenuifolia
    (b) sibthorpiana (Guss.) Tutin

11 Myrrhoides Heister ex Fabr.
   1 nodosa (L.) Cannon                     22      Hu      Baksay, 1956

12 Chaerophyllum L.
   1 aromaticum L.
   2 byzantinum Boiss.
   3 heldreichii Orph. ex Boiss.
   4 coloratum L.
   5 creticum Boiss. & Heldr.
   6 azoricum Trelease
   7 hirsutum L.                            22      He      Rohner, 1954
   8 elegans Gaudin                         22      He      Rohner, 1954
   9 villarsii Koch                         22      He      Rohner, 1954
  10 bulbosum L.
    (a) bulbosum
    (b) prescottii (DC.) Nyman
  11 aureum L.
  12 temulentum L.                          14      Au Da Ga   Böcher & Larsen, 1955
                                           22      Ho      Gadella &
                                                           Kliphuis, 1967a
```

13 Anthriscus Pers.
  1 sylvestris (L.) Hoffm.   16   Fe   V. Sorsa, 1962
  2 nitida (Wahlenb.) Garcke
  3 nemorosa (Bieb.) Sprengel
  4 fumarioides (Waldst. & Kit.) Sprengel
  5 cerefolium (L.) Hoffm.
  6 caucalis Bieb.   14   Su   Lövkvist, 1963
  7 tenerrima Boiss. & Spruner

14 Scandix L.
  1 stellata Banks & Solander
  2 australis L.
   (a) microcarpa (Lange) Thell.
   (b) brevirostris (Boiss. & Reuter) Thell.
   (c) grandiflora (L.) Thell.
   (d) australis
  3 pecten-veneris L.
   (a) brachycarpa (Guss.) Thell.
   (b) pecten-veneris   26   Su   Lövkvist, 1963
   (c) macrorhyncha (C.A. Meyer) Rouy &
      Camus

15 Myrrhis Miller
  1 odorata (L.) Scop.

16 Molopospermum Koch
  1 peloponnesiacum (L.) Koch

17 Coriandrum L.
  1 sativum L.   22   Lu   Gardé & Malheiros-
                                       Gardé, 1949

18 Bifora Hoffm.
  1 testiculata (L.) Roth
  2 radians Bieb.

19 Scaligeria DC.
  1 cretica (Miller) Boiss.

20 Smyrnium L.
  1 olusatrum L.   22   Br   Rutland, 1941
  2 orphanidis Boiss.
  3 apiifolium Willd.
  4 perfoliatum L.
  5 rotundifolium Miller

21 Bunium L.
  1 bulbocastanum L.
  2 alpinum Waldst. & Kit.
   (a) alpinum
   (b) corydalinum (DC.) Nyman
   (c) montanum (Koch) P.W. Ball
   (d) macuca (Boiss.) P.W. Ball
   (e) petraeum (Ten.) Rouy & Camus
  3 ferulaceum Sibth. & Sm.
  4 pachypodum P.W. Ball

22 Conopodium Koch
  1 majus (Gouan) Loret
  2 pyrenaeum (Loisel.) Miégeville
  3 bourgaei Cosson

```
 4 ramosum Costa
 5 capillifolium (Guss.) Boiss.
 6 thalictrifolium (Boiss.) Calestani
 7 bunioides (Boiss.) Calestani

23 Huetia Boiss.
    1 cynapioides (Guss.) P.W. Ball
     (a) cynapioides
     (b) macrocarpa (Boiss. & Spruner) P.W. Ball
     (c) divaricata (Boiss. & Heldr.) P.W. Ball
    2 cretica (Boiss. & Heldr.) P.W. Ball
    3 pumila (Sibth. & Sm.) Boiss. & Reuter

24 Muretia Boiss.
    1 lutea (Hoffm.) Boiss.

25 Pimpinella L.
    1 rigidula (Boiss. & Orph.) H. Wolff
    2 lutea Desf.
    3 gracilis (Boiss.) H. Wolff
    4 procumbens (Boiss.) H. Wolff
    5 anisum L.
    6 cretica Poiret
    7 peregrina L.
    8 tragium Vill.
     (a) tragium
     (b) lithophila (Schischkin) Tutin
     (c) polyclada (Boiss. & Heldr.) Tutin
     (d) depressa (DC.) Tutin
     (e) titanophila (Woronow) Tutin
    9 pretenderis (Heldr.) Orph. ex Halácsy
   10 villosa Schousboe
   11 anisoides Briganti
   12 serbica (Vis.) Bentham & Hooker fil.
      ex Drude
```

13 major (L.) Hudson	20	Hu	Baksay, 1957
14 siifolia Leresche			
15 bicknellii Briq.			
16 saxifraga L.	40	Su	Lövkvist, 1963

```
26 Aegopodium L.
```
1 podagraria L.	22	Su	Lövkvist, 1963
	c.44	Fe	V. Sorsa, 1962

```
27 Sium L.
```
1 latifolium L.	20	Ho	Gadella & Kliphuis, 1966
2 sisarum L.			

```
28 Berula Koch
```
1 erecta (Hudson) Coville	20	Su	Lövkvist, 1963

```
29 Crithmum L.
    1 maritimum L.

30 Sclerochorton Boiss.
    1 junceum (Sibth. & Sm.) Boiss.

31 Dethawia Endl.
```
1 tenuifolia (Ramond ex DC.) Godron	22	Ga	Favarger & Huynh, 1964

32 Seseli L.
  1 libanotis (L.) Koch   22   Ga   Larsen, 1954
    (a) libanotis
    (b) intermedium (Rupr.) P.W. Ball
  2 sibiricum (L.) Garcke
  3 condensatum (L.) Reichenb. fil.
  4 peucedanoides (Bieb.) Kos.-Pol.   22   Hu   Baksay, 1958
  5 gummiferum Pallas ex Sm.
    (a) gummiferum
    (b) crithmifolium (DC.) P.H. Davis
    (c) aegaeum P.H. Davis
  6 lehmannii Degen
  7 vayrehanum Font Quer
  8 intricatum Boiss.
  9 granatense Willk.
 10 peixoteanum Samp.
 11 nanum Dufour
 12 montanum L.
    (a) montanum
    (b) tommasinii (Reichenb. fil.) Arcangeli
    (c) polyphyllum (Ten.) P.W. Ball
 13 cantabricum Lange
 14 bocconi Guss.
 15 elatum L.
    (a) elatum
    (b) gouanii (Koch) P.W. Ball
    (c) osseum (Crantz) P.W. Ball   18   Hu   Baksay, 1958
    (d) austriacum (G. Beck) P.W. Ball
 16 annuum L.   16   Cz   Činčura & Hindakova,
    (a) annuum   1963
    (b) carvifolium (Vill.) P. Fourn.
 17 campestre Besser
 18 arenarium Bieb.
 19 tortuosum L.
 20 pallasii Besser   16   Hu   Baksay, 1956
 21 glabratum Willd. ex Schultes
 22 strictum Ledeb.
 23 rigidum Waldst. & Kit.
    (a) rigidum
    (b) peucedanifolium (Besser) Nyman
 24 rhodopeum Velen.
 25 malyi A. Kerner
 26 hippomarathrum Jacq.
    (a) hippomarathrum   20   Cz   Činčura & Hindakova,
    (b) hebecarpum (DC.) Drude   1963
 27 dichotomum Pallas ex Bieb.
 28 tomentosum Vis.
 29 globiferum Vis.
 30 leucospermum Waldst. & Kit.   22   Hu   Baksay, 1956
 31 degenii Urum.
 32 parnassicum Boiss. & Heldr.
 33 bulgaricum P.W. Ball
 34 gracile Waldst. & Kit.

33 Oenanthe L.
  1 globulosa L.
    (a) globulosa   22   Lu   Cook, 1968
    (b) kunzei (Willk.) Nyman
  2 lisae Moris
  3 fistulosa L.   22   Br   Cook, 1968
  4 pimpinelloides L.

```
 5 millefolia Janka
 6 tenuifolia Boiss. & Orph.
 7 silaifolia Bieb.                     22      Br      Rutland, 1941
 8 peucedanifolia Pollich
 9 lachenalii C.C. Gmelin               22      Br Lu   Cook, 1968
10 banatica Heuffel
11 crocata L.                           22      Br Lu   Cook, 1968
12 fluviatilis (Bab.) Coleman           22      Br      Cook, 1968
13 aquatica (L.) Poiret                 22      Br      Cook, 1968

34 Lilaeopsis E.L. Greene
   1 attenuata (Hooker & Arnott) Fernald

35 Aethusa L.
   1 cynapium L.
     (a) cynapium                       20      Au      Titz, 1966
     (b) agrestis (Wallr.) Dostál       20      Su      Lövkvist, 1963
     (c) cynapioides (Bieb) Nyman       20      Su      Lövkvist, 1963

36 Portenschlagiella Tutin
   1 ramosissima (Portenschl.) Tutin

37 Athamanta L.
   1 macedonica (L.) Sprengel
     (a) macedonica
     (b) albanica (Alston & Sandwith) Tutin
     (c) arachnoidea (Boiss. & Orph.) Tutin
         Seseli farinosum Quezel & Contandr. 22   Gr    Quezel &
   2 sicula L.                                         Contandriopoulos, 1965
   3 turbith (L.) Brot.
     (a) turbith
     (b) haynaldii (Borbás & Uechtr.) Tutin
     (c) hungarica (Borbás) Tutin
   4 cretensis L.                       22      It      Favarger, 1965
   5 densa Boiss. & Orph.
   6 cortiana Ferrarini

38 Grafia Reichenb.
   1 golaka (Hacq.) Reichenb.

39 Xatardia Meissner
   1 scabra (Lapeyr.) Meissner

40 Foeniculum Miller
   1 vulgare Miller
     (a) piperitum (Ucria) Coutinho
     (b) vulgare

41 Anethum L.
   1 graveolens L.

42 Kundmannia Scop.
   1 sicula (L.) DC.

43 Silaum Miller
   1 silaus (L.) Schinz & Thell.

44 Trochiscanthes Koch
   1 nodiflora (Vill.) Koch

45 Meum Miller
```

1 athamanticum Jacq.

46 Physospermum Cusson
1 cornubiense (L.) DC.
2 verticillatum (Waldst. & Kit.) Vis.

47 Conium L.
1 maculatum L.

48 Hladnikia Reichenb.
1 pastinacifolia Reichenb. 22 Ju Susnik, 1962

49 Pleurospermum Hoffm.
1 austriacum (L.) Hoffm. 22 Hu Baksay, 1958
2 uralense Hoffm.

50 Aulacospermum Ledeb.
1 isetense (Sprengel) Schischkin

51 Lecokia DC.
1 cretica (Lam.) DC.

52 Cachrys L.
1 sicula L.
2 libanotis L.
3 pungens Jan ex Guss.
4 cristata DC.
5 ferulacea (L.) Calestani
6 trifida Miller
7 alpina Bieb.
8 odontalgica Pallas

53 Heptaptera Margot & Reuter
1 triquetra (Vent.) Tutin
2 colladonioides Margot & Reuter
3 angustifolia (Bertol.) Tutin
4 macedonica (Bornm.) Tutin

54 Magydaris Koch ex Dc.
1 pastinacea (Lam.) Paol.
2 panacifolia (Vahl) Lange

55 Hohenackeria Fischer & C.A. Meyer
1 exscapa (Steven) Kos.-Pol.
2 polyodon Cosson & Durieu

56 Bupleurum L.
1 rotundifolium L.
2 lancifolium Hornem.
3 longifolium L.
 (a) longifolium
 (b) aureum (Hoffm.) Soó
4 stellatum L. 16 Alps Favarger, 1954
5 angulosum L.
6 flavum Forskål
7 gracile D'Urv. 14 Gr Snogerup, 1962
8 aira Snogerup 14 Gr Snogerup, 1962
9 apiculatum Friv.
10 glumaceum Sibth. & Sm. 16 Gr Snogerup, 1962
11 baldense Turra
 (a) baldense

```
    (b) gussonei (Arcangeli) Tutin
12 flavicans Boiss. & Heldr.
13 karglii Vis.
14 capillare Boiss. & Heldr.
15 fontanesii Guss. ex Caruel
16 praealtum L.
17 commutatum Boiss. & Balansa
   (a) commutatum
   (b) glaucocarpum (Borbás) Hayek
   (c) aequiradiatum (H. Wolff) Hayek
18 brachiatum C. Koch ex Boiss.
19 gerardi All.
20 trichopodum Boiss. & Spruner
21 affine Sadler
22 asperuloides Heldr. ex Boiss.
23 tenuissimum L.                       16      Hu      Pólya, 1948
24 semicompositum L.
25 petraeum L.                          14      It      Favarger, 1959a
26 ranunculoides L.
   (a) ranunculoides                    42      Ga      Favarger, 1965b
   (b) gramineum (Vill.) Hayek          14      It      Favarger & Huynh,
                                                        1965
27 bourgaei Boiss. & Reuter
28 multinerve DC.
29 falcatum L.
   (a) falcatum                         16      Hu      Baksay, 1956
   (b) cernuum (Ten.) Arcangeli
       subsp. dilatatum                 32      Hu      Baksay, 1956
30 elatum Guss.
31 rigidum L.
   (a) rigidum
   (b) paniculatum (Brot.) H. Wolff
32 spinosum Gouan
33 fruticescens L.
34 dianthifolium Guss.
35 barceloi Cosson ex Willk.
36 acutifolium Boiss.
37 foliosum Salzm. ex DC.
38 gibraltarium Lam.
39 fruticosum L.

57 Trinia Hoffm.
 1 glauca (L.) Dumort.
   (a) glauca
   (b) carniolica (A. Kerner ex Janchen)
 2 multicaulis (Poiret) Schischkin
 3 hispida Hoffm.
 4 ramosissima (Fischer ex Trev.) Koch  20      Hu      Baksay, 1958
 5 kitaibelii Bieb.
 6 muricata Godet
 7 dalechampii (Ten.) Janchen
 8 guicciardii (Boiss. & Heldr.) Drude
 9 crithmifolia (Willd.) H. Wolff

58 Cuminum L.
 1 cyminum L.

59 Apium L.
 1 graveolens L.
 2 nodiflorum (L.) Lag.                  22      Br      Rutland, 1941
 3 repens (Jacq.) Lag.                   22      Hu      Baksay, 1956
```

4 inundatum (L.) Reichenb. fil.   22   Hs   Gadella & Kliphuis,
                                                    1966
5 crassipes (Koch ex Reichenb.) Reichenb.
  fil.

60 Petroselinum Hill
  1 crispum (Miller) A.W. Hill   22   Br   Rutland, 1941
  2 segetum (L.) Koch

61 Ridolfia Moris
  1 segetum Moris

62 Sison L.
  1 amomum L.

63 Cicuta L.
  1 virosa L.   22   Hu   Baksay, 1956

64 Cryptotaenia DC.
  1 canadensis (L.) DC.

65 Lereschia Boiss.
  1 thomasii (Ten.) Boiss.   12   It   Grau, ined.

66 Ammi L.
  1 visnaga (L.) Lam.
  2 crinitum Guss.
  3 majus L.
  4 huntii H.C. Watson
  5 trifoliatum (H.C. Watson) Trelease

67 Ptychotis Koch
  1 saxifraga (L.) Loret & Barrandon

68 Ammoides Adanson
  1 pusilla (Brot.) Breistr.

69 Thorella Briq.
  1 verticillatinundata (Thore) Briq.

70 Falcaria Fabr.
  1 vulgaris Bernh.   22   Su   Löve & Löve, 1942

71 Carum L.
  1 carvi L.   20   Su   Håkansson, 1953
  2 multiflorum (Sibth. & Sm.) Boiss.
   (a) multiflorum
   (b) strictum (Griseb.) Tutin
  3 verticillatum (L.) Koch
  4 rigidulum (Viv.) Koch ex DC.
  5 heldreichii Boiss.

72 Stefanoffia H. Wolff
  1 daucoides (Boiss.) H. Wolff

73 Brachyapium (Baillon) Maire
  1 dichotomum (L.) Maire

74 Endressia Gay
  1 pyrenaica (Gay ex DC.) Gay
  2 castellana Coincy

75 Cnidium Cusson
  1 dubium (Schkuhr) Thell.   20   Hu   Baksay, 1958
  2 silaifolium (Jacq.) Simonkai
   (a) silaifolium
   (b) orientale (Boiss.) Tutin

76 Selinum L.
  1 carvifolia (L.) L.   22   Hu   Pólya, 1949
  2 pyrenaeum (L.) Gouan

77 Ligusticum L.
  1 mutellinoides (Crantz) Vill.   22   He   Favarger, 1954
  2 mutellina (L.) Crantz   22   He   Favarger, 1954
  3 corsicum Gay
  4 albanicum Jáv.
  5 lucidum Miller
   (a) lucidum
   (b) huteri (Porta & Rigo) O. Bolós
  6 ferulaceum All.
  7 scoticum L.

78 Cenolophium Koch ex DC.
  1 denudatum (Hornem.) Tutin

79 Conioselinum Hoffm.
  1 tataricum Hoffm.

80 Angelica L.
  1 palustris (Besser) Hoffm.
  2 sylvestris L.   22   Hu   Pólya, 1950
  3 archangelica L.
   (a) archangelica   22   Su   Vaarama, 1947
   (b) litoralis (Fries) Thell.   22   Fe   Vaarama, 1947
  4 hererocarpa Lloyd
  5 razulii Gouan
  6 laevis Gay ex Ave-Lall.
  7 angelicastrum (Hoffmanns. & Link)
     Coutinho
  8 pachycarpa Lange

81 Levisticum Hill
  1 officinale Koch

82 Palimbia Besser
  1 rediviva (Pallas) Thell.

83 Bonannia Guss.
  1 graeca (L.) Halácsy

84 Johrenia DC.
  1 distans (Griseb.) Halácsy

85 Capnophyllum Gaertner
  1 peregrinum (L.) Lange

86 Ferula L.
  1 communis L.
   (a) communis
   (b) glauca (L.) Rouy & Camus
  2 tingitana L.
  3 sadlerana Ledeb.   22   Hu   Baksay, 1956

```
  4 heuffelii Griseb. ex Heuffel
  5 nuda Sprengel
  6 orientalis L.
  7 tatarica Fischer ex Sprengel
  8 caspica Bieb.

87 Ferulago Koch
  1 nodosa (L.) Boiss.
  2 thyrsiflora (Sibth. & Sm.) Koch
  3 asparagifolia Boiss.
  4 campestris (Besser) Grec.
  5 lutea (Poiret) Grande
  6 capillaris (Link ex Sprengel) Coutinho
  7 granatensis Boiss.
  8 sylvatica (Besser) Reichenb.
  9 sartorii Boiss.

88 Eriosynaphe DC.
  1 longifolia (Fischer ex Sprengel) DC.

89 Opopanax Koch
  1 chironium (L.) Koch              22      Hs      Gardé & Malheiros-
                                                     Gardé, 1949
  2 hispidus (Friv.) Griseb.

90 Peucedanum L.
  1 officinale L.                    66      Hu      Pólya, 1949
    (a) officinale
    (b) stenocarpum (Boiss. & Reuter) Font
        Quer
  2 longifolium Waldst. & Kit.
  3 tauricum Bieb.
  4 ruthenicum Bieb.
  5 rochelianum Heuffel
  6 paniculatum Loisel.
  7 gallicum Latourr.
  8 aragonense Rouy & Camus
  9 coriaceum Reichenb.
 10 oligophyllum (Griseb.) Vandas
    (a) oligophyllum
    (b) aequiradium (Velen.) Tutin
 11 achaicum Halácsy
 12 carvifolia Vill.
 13 schottii Besser ex DC.           22      It      Favarger, 1959a
 14 vittijugum Boiss.
 15 obtusifolium Sibth. & Sm.
 16 arenarium Waldst. & Kit.
    (a) arenarium
    (b) neumayeri (Vis.) Stoj. & Stefanov
 17 aegopodioides (Boiss.) Vandas
 18 latifolium (Bieb.) DC.
 19 alsaticum L.                     22      Hu      Pólya, 1949
 20 venetum (Sprengel) Koch
 21 austriacum (Jacq.) Koch
 22 oreoselinum (L.) Moench
 23 lancifolium Lange
 24 palustre (L.) Moench
 25 cervaria (L.) Lapeyr.
 26 ostruthium (L.) Koch
 27 verticillare (L.) Koch ex DC.
 28 hispanicum (Boiss.) Endl.
```

```
  29 alpinum (Sieber ex Schultes) B.L.
       Burtt & P.H. Davis

  91 Pastinaca L.
       1 sativa L.
         (a) sativa                          22    Su    Lövkvist, 1963
         (b) sylvestris (Miller) Rouy & Camus
         (c) urens (Req. ex Godron) Čelak
         (d) divaricata (Desf.) Rouy & Camus
       2 latifolia (Duby) DC.
       3 lucida L.
       4 hirsuta Pančić

  92 Heracleum L.
       1 minimum Lam.
       2 austriacum L.                       22    Au    Polatschek, 1966
         (a) austriacum
         (b) siifolium (Scop.) Nyman
       3 orphanidis Boiss.
       4 carpaticum Porc.
       5 sphondylium L.                      22    Su    Håkansson, 1953
         (a) alpinum (L.) Bonnier & Layens
         (b) pyrenaicum (Lam.) Bonnier & Layens
         (c) transsilvanicum (Schur) Brummitt
         (d) montanum (Schleicher ex Gaudin) Briq.
         (e) sphondylium
         (f) verticillatum (Pančić) Brummitt
         (g) orsinii (Guss.) H. Neumayer
         (h) ternatum (Velen.) Brummitt
         (i) sibiricum (L.) Simonkai         22    Su    Håkansson, 1953
       6 ligusticifolium Bieb.
       7 stevenii Manden.
       8 pubescens (Hoffm.) Bieb.
       9 mantegazzianum Sommier & Levier

  93 Malabaila Hoffm.
       1 aurea (Sibth. & Sm.) Boiss.         22    Gr    Moore, ined.
       2 graveolens (Sprengel) Hoffm.
       3 involucrata Boiss. & Spruner

  94 Tordylium L.
       1 maximum L.
       2 officinale L.
       3 pestalozzae Boiss.
       4 apulum L.
       5 byzantinum (Aznav.) Hayek

  95 Laser Borkh.
       1 trilobum (L.) Borkh.

  96 Elaeoselinum Koch ex DC.
       1 asclepium (L.) Bertol.
         (a) asclepium
         (b) meoides (Desf.) Fiori
       2 tenuifolium (Lag.) Lange
       3 foetidum (L.) Boiss.
       4 gummiferum (Desf.) Tutin

  97 Guillonea Cosson
       1 scabra (Cav.) Cosson
```

98 Laserpitium L.
  1 siler L.
   (a) siler
   (b) garganicum (Ten.) Arcangeli
   (c) zernyi (Hayek) Tutin
  2 latifolium L.   22   Hu   Baksay, 1961
  3 longiradium Boiss.
  4 nestleri Soyer-Willemet
  5 krapfii Crantz
   (a) krapfii
   (b) gaudinii (Moretti) Thell.
  6 peucedanoides L.   22   It   Favarger, 1959a
  7 archangelica Wulfen
  8 nitidum Zanted.   22   It   Favarger, 1959a
  9 halleri Crantz
   (a) halleri
   (b) cynapiifolium (Viv. ex DC.) P.   22   Co   Contandriopoulos,
     Fourn.   1964a
  10 gallicum L.
  11 pseudomeum Orph.,Heldr. & Sart. ex Boiss.
  12 prutenicum L.
   (a) prutenicum
   (b) dufourianum (Rouy & Camus) Tutin
  13 hispidum Bieb.

99 Thapsia L.
  1 villosa L.
  2 maxima Miller
  3 garganica L.

100 Rouya Coincy
  1 polygama (Desf.) Coincy

101 Melanoselinum Hoffm.
  1 decipiens (Schrader & Wendl.) Hoffm.

102 Torilis Adanson
  1 nodosa (L.) Gaertner   22   Lu   Gardé & Malheiros-
                                                                 Gardé, 1949
  2 arvensis (Hudson) Link   12   Hu   Baksay, 1958
   (a) neglecta (Schultes) Thell.
   (b) arvensis
   (c) elongata (Hoffmanns. & Link) Cannon
   (d) purpurea (Ten.) Hayek
  3 japonica (Houtt.) DC.   16   Su   Lövkvist, 1963
  4 ucranica Sprengel
  5 tenella (Delile) Reichenb. fil.
  6 leptophylla (L.) Reichenb. fil

103 Astrodaucus Drude
  1 orientalis (L.) Drude
  2 littoralis (Bieb.) Drude

104 Turgeniopsis Boiss.
  1 foeniculacea (Fenzl) Boiss.

105 Caucalis L.
  1 platycarpos L.   20   Hs   Lorenzo-Andreu, 1951

106 Turgenia Hoffm.
  1 latifolia (L.) Hoffm.

107 Orlaya Hoffm.
1 kochii Heywood   16   It   Larsen, 1956a
2 grandiflora (L.) Hoffm.   20   Hu   Baksay, 1956
3 daucorlaya Murb.

108 Daucus L.
1 durieua Lange
2 muricatus (L.) L.
3 broteri Ten.
4 halophilus Brot.
5 aureus Desf.
6 guttatus Sibth. & Sm.
  (a) guttatus
  (b) zahariadii Heywood
7 involucratus Sibth. & Sm.
8 carota L.
  (a) carota   18   Ho   Gadella & Kliphuis,
                                                                     1966
  (b) maritimus (Lam.) Batt.
  (c) major (Vis.) Arcangeli
  (d) maximus (Desf.) Ball
  (e) sativus (Hoffm.) Arcangeli
  (f) gummifer Hooker fil.
  (g) commutatus (Paol.) Thell.
  (h) hispanicus (Gouan) Thell.
  (i) hispidus (Arcangeli) Heywood
  (j) gadecaei (Rouy & Camus) Heywood
  (k) drepanensis (Arcangeli) Heywood
  (l) rupestris (Guss.) Heywood
9 setifolius Desf.
10 crinitus Desf.

109 Pseudorlaya (Murb.) Murb.
1 pumila (L.) Grande   16   It   Larsen, 1956a
2 minuscula (Pau ex Font Quer) Lainz

110 Artedia L.
1 squamata L.

CXXX DIAPENSIACEAE

1 Diapensia L.
1 lapponica L.   12   Is   Löve & Löve, 1956

CXXXI PYROLACEAE

1 Pyrola L.
1 minor L.   46   Fe   V. Sorsa, 1962
2 media Swartz   92   Da   Hagerup, 1941
3 chlorantha Swartz   46   Da   Hagerup, 1941
4 rotundifolia L.
  (a) rotundifolia   46   Su   Knaben & Engelskjon,
                                                                    1968
  (b) maritima (Kenyon) E.F. Warburg   46   Da No   Knaben & Engelskjon,
                                                                    1968
5 grandiflora Radius   46   Is   Löve & Löve, 1956
6 norvegica Knaben   46   No   Knaben, 1950
7 carpatica J. Holub & Křísa   46   Cz   Holub & Křísa, 1971

153

2 Orthilia Rafin.
  1 secunda (L.) House
   (a) secunda   38   Da   Hagerup, 1941
   (b) obtusata (Turcz.) Böcher

3 Moneses Salisb.
  1 uniflora (L.) A. Gray   26   Da   Hagerup, 1941

4 Chimaphila Pursh
  1 umbellata (L.) W. Barton   26   Da   Knaben & Engelskjon,
                                         1968
  2 maculata (L.) Pursh

5 Monotropa L.
  1 hypopitys L.   16, 48 Da   Hagerup, 1944

CXXXII ERICACEAE

1 Erica L.
  1 sicula Guss.
  2 ciliaris L.   24   Br   Glanville, ined.
  3 mackaiana Bab.   24   Hb Hs Glanville, ined.
  4 tetralix L.   24   Da   Hagerup, 1928
  5 terminalis Salisb.
  6 cinerea L.   24   Fa   Hagerup, 1928
  7 australis L.
  8 umbellata L.
  9 arborea L.
 10 lusitanica Rudolphi
 11 vagans L.   24   Br   Maude, 1940
 12 manipuliflora Salisb.
 13 multiflora L.
 14 herbacea L.   24   Da   Hagerup, 1928
 15 erigena R. Ross   24   Hb   Glanville, ined.

2 Bruckenthalia Reichenb.
  1 spiculifolia (Salisb.) Reichenb.

3 Calluna Salisb.
  1 vulgaris (L.) Hull   16   Is   Löve & Löve, 1956b

4 Cassiope D. Don
  1 tetragona (L.) D. Don
  2 hypnoides (L.) D. Don   32   Is   Löve & Löve, 1961

5 Rhododendron L.
  1 luteum Sweet
  2 ponticum L.
   (a) ponticum
   (b) baeticum (Boiss. & Reuter) Hand.-
      Mazz.
  3 ferrugineum L.
  4 myrtifolium Schott & Kotschy
  5 hirsutum L.
  6 lapponicum (L.) Wahlenb.

6 Ledum L.
  1 palustre L.
   (a) palustre   c.52   Fe   V. Sorsa, 1962
   (b) groenlandicum (Oeder) Hultén

7 Rhodothamnus Reichenb.
  1 chamaecistus (L.) Reichenb.   24   Au   Polatschek, 1966

8 Kalmia L.
  1 angustifolia L.
  2 polifolia Wangenh.

9 Loiseleuria Desv.
  1 procumbens (L.) Desv.   24   Is   Löve & Löve, 1956

10 Phyllodoce Salisb.
   1 caerulea (L.) Bab.   24   Fe   Sorsa, 1963

11 Daboecia D. Don
   1 cantabrica (Hudson) C. Koch   24   Hb   Maude, 1939
   2 azorica Tutin & E.F. Warburg

12 Arbutus L.
   1 unedo L.   26   Hb   Sealy & Webb, 1950
   1 andrachne L.

13 Arctostaphylos Adanson
   1 uva-ursi (L.) Sprengel   52   Is   Löve & Löve, 1956
   2 alpinus (1.) Sprengel   26   Rs(N)   Sokolovskaya &
                                                            Strelkova, 1960

14 Gaultheria L.
   1 shallon Pursh

15 Pernettya Gaud.-Beaupré
   1 mucronata (L. fil.) Gaud.-Beaupré
     ex Sprengel

16 Andromeda L.
   1 polifolia L.   c.48   Fe   Sorsa, 1963

17 Chamaedaphne Moench
   1 calyculata (L.) Moench   22   Fe   Sorsa, 1966

18 Vaccinium L.
   1 oxycoccos (Hill) A. Gray   48   Da   Hagerup, 1940
                                           72   Da   Hagerup, 1940
   2 microcarpum (Turcz. ex Rupr.) Schmalh. 24   Is   Löve & Löve, 1956
   3 macrocarpon Aiton
   4 vitis-idaea L.   24   Is   Löve & Löve, 1956
     (a) vitis-idaea
     (b) minus (Loddiges) Hultén
   5 uliginosum L.
     (a) uliginosum   48   Fe   Rousi, 1967
     (b) microphyllum Lange   24   Is   Löve & Löve, 1956
   6 myrtillus L.   24   Br   I. & O. Hedberg, 1961
   7 arctostaphylos L.
   8 cylindraceum Sm.

CXXXIII EMPETRACEAE

1 Corema D. Don
  1 album (L.) D. Don   26   Lu   Hagerup, 1941a

2 Empetrum L.

1 nigrum L.
 (a) nigrum 26 Fe Sorsa, 1963
 (b) hermaphroditum (Hagerup) Böcher 52 Is Löve & Löve, 1956

CXXXIV MYRSINACEAE

1 Myrsine L.
 1 africana L.

CXXXV PRIMULACEAE

1 Primula L.
 1 vulgaris Hudson
  (a) vulgaris 22 Br Eaton, 1968
  (b) balearica (Willk.) W.W. Sm. &
   Forrest
  (c) sibthorpii (Hoffmanns.) W.W. Sm.
   & Forrest
 2 elatior (L.) Hill
  (a) elatior 22 Br Eaton, 1968
  (b) leucophylla (Pax) H.-Harrison ex
   W.W. Sm. & Fletcher
  (c) intricata (Gren. & Godron) Ludi
  (d) lofthousei (H.-Harrison) W.W. Sm.
   & Fletcher 22 Hs Valentine, 1978
  (e) pallasii (Lehm.) W.W. Sm. & Forrest
 3 veris L.
  (a) veris 22 Fe Sorsa, 1963
  (b) canescens (Opiz) Hayek ex Ludi
  (c) columnae (Ten.) Ludi
  (d) macrocalyx (Bunge) Ludi
 4 farinosa L.
  (a) farinosa 18 Br Su Davies, 1953
   36 Br Davies, 1953
  (b) exigua (Velen.) Hayek
 5 frondosa Janka
 6 longiscapa Ledeb.
 7 halleri J.F. Gmelin
 8 scotica Hooker 54 Br Dovaston, 1955
 9 scandinavica Bruun 72 Br Dovaston, 1955
10 stricta Hornem. 126 Is Löve & Löve, 1956
11 nutans Georgi 22 Su Bruun, 1932
12 egaliksensis Wormsk.
13 deorum Velen. 66 He Kress, 1963
14 glutinosa Wulfen 66 It Kress, 1963
   67 It Kress, 1963
15 spectabilis Tratt. 66 It Kress, 1963
16 glaucescens Moretti 66 It Kress, 1963
17 wulfeniana Schott 66 Au Kress, 1963
18 clusiana Tausch c.198 Au Kress, 1963
19 minima L. 66 67 68 Au It Kress, 1963
   69 70 73
20 kitaibeliana Schott 66 Ju Kress, 1963
21 integrifolia L. 66 c.68 Ga It Kress, 1963
   c.70
22 tyrolensis Schott 66 It Kress, 1963
23 allionii Loisel. 66 Ga Kress, 1963
24 palinuri Petagna 66 It Kress, 1963
25 latifolia Lapeyr. 64 65 66 He Kress, 1963
   67

26 marginata Curtis	62 63 ?124	Ga	Favarger & Huynh, 1964
	126 127 128	It	Kress, 1963
27 carniolica Jacq.	62 c.68	He	Kress, 1963
28 auricula L.	62	Po	Skalińska et al., 1959
	63 64 65 66	It	Kress, 1963
29 hirsuta All.	62 63 64 67	It	Kress, 1963
30 pedemontana Thomas ex Gaudin	62	It	Kress, 1963
31 apennina Widmer	62	It	Kress, 1963
32 daonensis (Leybold) Leybold	62 63 64	It	Kress, 1963
33 villosa Wulfen	62 63 64	It	Kress, 1963

2 Vitaliana Sesler
  1 primuliflora Bertol.
    (a) primuliflora
    (b) canescens O. Schwarz

(c) cinerea (Sund.) I.K. Ferguson	40	Ga	Favarger & Küpfer, 1968

    (d) assoana Lainz
    (e) praetutiana (Buser ex Sünd.) I.K. Ferguson

3 Androsace L.
  1 maxima L.
  2 septentrionalis L.

3 chaixii Gren. & Godron	34	Ga	Favarger, 1965b

  4 filiformis Retz.
  5 elongata L.

6 lactea L.	76	Po He	Favarger, 1962
7 obtusifolia All.	36	He	Favarger, 1962
	38	He	Favarger, 1958

  8 hedraeantha Griseb.
  9 carnea L.

(a) carnea	38	Ga	Favarger, 1958
(b) laggeri (Huet) Nyman	38	Ga	Favarger, 1958
(c) rosea (Jordan & Fourr.) Rouy	38	Ga	Favarger, 1958
(d) brigantiaca (Jordan & Fourr.) I.K. Ferguson	78	He	Favarger, 1958
10 pyrenaica Lam.	38	Ga/Hs	Kress, 1963
11 chamaejasme Wulfen	20	Po	Skalińska, 1959
12 villosa L.	20	He	Favarger, 1958
13 helvetica (L.) All.	40	He	Favarger, 1958
14 vandellii (Turra) Chiov.	40	Ga Hs	Kress, 1963
15 cylindrica DC.	40	Hs	Kress, 1963
16 mathildae Levier	40	It	Kress, 1963
17 pubescens DC.	40	He Hs	Kress, 1963
18 ciliata DC.	c.80	Hs	Kress, 1963
19 hausmannii Leybold	40	It	Kress, 1963
20 alpina (L.) Lam.	40	He	Favarger, 1958
21 wulfeniana Sieber ex Koch	40	Au	Kress, 1963
22 brevis (Hegetschw.) Cesati	40	It	Kress, 1963

4 Cortusa L.

1 matthioli L.	24	Po	Skalińska et al., 1959

5 Soldanella L.

1 pusilla Baumg.	40	Rm	Tarnavschi, 1948

  2 minima Hoppe

(a) minima	40	It	Favarger, 1965
(b) samnitica Cristofolini & Pignatti			
3 austriaca Vierh.			
4 alpina L.	40	He	Larsen, 1954
5 carpatica Vierh.	40	Po	Skalińska, 1950
6 pindicola Hausskn.			
7 hungarica Simonkai			
(a) hungarica			
(b) major (Neilr.) S. Pawl.	40	Po	Skalińska, 1950
8 dimoniei Vierh.			
9 montana Willd.	40	Po	Skalińska, 1950
10 villosa Darracq			

6 Hottonia L.
1 palustris L.	20	Po	Skalińska et al., 1961

7 Cyclamen L.
1 hederifolium Aiton	24	Bu	Kuzmanov & Kozuharov, ined.
2 graecum Link	78-80	Gr	Glasau, 1939
3 purpurascens Miller	34	Au	Glasau, 1939
4 repandum Sibth. & Sm.			
5 creticum Hildebr.	24	Cr	Glasau, 1939
6 balearicum Willk.	18	Bl	Glasau, 1939
	20	Bl	Kress, 1963
7 coum Miller			
8 persicum Miller			

8 Lysimachia L.
1 nemorum L.	16	Ho	Gadella & Kliphuis, 1963
	18	Po	Skalińska et al., 1964
	28	Ge	Wulff, 1938
2 serpyllifolia Schreber			
3 vulgaris L.	56 84	Ho	Gadella & Kliphuis, 1966, 1968
4 terrestris (L.) Britton, E.E. Sterns & Poggenb.			
5 ciliata L.			
6 nummularia L.	32 43 45	Ho	Gadella & Kliphuis, 1963, 1967a
	36	Po	Skalińska et al., 1964
7 punctata L.			
8 minoricensis Rodr.			
9 ephemerum L.			
10 clethroides Duby			
11 atropurpurea L.			
12 dubia Aiton			
13 thyrsiflora L.	54	Ho	Gadella & Kliphuis, 1963

9 Trientalis L.
1 europaea L.	c.160	Is	Löve & Löve, 1956

10 Asterolinon Hoffmanns. & Link
1 linum-stellatum (L.) Duby

11 Glaux L.
1 maritima L.	30	Is	Löve & Löve, 1956

12 Anagallis L.

1 minima (L.) E.H.L. Krause	22	Da	Hagerup, 1941b
2 crassifolia Thore			
3 tenella (L.) L.	22	Au	Haffner, 1950
4 arvensis L.	40	Gr	Kollman & Feinbrun, 1968
5 foemina Miller	40	?Br	Marsden-Jones & Weiss, 1960
6 monelli L.	22	Hs	D.M. Moore, ined.

13 Samolus L.

1 valerandi L.	?24	Ge	Wulff, 1937a
	26	Ho	Gadella & Kliphuis, 1968

14 Coris L.

1 monspeliensis L.	18	It	Kress, 1963
	56	Hs	Gadella et al., 1966
2 hispanica Lange			

CXXXVI PLUMBAGINACEAE

1 Plumbago L.

1 europaea L.	12	Rm	Tarnavschi, 1938

2 Ceratostigma Bunge
   1 plumbaginoides Bunge

3 Acantholimon Boiss.
   1 androsaceum (Jaub. & Spach) Boiss.

4 Armeria Willd.

1 maritima (Miller) Willd.			
(a) maritima	18	Is	Löve & Löve, 1956
(b) sibirica (Turcz. ex Boiss.) Nyman			
(c) miscella (Merino) Malagarriga			
(d) elongata (Hoffm.) Bonnier	18	Br	Baker, 1959
(e) halleri (Wallr.) Rothm.	18	Ge	Griesinger, 1937
(f) purpurea (Koch) A. & D. Löve	18	Ge	Griesinger, 1937
(g) alpina (Willd.) P. Silva	18	Ju	Lovka et al., 1971
	36	Europe	Donadille, 1967
(h) barcensis (Simonkai) P. Silva			
2 canescens (Host) Boiss.			
(a) canescens	18	Europe	Donadille, 1967
(b) nebrodensis (Guss.) P. Silva	18	Europe	Donadille, 1967
3 undulata (Bory) Boiss.			
4 rumelica Boiss.	18	Europe	Donadille, 1967
5 sancta Janka			
6 vandasii Hayek			
7 denticulata (Bertol.) DC.			
8 alliacea (Cav.) Hoffmans. & Link	18	Europe	Donadille, 1967
9 ruscinonensis Girard	18	Ga	Donadille, 1967
10 pseudarmeria (Murray) Mansfeld	18	Lu	Donadille, 1967
11 vestita Willk.			
12 langei Boiss. ex Lange			
(a) langei			
(b) daveaui (Coutinho) P. Silva			
13 villosa Girard	18	Hs	Phillips, 1938
14 transmontana (Samp.) Lawrence			

```
15 morisii Boiss.
16 colorata Pau
17 macropoda Boiss.
18 filicaulis (Boiss.) Boiss.            18    Ga/Hs    Donadille, 1967
19 trachyphylla Lange
20 littoralis Willd.                     18    Europe   Donadille, 1967
21 duriaei Boiss.
22 arcuata Welw. ex Boiss. & Reuter
23 girardii (Bernis) Litard.            18    Europe   Donadille, 1967
24 eriophylla Willk.
25 humilis (Link) Schultes
   (a) humilis                           18    Lu       Fernandes, 1950
   (b) odorata (Samp.) P. Silva
26 sardoa Sprengel                       18    Sa       Donadille, 1967
27 leucocephala Salzm. ex Koch           18    Co       Contandriopoulos,
28 multiceps Wallr.                      18    Co          1962
29 soleirolii (Duby) Godron              18    Co       Donadille, 1967
30 juniperifolia (Vahl) Hoffmanns. & Link 18  Hs       Sugiura, 1938
31 splendens (Lag. & Rodr.) Webb
   (a) splendens                         18    Hs       Donadille, 1967
   (b) bigerrensis (C. Vicioso & Beltran)
       P. Silva
32 pubigera (Desf.) Boiss.
33 berlengensis Daveau
34 welwitschii Boiss.                    18    Lu       Donadille, 1967
35 pungens (Link) Hoffmanns. & Link      18    Europe   Donadille, 1967
36 pinifolia (Brot.) Hoffmanns. & Link   18    Lu       Donadille, 1967
37 rouyana Daveau                        18    Europe   Donadille, 1967
38 macrophylla Boiss. & Reuter           18    Hs/Lu    Sugiura, 1944
39 velutina Welw. ex Boiss. & Reuter
40 gaditana Boiss.
41 hirta Willd.
42 hispalensis Pau
43 cariensis Boiss.


5 Limonium Miller
  1 sinuatum (L.) Miller
  2 bonduellei (Lestib.) O. Kuntze
  3 thouinii (Viv.) O. Kuntze
  4 brassicifolium (Webb & Berth.) O.
    Kuntze
  5 ferulaceum (L.) O. Kuntze
  6 diffusum (Pourret) O. Kuntze
  7 caesium (Girard) O. Kuntze            16
  8 insigne (Cosson) O. Kuntze            18    Hs       Erben, 1978
    (a) insigne
    (b) carthaginiense Pignatti
  9 latifolium (Sm.) O. Kuntze
 10 asterotrichum (Salmon) Salmon
 11 sareptanum (A. Becker) Gams
 12 bungei (Claus) Gamajun.
 13 gmelinii (Willd.) O. Kuntze           18    Hu       Borhidi, 1968
                                          27    Rs(E)    Aleskoorskij, 1929
                                          36    Hu       Polya, 1948

 14 meyeri (Boiss.) O. Kuntze
 15 hirsuticalyx Pignatti
 16 tomentellum (Boiss.) O. Kuntze
 17 vulgare Miller
    (a) vulgare                           36    Lu       Rodrigues, 1953
    (b) serotinum (Reichenb.) Gams
 18 humile Miller                         36    Br       Choudhuri, 1942
```

19 bellidifolium (Gouan) Dumort.   18   Rs(E)   Aleskoovskij, 1929
20 cancellatum (Bernh. ex Bertol.) O.
   Kuntze
21 vestitum (Salmon) Salmon
22 johannis Pignatti
23 cordatum (L.) Miller
24 furfuraceum (Lag.) O. Kuntze   18   Hs   Erben, 1978
25 lucentinum Pignatti & Freitag
26 dichotomum (Cav.) O. Kuntze
27 calcarae (Tod. ex Janka) Pignatti
28 catalaunicum (Willk. & Costa) Pignatti
   (a) catalaunicum
   (b) procerum (Willk.) Pignatti
   (c) viciosoi (Pau) Pignatti
29 articulatum (Loisel.) O. Kuntze
30 japygicum (Groves) Pignatti
31 minutum (L.) Fourr.
32 caprariense (Font Quer & Marcos)
   Pignatti
   (a) caprariense
   (b) multiflorum Pignatti
33 parvifolium (Tineo) Pignatti
34 acutifolium (Reichenb.) Salmon
35 aragonense (Debeaux) Pignatti
36 cosyrense (Guss.) O. Kuntze
37 pontium Pignatti
38 multiforme (U. Martelli) Pignatti
39 bocconei (Lojac.) Litard.
40 anfractum (Salmon) Salmon
41 melium (Nyman) Pignatti
42 tremolsii (Rouy) P. Fourn.   18   Hs   Erben, 1978
43 calaminare Pignatti ex Pignatti   18   Hs   Erben, 1978
44 tenuiculum (Tineo ex Guss.) Pignatti
45 hermaeum (Pignatti) Pignatti
46 remotispiculum (Lacaita) Pignatti
47 minutiflorum (Guss.) O. Kuntze
   (a) minutiflorum
   (b) balearicum Pignatti
48 inarimense (Guss.) Pignatti
   (a) inarimense
   (b) ebusitanum (Font Quer) Pignatti
49 tenoreanum (Guss.) Pignatti
50 laetum (Nyman) Pignatti
51 gougetianum (Girard) O. Kuntze
52 albidum (Guss.) Pignatti
53 frederici (W. Barbey) Rech. fil.
54 panormitanum (Tod.) Pignatti
55 graecum (Poiret) Rech. fil.
   (a) graecum
   (b) divaricatum (Rouy) Pignatti
56 oleifolium Miller
   (a) oleifolium
   (b) dictyocladum Arcangeli
   (c) sardoum (Pignatti) Pignatti)
   (d) pseudodictyocladum (Pignatti) Pignatti
57 carpathum (Rech. fil.) Rech. fil.
58 emarginatum (Willd.) O. Kuntze
59 duriusculum (Girard) Fourr.
60 ramosissimum (Poiret) Maire
   (a) confusum (Gren. & Godron) Pignatti
   (b) provinciale (Pignatti) Pignatti

```
    (c) tommasinii (Pignatti) Pignatti
    (d) siculum Pignatti
    (e) doerfleri (Halácsy) Pignatti
 61 ocymifolium (Poiret) O. Kuntze
 62 densissimum (Pignatti) Pignatti
 63 parvibracteatum Pignatti
 64 girardianum (Guss.) Fourr.              35      Ga      Baker, 1955, 1971
 65 densiflorum (Guss.) O. Kuntze           26      Si      Dolcher &
 66 dufourei (Girard) O. Kuntze                             Pignatti, 1971
 67 binervosum (G.E. Sm.) Salmon         34-36      Br      Baker, 1955, 1971
 68 recurvum Salmon                         27      Br      Baker, 1955, 1971
 69 transwallianum (Pugsley) Pugsley        27      Br      Baker, 1955, 1971
                                            35      Hb      Baker, 1955, 1971
 70 paradoxum Pugsley                       33      Br      Baker, 1955, 1971
 71 salmonis (Sennen & Elias) Pignatti
 72 auriculae-ursifolium (Pourret) Druce    25      Ga      Baker, 1955, 1971
                                            26
    (a) auriculae-ursifolium
    (b) lusitanicum Pignatti
    (c) multiflorum (Pignatti) Pignatti
 73 majoricum Pignatti
 74 ovalifolium (Poiret) O. Kuntze          16      Hs      Erben, 1978
    (a) gallicum Pignatti
    (b) lusitanicum Pignatti
 75 biflorum (Pignatti) Pignatti
 76 lausianum Pignatti
 77 delicatulum (Girard) O. Kuntze          27      Hs      Pignatti, ined.
    (a) delicatulum
    (b) tournefortii Pignatti
    (c) valentinum Pignatti
 78 gibertii (Sennen) Sennen             25-27      Hs      Pignatti, ined.
 79 costae (Willk.) Pignatti
 80 sibthorpianum (Guss.) O. Kuntze
 81 eugeniae Sennen                         16      Hs      Pignatti, ined.
 82 coincyi Sennen                          16      Hs      Pignatti, ined.
 83 album (Coincy) Sennen                   18      Hs      Pignatti, ined.
 84 cymuliferum (Boiss.) Sauvage & Vindt    16      Hs      Erben, 1978
 85 supinum (Girard) Pignatti            26-27      Hs
 86 suffruticosum (L.) O. Kuntze
 87 echioides (L.) Miller                   18      Ga Hs   Erben, 1978

 6 Goniolimon Boiss.
    1 elatum (Fischer ex Sprengel) Boiss.
    2 speciosum (L.) Boiss.
    3 heldreichii Halácsy
    4 dalmaticum (C. Presl) Reichenb. fil.
    5 tataricum (L.) Boiss.
    6 tauricum Klokov
    7 rubellum (S.G. Gmelin) Klokov & Grossh.
    8 graminifolium (Aiton) Boiss.
    9 collinum (Griseb.) Boiss.
   10 besseranum (Schultes ex Reichenb.)      36      Rm      Tarnavschi, 1938
      Kusn.
   11 sartorii Boiss.

 7 Limoniastrum Heister ex Fabr.
    1 monopetalum (L.) Boiss.

 8 Psylliostachys (Jaub. & Spach) Nevski
    1 spicata (Willd.) Nevski
```

CXXXVII EBENACEAE

1 Diospyros L.
  1 lotus L.

CXXXVIII STYRACACEAE

1 Styrax L.
  1 officinalis L.

CXXXIX OLEACEAE

1 Jasminum L.
  1 nudiflorum Lindley
  2 officinale L.
  3 fruticans L.
  4 humile L.

2 Fontanesia Labill.
  1 philliraeoides Labill.

3 Forsythia Vahl
  1 europaea Degen & Bald.

4 Fraxinus L.
  1 ornus L.
  2 pennsylvanica Marshall
  3 excelsior L.
    (a) excelsior
    (b) coriariifolia (Scheele) E. Murray
  4 angustifolia Vahl
    (a) angustifolia
    (b) oxycarpa (Bieb. ex Willd.) Franco &
        Rocha Afonso
  5 pallisiae Wilmott

3 excelsior L.	46	Ho	Gadella & Kliphuis,
(a) excelsior	46	Ho	1963

5 Syringa L.
  1 vulgaris L.
  2 josikaea Jacq. fil. ex Reichenb.

6 Ligustrum L.
  1 vulgare L.   46   Ho   Gadella & Kliphuis, 1966

7 Olea L.
  1 europaea L.   46   It   Breviglieri & Battaglia, 1954

8 Phillyrea L.
  1 angustifolia L.
  2 latifolia L.   46   It   Armenise, 1958

9 Picconia DC.
  1 azorica (Tutin) Knobl.

CXL GENTIANACEAE

1 Cicendia Adanson

1 filiformis (L.) Delarbre	26	Ga	Favarger, 1960

2 Exaculum Caruel

1 pusillum (Lam.) Caruel	20	Ga	Favarger, 1969

3 Blackstonia Hudson
1 perfoliata (L.) Hudson

(a) perfoliata	40	He Gr Ga It	Zeltner, 1966
(b) serotina (Koch ex Reichenb.) Vollmann	40	Ga Gr Hs Cz	Zeltner, 1966
(c) imperfoliata (L. fil.) Franco & Rocha Afonso	20	Ga	Zeltner, 1962
(d) grandiflora (Viv.) Maire			

4 Centaurium Hill

1 scilloides (L. fil.) Samp.	20	Br Lu Hs	Zeltner, 1970
2 erythraea Rafn			
(a) erythraea	40	Br Ga Hs It Si	Zeltner, 1962, 1970
(b) turcicum (Velen.) Melderis			
(c) rumelicum (Velen.) Melderis	20	Ga It Gr Sa Co	Zeltner, 1970
(d) grandiflorum (Biv.) Melderis			
(e) rhodense (Boiss. & Reuter) Melderis	40	Si It Sa	Zeltner, 1970
(f) majus (Hoffmanns. & Link) Melderis	20	Hs Lu It Ga Sa	Zeltner, 1963 Zeltner, 1970
3 suffruticosum (Griseb.) Ronniger	20	Hs	Zeltner, 1970
4 chloodes (Brot.) Samp.	40	Lu	Zeltner, 1970
5 littorale (D. Turner) Gilmour			
(a) littorale	38	Ge	Wulff, 1937a
	40	Ge No	Zeltner, 1970
	c.56	Br	Warburg, 1939
(b) uliginosum (Waldst. & Kit.) Melderis	40	Be Br Ho Su	Zeltner, 1970
6 favargeri Zeltner	20	Ga	Zeltner, 1966
7 triphyllum (W.L.E. Schmidt) Melderis	20	Hs	Zeltner, 1967
8 rigualii Esteve			
9 linariifolium (Lam.) G. Beck	20	Hs	Zeltner, 1963
10 microcalyx (Boiss. & Reuter) Ronniger			
11 pulchellum (Swartz) Druce	36	Hs Ga Gr Ju Au Br Be Ho	Zeltner, 1962, 1963, 1966
12 tenuiflorum (Hoffmanns. & Link) Fritsch			
(a) tenuiflorum	40	Gr Lu Sa	Zeltner, 1966
(b) acutiflorum (Schott) Zeltner	20	Hs Ga It Co Si	Zeltner, 1961
13 spicatum (L.) Fritsch	22	Hs Si Sa Ga Lu	Zeltner, 1970
14 maritimum (L.) Fritsch	20	Co Ga Sa	Zeltner, 1970

5 Gentiana L.

1 lutea L.			
(a) lutea	40	Ga	Favarger, 1949b
(b) symphyandra (Murb.) Hayek			
2 punctata L.	40	Po	Skalińska, 1952
3 pannonica Scop.	40	He	Favarger, 1965b
4 purpurea L.	40	No	Knaben & Engelskjon, 1967
5 burseri Lapeyr.			

(a) burseri	40	Ga	Favarger & Kupfer, 1968
(b) villarsii (Griseb.) Rouy	c.40	Ga	Favarger & Kupfer, 1968
6 asclepiadea L.	36	Po	Skalińska, 1950
7 pneumonanthe L.	26	Hu	Pólya, 1950
8 frigida Haenke	24	Po	Skalińska, 1950
9 froelichii Jan ex Reichenb.	42	Ju	Favarger, 1965b
10 cruciata L.	52	Po	Skalińska, 1952
(a) cruciata			
(b) phlogifolia (Schott & Kotschy) Tutin			
11 decumbens L. fil.			
12 prostrata Haenke	c.36	He	Favarger, 1952
13 pyrenaica L.	26	Ge	Favarger & Kupfer, 1968
14 boryi Boiss.			
15 clusii Perr. & Song.	36	Po	Skalińska, 1950
16 occidentalis Jakowatz			
17 ligustica R. de Vilmorin & Chopinet			
18 acaulis L.	36	Ju	Lovka et al., 1971
19 alpina Vill.	36	Ga	Favarger & Kupfer, 1968
20 dinarica G. Beck			
21 angustifolia Vill.	36	Ga	Favarger, 1965
22 verna L.			
(a) verna	28	Po	Skalińska, 1950
(b) pontica (Soltok.) Hayek			
(c) tergestina (G. Beck) Hayek			
23 brachyphylla Vill.			
(a) brachyphylla	28	He	Favarger, 1965
(b) favratii (Rittener) Tutin	28 30 + 2B	He	Favarger, 1965
	32	Ga	Favarger, 1965
24 pumila Jacq.	20	Au	Favarger, 1965
(a) pumila			
(b) delphinensis (Beauverd) P. Fourn.			
25 bavarica L.	30	It	Favarger, 1965
26 rostanii Reuter ex Verlot	30	Ga	Favarger, 1969
27 terglouensis Hacq.	40	It	Favarger, 1965
(a) tergloensis			
(b) schleicheri (Vacc.) Tutin			
28 nivalis L.	14	Po	Skalińska et al., 1964
29 utriculosa L.			

6 Gentianella Moench

1 tenella (Rottb.) Börner	10	No	Knaben, 1950
2 nana (Wulfen) Pritchard	30	Au	Favarger, 1965
3 ciliata (L.) Borkh.			
(a) ciliata	44	Po	Skalińska, 1959
(b) doluchanovii (Grossh.) Pritchard			
4 detonsa (Rottb.) G. Don fil.	44	Is	Löve & Löve, 1956
5 barbata (Froelich) Berchtold & J. Presl			
6 crispata (Vis.) J. Holub			
7 columnae (Ten.) J. Holub			
8 campestris (L.) Börner			
(a) campestris	36	Is	Löve & Löve, 1956
(b) baltica (Murb.) Tutin			
9 hypericifolia (Murb.) Pritchard	36	Ga	Favarger & Kupfer, 1968
10 amarella (L.) Börner	36	Is	Löve & Löve, 1956

```
    (a) amarella
    (b) septentrionalis (Druce) Pritchard
11 uliginosa (Willd.) Börner              c.54    Da    Holmen, 1961
12 anglica (Pugsley) E.F. Warburg
    (a) anglica
    (b) cornubiensis Pritchard
13 bulgarica (Velen.) J. Holub
14 ramosa (Hegetschw.) J. Holub
15 pilosa (Wettst.) J. Holub
16 engadinensis (Wettst.) J. Holub         36      He    Favarger &
                                                         Huynh, 1964
17 anisodonta (Borbás) Á. & D. Löve        36      It    Favarger, 1965b
18 aspera (Hegetschw. & Heer) Dostál ex
    Skalický
19 germanica (Willd.) E.F. Warburg         36      Ge    Reese, 1961
20 austriaca (A. & J. Kerner) J. Holub
21 lutescens (Velen.) J. Holub             36      Po    Skalińska, 1950
22 aurea (L.) H. Sm.                       36      Is    Löve & Löve, 1956


7 Lomatogonium A. Braun
   1 carinthiacum (Wulfen) Reichenb.       40      Au    Fürnkranz, 1965
   2 rotatum (L.) Fries ex Fernald         10      Is    Löve & Löve, 1956


8 Swertia L.
   1 perennis L.                           28      Po    Skalińska,et al.,
                                                         1959


9 Halenia Borkh.
   1 corniculata (L.) Cornaz


CXLI MENYANTHACEAE

1 Menyanthes L.
   1 trifoliata L.                         54      Su    Palmgren, 1943

2 Nymphoides Séguier
   1 peltata (S.G. Gmelin) O. Kuntze       54      Ho    Gadella & Kliphuis,
                                                         1965


CXLII APOCYNACEAE

1 Nerium L.
   1 oleander L.

2 Trachomitum Woodson
   1 venetum (L.) Woodson
   2 sarmatiense Woodson
   3 tauricum (Pobed.) Pobed.

3 Rhazya Decne
   1 orientalis (Decne) A.DC.

4 Vinca L.
   1 minor L.                              46      Br    Rutland, 1941
   2 herbacea Waldst. & Kit.
   3 difformis Pourret                     c.46    Lu    K. Jones, unpub.
    (a) difformis
    (b) sardoa Stearn
   4 balcanica Pénzes
```

```
5 major L.                              92      Br      Rutland, 1941

CXLIII ASCLEPIADACEAE

1 Periploca L.
  1 graeca L.                           24      It      Lopane, 1951
  2 laevigata Aiton

2 Araujia Brot.
  1 sericifera Brot.

3 Gomphocarpus R. Br.
  1 fruticosus (L.) Aiton               16      Hs      Björkqvist et al.,
                                                        1969

4 Asclepias L.
  1 syriaca L.

5 Cynanchum L.
  1 acutum L.                           18      It      Francini, 1927

6 Vincetoxicum N.M. Wolf
   1 canescens (Willd.) Decne
   2 speciosum Boiss. & Spruner
   3 nigrum (L.) Moench
   4 scandens Sommier & Levier
   5 huteri Vis. & Ascherson
   6 juzepczukii (Pobed.) Privalova
   7 rossicum (Kleopow) Barbarich
   8 schmalhausennii (Kusn.) Markgraf
   9 fuscatum (Hornem.) Reichenb. fil.
  10 pannonicum (Borhidi) J. Holub      44      Hu      Borhidi, 1968
  11 hirundinaria Medicus               22      Cz Da He  Böcher & Larsen,
                                                Hu It Sa  1955

    (a) lusitanicum Markgraf
    (b) intermedium (Loret & Barrandon)
          Markgraf
    (c) nivale (Boiss. & Heldr.) Markgraf
    (d) adriaticum (G. Beck) Markgraf
    (e) stepposum (Pobed.) Markgraf
    (f) cretaceum (Pobed.) Markgraf
    (g) jailicola (Juz.) Markgraf
    (h) contiguum (Koch) Markgraf
    (i) hirundinaria

7 Cionura Griseb.
  1 erecta (L.) Griseb.

8 Caralluma R. Br.
  1 europaea (Guss.) N.E. Br.
  2 munbyana (Decne) N.E. Br.


CXLIV RUBIACEAE

1 Putoria Pers.
  1 calabrica (L. fil.) DC.             22      Al      Strid, 1971

2 Sherardia L.
  1 arvensis L.                         22      Po      Skalińska et al.,
                                                        1964
```

3 Crucianella L.
  1 maritima L.   22   Lu   Rodrigues, 1953
  2 macrostachya Boiss.
  3 bithynica Boiss.
  4 graeca Boiss.   22   Gr   Ehrendorfer, ined.
  5 angustifolia L.   22   Bu Hs   Ehrendorfer, ined.
  6 imbricata Boiss.
  7 latifolia L.   44   Bl   Dahlgren et al.,
                                                        1971
  8 patula L.

4 Asperula L.
  1 aristata L. fil.
   (a) scabra (J. & C. Presl) Nyman   20
                                      40
   (b) nestia (Rech. fil.) Ehrend. &
      Krendl   40
   (c) thessala (Boiss. & Heldr.) Hayek
   (d) oreophila (Briq.) Hayek   20
   (e) condensata (Heldr. ex Boiss.)
      Ehrend. & Krendl   20
  2 wettsteinii Adamović
  3 tenella Heuffel ex Degen
  4 crassifolia L.
  5 calabra (Fiori) Ehrend. & Krendl
  6 staliana Vis.
  7 garganica Huter, Porta & Rigo ex
    Ehrend. & Krendl
  8 suffruticosa Boiss. & Heldr.
  9 ophiolithica Ehrend.
 10 idaea Halácsy
 11 suberosa Sibth. & Sm.
 12 gussonii Boiss.
 13 pumila Moris
 14 oetaea (Boiss.) Heldr. ex Halácsy
 15 nitida Sibth. & Sm.
 16 lutea Sibth. & Sm.
   (a) lutea
   (b) rigidula (Halácsy) Ehrend.
   (c) euboea Ehrend.
   (d) mungieri (Boiss. & Heldr.) Ehrend.
     & Krendl
 17 abbreviata (Halácsy) Rech. fil.
 18 pulvinaris (Boiss.) Heldr. ex Boiss.
 19 boissieri Heldr. ex Boiss.   22   Gr   Faure & Pietrera,
                                                        1969
 20 neilreichii G. Beck   20   Rs(K)   Fagerlind, 1934
 21 beckiana Degen
 22 neglecta Guss.
 23 rupicola Jordan
 24 pyrenaica L.
 25 cretacea Willd.
 26 supina Bieb.   20   Rs(K)   Ehrendorfer, ined.
 27 tephrocarpa Czern. ex M. Popov &
    Chrshan.
 28 exasperata V. Krecz. ex Klokov
 29 petraea V. Krecz. ex Klokov
 30 occidentalis Rouy
 31 cynanchica L.   20   Cz   Mičieta, 1981
                                      40
 32 rumelica Boiss.

33 danilewskiana Basiner
34 laevissima Klokov
35 graveolens Bieb. ex Schultes &
   Schultes fil.
36 diminuta Klokov
37 leiograveolens M. Popov & Chrshan.
38 savranica Klokov
39 setulosa Boiss.
40 littoralis Sibth. & Sm.

41 incana Sibth. & Sm.	44	Cr	Ehrendorfer, ined.
42 taygetea Boiss. & Heldr.			
43 rupestris Tineo			
44 hirsuta Desf.	22 44	Hs	Ehrendorfer & Ernet, ined.
45 taurica Pacz.			
46 arcadiensis Sims	22	Gr	Faure & Pietrera, 1969
47 doerfleri Wettst.	22	Ju	Ehrendorfer & Ernet, ined.
48 hirta Ramond	22	Ga	Favarger & Küpfer, 1968
49 hercegovina Degen			
50 capitata Kit. ex Schultes	22	Rm	Ehrendorfer &
51 hexaphylla All.	22	Ga	Ernet, ined.
52 taurina L.	22	Hu	Baksay, 1956
(a) taurina			
(b) leucanthera (G. Beck) Hayek			
53 involucrata Wahlenb.			
54 laevigata L.	44		error
55 tinctoria L.	22	Cz	Holub et al., 1970
	44	Po	Skalińska et al., 1959
56 arvensis L.			
57 rigida Sibth. & Sm.	22	Cr	Ehrendorfer, ined.
58 tournefortii Sieber ex Sprengel	22	Cr	Faure & Pietrera, 1969
59 muscosa Boiss. & Heldr.	22	Gr	Faure & Pietrera, 1969
60 baenitzii Heldr. ex Boiss.			
61 chlorantha Boiss. & Heldr.	22	Gr	Faure & Pietrera, 1969
62 scutellaris Vis.	22	Ju	Ehrendorfer &
63 baldaccii (Halácsy) Ehrend.			Ernet, ined.
64 saxicola Ehrend.			
65 boryana (Walpers) Ehrend.			
66 purpurea (L.) Ehrend.			
(a) purpurea	22	It	Krendl &
(b) apiculata (Sibth. & Sm.) Ehrend.	22	Gr	Ehrendorfer, ined.

5 Galium L.

1 paradoxum Maxim.			
2 rotundifolium L.	22	Po	Skalińska et al., 1964
3 scabrum L.	22	Co	Ehrendorfer
4 baillonii Brandza	22	Rm	&
5 broterianum Boiss. & Reuter	22	Lu	Krendl, ined.
6 boreale L.	44	Po	Skalińska et al., 1961
	55	Po	Skalińska et al., 1961
	66	Po	Skalińska et al., 1961

7 rubioides L.	66		error
8 fruticosum Willd.	22	Cr	Ehrendorfer, ined.
9 ephedroides Willk.			
10 odoratum (L.) Scop.	44	Po	Piotrowicz, 1958
11 triflorum Michx			
12 rivale (Sibth. & Sm.) Griseb.	66	Ju	Ehrendorfer, ined.
13 uliginosum L.	22	Ho	Gadella & Kliphuis, 1963
	44	Ho	Gadella & Kliphuis, 1963
14 viridiflorum Boiss. & Reuter			
15 debile Desv.	24	Bu	Ančev, 1974
16 palustre L.	24	Po	Skalińska et al., 1961
	48	Ju Rm	Teppner et al., 1976
17 elongatum C. Presl	96	Au Da Ge	Teppner et al., 1976
	144	Au	Teppner et al., 1976
18 trifidum L.	24	Au	Puff, 1976
brevipes	24	Is	Löve & Löve, 1956
19 saturejifolium Trev.			
20 boissieranum Ehrend. & Krendl			
21 baeticum (Rouy) Ehrend. & Krendl			
22 concatenatum Cosson			
23 humifusum Bieb.			
24 maritimum L.			
25 tunetanum Lam.			
26 verum L.			
(a) verum	22	Bu	Ančev, 1974
	44	Cz	Májovský et al., 1970
(b) wirtgenii (F.W. Schultz) Oborny	22	Ge	Fritsch, 1973
27 x pomeranicum Retz.			
28 thymifolium Boiss. & Heldr.			
29 kerneri Degen & Dörfler	22	Bu	Markova &
30 degenii Bald. ex Degen			Ančev, 1973
31 erythrorrhizon Boiss. & Reuter			
32 pulvinatum Boiss.			
33 litorale Guss.			
34 arenarium Loisel.	66	Ga	Krendl, ined.
35 firmum Tausch	22	Ju	Ehrendorfer & Krendl, ined.
36 heldreichii Halácsy	22	Gr	Ehrendorfer, 1974
37 protopycnotrichum Ehrend. & Krendl	22	Bu Ju Tu	Krendl, ined.
38 mollugo L.	22	Au It Ju	Krendl, 1967
39 album Miller			
(a) album	44	Bu	Ančev, 1974
(b) pycnotrichum (H. Braun) Krendl	44	Bu	Ančev, 1974
(c) prusense (C. Koch) Ehrend. & Krendl	44	Gr Tu	Ehrendorfer &
40 reiseri Halacsy	22	Gr	Krendl, ined.
41 fruticescens Cav.			
42 corrudifolium Vill.	22	Ju It	Krendl, 1967
43 truniacum (Ronniger) Ronniger	22	Au	Krendl, 1967
44 montis-arerae Merxm. & Ehrend.	22	It	Krendl, 1967
45 meliodorum (G. Beck) Fritsch	44	Au	Krendl, 1967
46 lucidum All.	44	Bu	Ančev, 1974
47 bernardii Gren. & Godron			
48 aetnicum Biv.	44	It Si	Krendl, 1976
49 cinereum All.	44	Ga	Krendl, 1976
50 crespianum J.J. Rodr.			
51 mirum Rech. fil.	22	Bu	Ančev, 1974
52 scabrifolium (Boiss.) Hausskn.	22 44	Bu	Ančev, 1974

53 peloponnesiacum Ehrend. & Krendl	44	Gr	Krendl, 1974
54 asparagifolium Boiss. & Heldr.			
55 melanantherum Boiss.	22	Gr	Krendl, ined.
56 rhodopeum Velen.	22 44	Bu	Ančev, 1974
57 incurvum Sibth. & Sm.	44	Cr	Ehrendorfer, ined.
58 flavescens Borbas	44	Bu	Ančev, 1974
59 pruinosum Boiss.			
60 glaucophyllum E. Schmid			
61 murcicum Boiss. & Reuter			
62 octonarium (Klokov) Pobed.	22	Bu	Krendl, ined.
63 glaucum L.	22	Au	Ernet &
	44	Au Cz Ga	Ehrendorfer, ined.
64 biebersteinii Ehrend.			
65 xeroticum (Klokov) Pobed.			
66 volhynicum Pobed.			
67 moldavicum (Dobrescu) Franco			
68 kitaibelianum Schultes & Schultes fil.	22	Rm	Ehrendorfer, 1975
69 pseudaristatum Schur	22	Bu	Ančev, 1974
70 aristatum L.	22	Ga It	Ehrendorfer, 1975
71 abaujense Borbas	44	Cz	Ehrendorfer, 1975
72 polonicum Blocki	44	Po	Piotrowicz, 1958
73 laconicum Boiss. & Heldr.	22	Gr	Faure & Pietrera, 1969
74 procurrens Ehrend.	22	Bu	Ehrendorfer & Ančev, 1975
75 laevigatum L.	44	Ju	Ehrendorfer, 1975
76 schultesii Vest	44	Ju	Ehrendorfer, 1975
	66	Cz	Majovsky et al., 1970
77 sylvaticum L.	22	Ho	Gadella & Kliphuis, 1963
78 longifolium (Sibth. & Sm.) Griseb.	22	Tu	Ehrendorfer, 1975
79 bulgaricum Velen.	22	Bu	Ančev, ined.
80 saxosum (Chaix) Breistr.	22	Ga	Ehrendorfer, ined.
81 cometerhizon Lapeyr.			
82 incanum Sibth. & Sm.			
(a) incanum	44	Gr	Faure & Pietrera, 1969
(b) creticum Ehrend.			
83 cyllenium Boiss. & Heldr.	22	Gr	Faure & Pietrera, 1969
84 palaeoitalicum Ehrend.	20	It	Ehrendorfer, 1974
85 pyrenaicum Gouan	22	Ga Hs	Ehrendorfer, ined.
	44	Ga	Ehrendorfer, ined.
86 corsicum Sprengel	22 44	Co	Ehrendorfer, ined.
87 obliquum Vill.	22	Ga	Ehrendorfer, 1955
	44	It Ga	Ehrendorfer, 1955
88 rubrum L.	88	It	Ehrendorfer, 1955
89 x centroniae Cariot	88	It	Ehrendorfer, 1955
90 x carmineum Beauverd			
91 balearicum Briq.			
92 valentinum Lange			
93 rosellum (Boiss.) Boiss. & Reuter	22	Hs	Küpfer, 1968
94 helodes Hoffmanns. & Link	22	Lu	Ehrendorfer, ined.
95 rivulare Boiss. & Reuter	22	Hs	Ehrendorfer, ined.
96 asturiocantabricum Ehrend.	88	Hs	Ehrendorfer, ined.
97 papillosum Lapeyr.	22	Hs	Ehrendorfer, ined.
	44	Ga Hs	Ehrendorfer, ined.
98 pinetorum Ehrend.	22 44	Hs	Ehrendorfer, ined.
99 marchandii Roemer & Schultes	88	Ga Hs	Ehrendorfer, ined.

100 nevadense Boiss. & Reuter	22	Hs	Küpfer, 1969
101 timeroyi Jordan	22	Ga	Ehrendorfer, 1962
102 fleurotii Jordan	44	Ga	Ehrendorfer, 1962
	88	Ga	Ehrendorfer, 1962
103 pumilum Murray	66	Europe	Ehrendorfer, 1962
	88	Ge	Ehrendorfer, 1962
104 valdepilosum H. Braun	22	Au	Ehrendorfer, 1962
	44	Da	Ehrendorfer, 1962
105 suecicum (Sterner) Ehrend.	22	Su	Ehrendorfer, 1962
106 oelandicum (Sterner & Hyl.) Ehrend.	22	Su	Ehrendorfer, 1962
107 cracoviense Ehrend.	22	Po	Piotrowicz, 1958
108 sudeticum Tausch			
109 sterneri Ehrend.	22	Br	Ehrendorfer, 1962
	44	Da No	Ehrendorfer, 1962
110 normanii O.C. Dahl	44	Is	Löve & Löve, 1956
111 austriacum Jacq.	22	Au	Titz, 1964
	44	Au	Titz, 1964
112 anisophyllon Vill.	22	Gr	Faure & Pietrara, 1969
	44	Po	Skalińska et al., 1964
	66	Au Ge	Ehrendorfer, 1963
	88	He It	Ehrendorfer, 1963
	110	It	Ehrendorfer, 1963
113 pseudohelveticum Ehrend.	44	Ga	Ehrendorfer, ined.
114 megalospermum All.	22	Au Ga He	Ehrendorfer, ined.
	44	Ga	Ehrendorfer, ined.
115 pusillum L.	22 88	Ga	Ehrendorfer, ined.
116 brockmannii Briq.	22	Hs	Ehrendorfer, ined.
117 idubedae (Pau ex Debeaux) Pau ex Ehrend.	22	Hs	Ehrendorfer, ined.
118 saxatile L.	22	Lu	Kliphuis, 1967
	44	Ho	Kliphuis, 1967
119 tendae Reichenb. fil.	22	Ga	Ehrendorfer, ined.
120 magellense Ten.	22	It	Ehrendorfer, 1963
121 margaritaceum A. Kerner	22	It	Ehrendorfer, ined.
122 baldense Sprengel	22	It	Ehrendorfer, 1953
123 noricum Ehrend.	44	Au	Ehrendorfer, 1953
124 demissum Boiss.	22	Bu	Ančev, 1974
125 stojanovii Degen	22	Bu	Ančev, 1974
126 cespitosum Lam.	22	Ga	Ehrendorfer, ined.
127 graecum L.			
(a) graecum	22	Cr	Ehrendorfer, ined.
(b) pseudocanum Ehrend.			
128 canum Req. ex DC.			
129 setaceum Lam.	22	Cr	Ehrendorfer, ined.
(a) setaceum			
(b) decaisnei (Boiss.) Ehrend.			
130 monachinii Boiss. & Heldr.	22	Cr	Ehrendorfer, ined.
131 spurium L.	20	Bu	Ančev, 1974
132 aparine L.	42	Au	Ehrendorfer, ined.
	44	Ho	Kliphuis, 1962b
	48	Au	Ehrendorfer, ined.
	62	Au	Ehrendorfer, ined.
	66	Bu	Ančev, 1974
	68	Au	Ehrendorfer, ined.
133 tricornutum Dandy	44	Au Ga It	Ehrendorfer, ined.
134 verrucosum Hudson	22	Ju	Ehrendorfer, ined.
135 intricatum Margot & Reuter			
136 capitatum Bory & Chaub.			
137 incrassatum Halácsy			

138 viscosum Vahl			
139 parisiense L.	44	Ga Hs Ju	Ehrendorfer, ined.
	66	Bu	Ancev, 1974
140 divaricatum Pourret ex Lam.	44	Ga Lu	Ehrendorfer, ined.
141 tenuissimum Bieb.	22	Asia	
	44	Bu	Ančev, 1974
142 minutulum Jordan			
143 recurvum Req. ex DC.			
144 verticillatum Danth.	22	Bu	Ančev, 1974
	44	Gr	Ehrendorfer, ined.
145 murale (L.) All.	44 Bot.	Gard.	Fagerlind, 1934

6 Callipeltis Steven
  1 cucullaris (L.) Rothm.

7 Cruciata Miller

1 laevipes Opiz	22	Cz	Májovský et al., 1970
2 taurica (Pallas ex Willd.) Ehrend.			
(a) euboea (Ehrend.) Ehrend.			
(b) taurica	44	Rs(K)	Ehrendorfer, ined.
3 glabra (L.) Ehrend.	22	It	Ehrendorfer, ined.
	44	Bu	Ančev, 1976
4 balcanica Ehrend.			
5 pedemontana (Bellardi) Ehrend.	18	Cz	Májovský et al., 1970

8 Valantia L.

1 aprica (Sibth. & Sm.) Boiss. & Heldr.	22	Gr	Faure & Pietrera, 1969
2 hispida L.	18	Bl	Dahlgren et al., 1971
3 muralis L.	18	Bl	Dahlgren et al., 1971

9 Rubia L.

1 peregrina L.	44	Bl	Dahlgren et al., 1971
angustifolia L.	66	Bl	Cardona, 1973
2 tenuifolia D'Urv.			
3 tatarica (Trev.) Friedrich Schmidt Petrop.			
4 tinctorum L.			

CXLV POLEMONIACEAE

1 Polemonium L.

1 caeruleum L.	18	No	Knaben & Engelskjon, 1967
2 acutiflorum Willd.	18	No	Knaben & Engelskjon, 1967
3 boreale Adams	18	No	Laane, 1966

2 Collomia Nutt.
  1 grandiflora Douglas ex Lindley

CXLVI CONVOLVULACEAE

1 Cuscuta L.
  1 australis R. Br.
   (a) tinei (Insenga) Feinbrun
   (b) cesattiana (Bertol.) Feinbrun
  2 campestris Yuncker
  3 suaveolens Ser.
  4 gronovii Willd.
  5 pedicellata Ledeb.
  6 europaea L.
  7 pellucida Butkov
  8 epilinum Weihe
  9 palaestina Boiss.
  10 atrans Feinbrun
  11 triumvirati Lange
  12 epithymum (L.) L.   14   Su   Ehrenberg, 1945
   (a) epithymum
   (b) kotschyi (Desmoulins) Arcangeli
  13 brevistyla A. Braun ex A. Richard
  14 planiflora Ten.
  15 approximata Bab.
   (a) approximata
   (b) macranthera (Heldr. & Sart. ex Boiss.)
       Feinbrun & W. Greuter
   (c) episonchum (Webb & Berth.) Feinbrun
  16 lupuliformis Krocker
  17 monogyna Vahl

2 Cressa L.
  1 cretica L.

3 Dichondra J.R. & G. Forster
  1 micrantha Urban

4 Calystegia R. Br.
  1 soldanella (L.) R. Br.   22   Lu   Rodriguez, 1953
  2 sepium (L.) R. Br.   22   Ho   Gadella & Kliphuis,
                                                         1967
   (a) sepium   22   Cz   Holub et al., 1970
   (b) roseata Brummitt   22   Br   Brummitt, 1973
   (c) americana (Sims) Brummitt
   (d) spectabilis Brummitt
  3 silvatica (Kit.) Griseb.   22   Br   Brummitt, 1973
  4 pulchra Brummitt & Heywood   22   Cz   Holub et al., 1970

5 Convolvulus L.
  1 persicus L.
  2 boissieri Steudel
   (a) boissieri
   (b) compactus (Boiss.) Stace
  3 libanoticus Boiss.
  4 dorycnium L.

5 cneorum L.
6 oleifolius Desr.
7 lanuginosus Desr.
8 calvertii Boiss.
9 holosericeus Bieb.
10 lineatus L.
11 cantabrica L.   30   Hu   Baksay, 1958
12 humilis Jacq.
13 tricolor L.
  (a) tricolor
  (b) cupanianus (Sa'ad) Stace
14 meonanthus Hoffmanns. & Link
15 pentapetaloides L.
16 siculus L.
  (a) siculus   44   Bl   Dahlgren et al.,
                                                                1971
  (b) agrestis (Schweinf.) Verdcourt
17 sabatius Viv.
18 valentinus Cav.
19 arvensis L.   50   Da   Hagerup, 1941b
20 farinosus L.
21 betonicifolius Miller
22 scammonia L.
23 althaeoides L.
  (a) althaeoides
  (b) tenuissimus (Sibth. & Sm.) Stace

6 Ipomoea L.
  1 stolonifera (Cyr.) J.F. Gmelin
  2 sagittata Poiret
  3 acuminata (Vahl) Roemer & Schultes
  4 purpurea Roth
  5 batatas (L.) Lam.

CXLVII HYDROPHYLLACEAE

1 Nemophila Nutt.
  1 menziesii Hooker & Arnott

2 Phacelia Juss.
  1 tanacetifolia Bentham

3 Wigandia Kunth
  1 caracasana Kunth

CXLVIII BORAGINACEAE

1 Argusia Boehmer
  1 sibirica (L.) Dandy   26   Rm   Tarnavschi, 1948

2 Heliotropium L.
  1 europaeum L.
  2 ellipticum Ledeb.
  3 lasiocarpum Fischer & C.A. Meyer
  4 dolosum De Not.
  5 suaveolens Bieb.
  (a) suaveolens
  (b) bocconei (Guss.) Brummitt
  6 halacsyi Reidl

```
  7 hirsutissimum Grauer
  8 micranthos (Pallas) Bunge
  9 arguzioides Kar. & Kir.
 10 curassavicum L.
 11 amplexicaule Vahl
 12 supinum L.

3 Arnebia Forskål
  1 decumbens (Vent.) Cosson & Kralik

4 Lithospermum L.
  1 officinale L.                         28       Ho      Gadella & Kliphuis,
                                                          1966

5 Neatostema I.M. Johnston
  1 apulum (L.) I.M. Johnston             28       Bl      Dahlgren et al.,
                                                          1971

6 Buglossoides Moench
  1 purpurocaerulea (L.) I.M. Johnston    16       Ge      Reese, 1952
  2 gastonii (Bentham) I.M. Johnston
  3 calabra (Ten.) I.M. Johnston          20       It      Grau, 1968b
  4 arvensis (L.) I.M. Johnston        28 42       Ga      Grau, 1968b
                                                   It      Gadella & Kliphuis,
                                                          1970
    (a) arvensis                          28       Scand.  Löve & Löve, 1944

    (b) permixta (Jordan ex F.W. Schultz)
        R. Fernandes                      28               Grau, 1968b
    (c) gasparrinii (Heldr. ex Guss.)
        R. Fernandes
    (d) sibthorpiana (Griseb.) R. Fernandes
  5 minima (Moris) R. Fernandes
  6 glandulosa (Velen.) R. Fernandes
  7 tenuiflora (L. fil.) I.M. Johnston

7 Lithodora Griseb.
  1 rosmarinifolia (Ten.) I.M. Johnston   26       Si      Fürnkranz, 1967
  2 fruticosa (L.) Griseb.                28       Hs      Lorenzo-Andreu, 1951
  3 hispidula (Sibth. & Sm.) Griseb.
  4 zahnii (Heldr. ex Halácsy) I.M. Johnston
  5 oleifolia (Lapeyr.) Griseb.
  6 nitida (Ern) R. Fernandes             50       Hs      Grau, 1971
  7 diffusa (Lag.) I.M. Johnston
    (a) diffusa                           24       Hs      Grau, 1968b
    (b) lusitanica (Samp.) P. Silva &
        Rozeira

8 Macrotomia DC.
  1 densiflora (Ledeb. ex Nordm.) Macbride

9 Onosma L.
  1 frutescens Lam.                       14       Gr      Grau, 1964a
  2 graeca Boiss.                         14       Gr      Grau, 1968b
  3 taygetea Boiss. & Heldr.
  4 visianii G.C. Clementi                18       Au      Teppner, 1971
  5 rhodopea Velen.
  6 setosa Ledeb.
  7 tricerosperma Lag.
    (a) tricerosperma
    (b) hispanica (Degen & Hervier) P.W. Ball
```

(c) granatensis (Dubeaux & Degen) Stroh
8 simplicissima L.
9 polyphylla Ledeb.
10 tinctoria Bieb.
11 propontica Aznav.
12 bubanii Stroh
13 fastigiata (Br.-Bl.) Br.-Bl. ex Lacaita

14 vaudensis Gremli	20	He	Teppner, 1971
15 arenaria Waldst. & Kit.	12	Hu	Baksay, 1957b
		Hu	Pólya, 1950
(a) arenaria	20	Au It	Teppner, 1971
(b) fallax (Borbás) Jáv.	26	Sb Ju	Teppner, 1971
16 austriaca (G. Beck) Fritsch	26	Au	Teppner, 1971
17 pseudarenaria Schur			
18 tridentina Wettst.	28	It	Grau, 1964a
19 helvetica (A. DC.) Boiss.	26 28	It	Grau, 1964a
		Ga It	Favarger, 1971
20 lucana Lacaita			
21 thracica Velen.			
22 echioides L.	14	It	Grau, 1968b
23 tornensis Jáv.	14	Cz	Baksay, 1957b
24 heterophylla Griseb.			
25 elegantissima Rech. fil. & Goulimy			
26 euboica Rech. fil.			
27 mattirolii Bald.			
28 montana Sibth. & Sm.	14	Gr	Grau, 1968b
29 erecta Sibth. & Sm.			
30 taurica Pallas ex Willd.	14	Rs(K)	Teppner, 1971
31 leptantha Heldr.			
32 stellulata Waldst. & Kit.	22	Ju	Teppner, 1971
33 spruneri Boiss.			

10 Cerinthe L.
  1 minor L.
  2 glabra Miller
  (a) minor

(b) tenuiflora (Bertol.) Domac	16	Co	Contandriopoulos, 1964a
(c) smithiae (A. Kerner) Domac			
3 major L.	16	Hs	Gadella et al., 1966

  4 retorta Sibth. & Sm.

11 Moltkia Lehm.

1 suffruticosa (L.) Brand	16	It	Grau, 1968b
2 petraea (Tratt.) Griseb.			
3 doerfleri Wettst.			

12 Alkanna Tausch

1 orientalis (L.) Boiss.	28	Gr	Grau, 1968b
2 graeca Boiss. & Spruner	30	Gr	Grau, 1968b
(a) graeca			
(b) baeotica (DC.) Nyman			
3 calliensis Heldr. ex Boiss.			
4 primuliflora Griseb.			
5 stribrnyi Velen.			
6 methanaea Hausskn.			
7 tinctoria (L.) Tausch	30	It	Grau, 1968b
8 corcyrensis Hayek			
9 lutea DC.	28	Bl	Dahlgren et al., 1971

10 pelia (Halácsy) Rech. fil.
11 pindicola Hausskn.
12 sandwithii Rech. fil.
13 noneiformis Griseb.
14 pulmonaria Griseb.
15 scardica Griseb.
16 sieberi DC.
17 sartoriana Boiss. & Heldr.

13 Halacsya Dörfler
  1 sendtneri (Boiss.) Dörfler   22   Ju   Sušnik, 1967

14 Echium L.
  1 albicans Lag. & Rodr.
  2 angustifolium Miller
  3 humile Desf.
  4 asperrimum Lam.   14   Ga   Delay, 1969
  5 italicum L.

                         32   Hs   Gadella et al., 1966
                         16   Cz   Májovský et al., 1970

  6 lusitanicum L.
    (a) lusitanicum
    (b) polycaulon (Boiss.) P. Gibbs
  7 flavum Desf.
  8 boissieri Steudel
  9 russicum J.F. Gmelin   24   Hu   Pólya, 1950
10 vulgare L.   16 32 Ho   Gadella & Kliphuis, 1966
11 plantagineum L.   16   Co   Litardière, 1943
12 sabulicola Pomel
13 creticum L.
    (a) creticum
    (b) coincyanum (Lacaita) R. Fernandes
    (c) algarbiense R. Fernandes   16   Lu   Fernandes, 1969
14 tuberculatum Hoffmanns. & Link 16   Lu   Rodriguez, 1953
15 gaditanum Boiss.
16 rosulatum Lange
17 parviflorum Moench   16   Bl   Dahlgren, et al., 1971

18 arenarium Guss.

15 Pulmonaria L.
  1 obscura Dumort.   14   Hu   Pólya, 1949
  2 officinalis L.   16 Ge Ju Au Merzmuller & It   Grau, 1969
                         14   Ga   Larsen, 1954

  3 rubra Schott   14   Rm   Merxmüller & Grau, 1969
  4 filarszkyana Jáv.   14   Rm   Tarnavschi, 1935
  5 stiriaca A. Kerner   18   Au Ju Wolkinger, 1966
  6 mollis Wulfen ex Hornem.   18   Au   Merxmüller & Grau, 1969

                         14
                         28   Rm   Tarnavschi, 1935
  7 vallarsae A. Kerner   22   It   Merxmüller & Grau, 1969
  8 affinis Jordan   22   Ga It Merxmüller & Grau, 1969

9 saccharata Miller	22	It	Merxmüller & Grau, 1969
10 montana Lej.	22	Ge	Merxmüller & Grau, 1969
	18	Ge	Merxmüller & Grau, 1969
11 longifolia (Bast.) Boreau	14	Ga	Merxmüller & Grau, 1969
12 angustifolia L.	14	Ga It	Merxmüller & Grau, 1969
13 visianii Degen & Lengyel	20	Au He It Ju	Merxmüller & Grau, 1969
14 kerneri Wettst.	26	Au	Merxmüller, 1970

16 Nonea Medicus
1 lutea (Desr.) DC.	14	Rm	Gusuleac & Tarnavschi, 1935
2 obtusifolia (Willd.) DC.			
3 pulla (L.) DC.	18	Rm	Tarnavschi & Lungeneau, 1970
	20	Au	Furnkranz, 1967

4 pallens Petrović
5 caspica (Willd.) G. Don fil.
6 versicolor (Steven) Sweet
7 ventricosa (Sibth. & Sm.) Griseb.
8 micrantha Boiss. & Reuter
9 vesicaria (L.) Reichenb.

17 Elizaldia Willk.
 1 calycina (Roemer & Schultes) Maire

18 Symphytum L.
1 officinale L.	24 + 4B	It	Gadella & Kliphuis, 1970
	24 40 48 26 36	Ho	Gadella & Kliphuis 1967
	54	Bu	Markova & Ivanova, 1970
(a) officinale			
(b) uliginosum (A. Kerner) Nyman			
2 asperum Lepechin			
3 x uplandicum Nyman	36	Fe	Vaarama, 1948
4 tuberosum L.			
(a) tuberosum	44	It	Grau, 1968b
(b) nodosum (Schur) Soó	96	Ge	Grau, 1968b
	100	Ju	
	18	Rm	Tarnavschi, 1948
	72	Bu	Markova & Ivanova, 1970
5 gussonei F.W. Schultz			
6 cordatum Waldst. & Kit. ex Willd.	18	Rm	Tarnavschi, 1948
7 naxicola Pawl.	30	Gr	Runemark, 1967
8 ibiricum Steven			
9 davisii Wickens			
10 cycladense Pawl.	30	Gr	Pawlowski, unpub.
11 tauricum Willd.	18	Rm	Tarnavschi, 1948
12 orientale L.			
13 bulbosum C. Schimper			
14 ottomanum Friv.	20	Bu	Markova & Ivanova, 1970

19 Procopiania Guşuleac
 1 cretica (Willd.) Guşuleac
 2 insularis Pawl.   28   Cr   Runemark, 1967a
 3 circinalis (Runemark) Pawl.   28   Cr   Runemark, 1967a

20 Brunnera Steven
 1 macrophylla (Adams) I.M. Johnston

21 Anchusa L.
 1 leptophylla Roemer & Schultes
 2 gmelinii Ledeb.
 3 ochroleuca Bieb.
 4 cespitosa Lam.
 5 undulata L.
  (a) undulata
  (b) hybrida (Ten.) Coutinho
 6 sartorii Heldr. ex Guşuleac
 7 granatensis Boiss.
 8 calcarea Boiss.
 9 crispa Viv.   16   Co   Contandriopoulos,
                                                                  1962
 10 officinalis L.   16 + 0-2B   Au   Furnkranz, 1967

 11 azurea Miller
 12 aggregata Lehm.
 13 thessala Boiss. & Spruner   24   Rm   Tarnavschi, 1948
 14 pusilla Guşuleac
 15 macrosyrinx Rech. fil.
 16 macedonica Degen & Dörfler
 17 stylosa Bieb.
 18 spruneri Boiss.
 19 barrelieri (All.) Vitman
 20 serpentinicola Rech. fil.
 21 aegyptiaca (L.) DC.
 22 arvensis (L.) Bieb.   48   It   Gadella & Kliphuis,
                                                                  1970
  (a) arvensis   48   Su   Lövkvist, 1963
  (b) orientalis (L.) Nordh.
 23 cretica Miller   16   It   Grau, 1968b
 24 variegata (L.) Lehm.   16   Gr   Grau, 1968b

22 Pentaglottis Tausch
 1 sempervirens (L.) Tausch ex L.H.
   Bailey

23 Borago L.
 1 officinalis L.   16   Rm   Tarnavschi, 1948
 2 pygmaea (DC.) Chater & W. Greuter   32   Co   Contandriopoulos,
                                                                  1962

24 Trachystemon D. Don
 1 orientalis (L.) G. Don fil.   56   Bu   Markova, 1970

25 Mertensia Roth
 1 maritima (L.) S.F. Gray   24   Is   Löve & Löve, 1956

26 Amsinckia Lehm.
 1 lycopsoides (Lehm.) Lehm.
 2 calycina (Moris) Chater

27 Asperugo L.
  1 procumbens L.

	48	Cz	Májovský et al., 1970

28 Rochelia Reichenb.
  1 disperma (L. fil.) C. Koch
    (a) disperma
    (b) retorta (Pallas) E. Kotejowa

29 Myosotis L.

1 incrassata Guss.	24	Ga	Merxmüller & Grau, 1963
2 cadmea Boiss.	24	Gr	Grau, 1968a
3 ucrainica Czern.			
4 pusilla Loisel.	24	Co Ga	Grau, 1968a
5 litoralis Steven ex Bieb.			
6 arvensis (L.) Hill			
(a) arvensis	52	Ho	Gadella & Kliphuis, 1967a
(b) umbrata (Rouy) O. Schwarz	66	Ga Ge Ju No Su	Grau, 1968a
7 ramosissima Rochel			
(a) ramosissima	48	Is	Löve & Löve, 1956
(b) globularis (Samp.) Grau			
8 ruscinonensis Rouy	?36 48	Ga	Blaise, 1966
9 discolor Pers.			
(a) discolor	72	Ga Ge Hs Lu	Grau, 1968a
	64	Ho	Gadella & Kliphuis, 1968
(b) dubia (Arrondeau) Blaise	24	Ga	Blaise, 1970
10 congesta R.J. Shuttlew. ex Albert & Reyn.	24 ?32	Gr	Grau, 1970
	48	Gr	Blaise, 1970
11 balbisiana Jordan			
12 persoonii Rouy	48	Hs	Grau, 1968a
13 stricta Link ex Roemer & Schultes	36	Ga	Blaise, 1966
	48	Bu Ga Ge Ju	Grau, 1968a
14 minutiflora Boiss. & Reuter	48	Hs	Grau, 1968a
15 speluncicola (Boiss.) Rouy	24	Ga	Merxmüller & Grau, 1963
16 refracta Boiss.			
(a) refracta	44	Cr Gr Hs	Grau, 1968a
(b) paucipilosa Grau	20	Gr	Grau, 1968a
(c) aegagrophila W. Grueter & Grau			
17 sylvatica Hoffm.			
(a) sylvatica	18 20	Au Ge	Grau, 1964b
(b) subarvensis Grau	18	It Ju	Grau, 1964b
(c) cyanea (Boiss. & Heldr.) Vestergren	18 20	Gr	Grau, 1964b
(d) elongata (Strobl) Grau	20 22	Si	Grau, 1964b
18 soleirolii Gren. & Godron	18	Co	Blaise, 1969
19 decumbens Host Host			
(a) decumbens	32	Au Ga He It Su	Grau, 1964b
(b) teresiana (Sennen) Grau	32 34	Ga Hs	Grau, 1964b
(c) variabilis (M. Angelis) Grau	32	Au	Grau, 1964b
(d) kerneri (Dalla Torre & Sarnth.) Grau	32	Au	Grau, 1964b
(e) florentina Grau	28	It	Grau, 1970
20 latifolia Poiret			
21 alpestris F.W. Schmidt	24 48 72 70	Au Ga Ge He It	Grau, 1964b

22 gallica Vestergren	24	Ga	Merxmüller & Grau, 1963
23 ambigens (Béguinot) Grau	24	It	Grau, 1964
24 stenophylla Knaf	48	Au	Merxmüller & Grau, 1963
25 asiatica (Vestergren) Schischkin & Serg.			
26 alpina Lapeyr.	24	Ga Hs	Grau, 1964b
27 corsicana (Fiori) Grau	24	Co	Contandriopoulos, 1966
28 suaveolens Waldst. & Kit. ex Willd.	24 48	Gr	Grau, 1964b
29 lithospermifolia (Willd.) Hornem.			
30 secunda A. Murray	24 48	Lu	Fernandes & Leitao, 1969
31 welwitschii Boiss. & Reuter	24	Lu	Grau, 1967a
32 stolonifera (DC.) Gay ex Leresche & Levier	24	Hs	Merxmüller & Grau, 1963
33 sicula Guss.	46 46+2	Ga Sa	Grau, 1967a
34 debilis Pomel	48	Lu	Grau, 1967a
35 laxa Lehm.			
(a) caespitosa (C.F. Schultz) Hyl. ex Nordh.			
(b) baltica (Sam.) Hyl. ex Nordh.	?84 88	Su Fe	Merxmüller & Grau, 1963
36 scorpioides L.	66	Au Ge	Merxmüller & Grau, 1963
	?64	Hu	Pólya, 1950
37 rehsteineri Wartm.	22	Ge	Merxmüller & Grau, 1963
38 nemorosa Besser	22 44	Cz	Holub et al., 1971
39 lamottiana (Br.-Bl. ex Chassagne) Grau	44	Ga Hs	Grau, 1970
40 azorica H.C. Watson			
41 sparsiflora Mikan ex Polh	18	Au	Merxmüller & Grau, 1963

30 Eritrichium Schrader ex Gaudin
   1 pectinatum (Pallas) DC.
   2 nanum (L.) Schrader ex Gaudin

(a) nanum	46	He	Favarger, 1965b

   (b) jankae (Simonkai) Jáv.
   3 villosum (Ledeb.) Bunge
   4 aretioides (Cham.) DC.

31 Lappula Gilib.

1 deflexa (Wahlenb.) Garcke	24	No	Knaben & Engelskjon, 1967

   2 spinocarpos (Forskål) Ascherson ex
     O. Kuntze
   3 echinophora (Pallas) O. Kuntze
   4 marginata (Bieb.) Gurke

5 squarrosa (Retz.) Dumort	48	Hu	Baksay, 1956

   (a) squarrosa
   (b) heteracantha (Ledeb.) Chater
   6 barbata (Bieb.) Gurke

32 Trigonotis Steven
   1 peduncularis (Trev.) Bentham ex Baker

33 Omphalodes Miller

1 scorpioides (Haenke) Schrank	24	Ge	Grau, 1967b

```
 2 nitida Hoffmanns. & Link              24      Lu      Grau, 1967b
 3 verna Moench                          48      It Lu   Grau, 1967b
 4 luciliae Boiss.
 5 linifolia (L.) Moench                 28      Lu      Grau, 1967b
 6 kuzinskyanae Willk.                   28      Lu      Grau, 1967b
 7 littoralis Lehm.
 8 brassicifolia (Lag.) Sweet

34 Cynoglossum L.
 1 officinale L.                         24      Cz      Májovský et al.,
                                                          1970
                                         28      Cz      Murín & Váchová,
                                                          1967



 2 dioscoridis Vill.
 3 columnae Ten.
 4 germanicum Jacq.                      24      Cz      Májovský et al.,
                                                          1970
 5 creticum Miller                       24      Ga      Delay, 1970
 6 sphacioticum Boiss. & Heldr.
 7 hungaricum Simonkai                   24      Hu      Baksay, 1958
 8 nebrodense Guss.
 9 cheirifolium L.
10 magellense Ten.
11 clandestinum Desf.

35 Solenanthus Ledeb.
 1 reverchonii Degen.
 2 biebersteinii DC.
 3 apenninus (L.) Fischer & C.A. Meyer
 4 scardicus Bornm.
 5 stamineus (Desf.) Wettst.
 6 albanicus (Degen & Bald.) Degen & Bald.

36 Mattiastrum (Boiss.) Brand
 1 lithospermifolium (Lam.) Brand

37 Rindera Pallas
 1 tetraspis Pallas
 2 umbellata (Waldst. & Kit.) Bunge
 3 graeca (A. DC.) Boiss. & Heldr.


CXLIX VERBENACEAE

1 Vitex L.
 1 agnus-castus L.

2 Lantana L.
 1 camara L.

3 Verbena L.
 1 rigida Sprengel
 2 bonariensis L.
 3 officinalis L.                        14      Hs      Björkqvist et al.,
                                                          1969
 4 supina L.
```

4 Lippia L.
　1 nodiflora (L.) Michx
　2 canescens Kunth
　3 triphylla (L'Hér.) O. Kuntze

CL CALLITRICHACEAE

1 Callitriche L.
　1 hermaphroditica L.　　　　　　　6　　Is　　Löve & Löve, 1956
　2 truncata Guss.
　(a) truncata
　(b) fimbriata Schotsman
　(c) occidentalis (Rouy) Schotsman　6　　Lu　　Schotsman, 1961a
　3 pulchra Schotsman　　　　　　　8　　Cr　　Schotsman, 1969
　4 lusitanica Schotsman　　　　　　8　　Lu　　Schotsman, 1967
　5 stagnalis Scop.　　　　　　　　10　Br Ga He Lu Schotsman, 1961b
　6 obtusangula Le Gall　　　　　　10　Co Ga Ge Lu Schotsman, 1961b
　7 cophocarpa Sendtner　　　　　　10　　He　　Schotsman, 1961b
　8 platycarpa Kütz.　　　　　　　20　　He　　Schotsman, 1961b
　9 palustris L.　　　　　　　　　20　　Is　　Löve & Löve, 1956
　10 hamulata Kütz. ex Koch　　　　38　　He　　Schotsman, 1961
　　　　　　　　　　　　　　　　　?40　　Is　　Löve & Löve, 1956
　11 brutia Petagna　　　　　　　　28　　Lu　　Schotsman, 1961c
　12 peploides Nutt.
　13 terrestris Rafin.

CLI LABIATAE

1 Ajuga L.
　1 orientalis L.　　　　　　　　32　　Gr　　Strid, 1965
　2 genevensis L.　　　　　　　　32　　Ge　　Scheerer, 1940
　3 pyramidalis L.　　　　　　　32　　Fe　　Sorsa, 1963
　4 reptans L.　　　　　　　　　32　　Bu　　Markova & Ivanova,
　　　　　　　　　　　　　　　　　　　　　　1970
　5 tenorii C. Presl
　6 salicifolia (L.) Schreber
　(a) salicifolia
　(b) bassarabica (Savul. & Zahar.)
　　　P.W. Ball
　7 laxmannii (L.) Bentham　　　62　　Bu　　Markova & Ivanova,
　　　　　　　　　　　　　　　　　　　　　　1971
　8 piskoi Degen & Bald.
　9 iva (L.) Schreber　　　　　c.86　　Bl　　Dahlgren et al.,
　　　　　　　　　　　　　　　　　　　　　　1971
　10 chamaepitys (L.) Schreber
　(a) chamaepitys　　　　　　　28　　Br　　Rutland, 1941
　(b) chia (Schreber) Arcangeli　30　　Gr　　Strid, 1965

2 Teucrium L.
　1 fruticans L.
　2 brevifolium Schreber　　　　30　　Gr　　Strid, 1965
　3 aroanium Orph. ex Boiss.
　4 pseudochamaepitys L.
　5 campanulatum L.
　6 arduini L.
　7 lamifolium D'Urv.
　8 halacsyanum Heldr.　　　　　32　　Gr　　Quezel & Contand-
　　　　　　　　　　　　　　　　　　　　　　riopoulos, 1966
　9 francisci-werneri Rech. fil.

10 heliotropifolium W. Barbey
11 scorodonia L.
  (a) scorodonia 32 Ho Gadella & Kliphuis, 1963

                                           ?34 Br Rutland, 1941
  (b) euganeum (Vis.) Arcangeli
  (c) baeticum (Boiss. & Reuter) Tutin
12 massiliense L. 32 Co Contandriopoulos, 1962

13 salviastrum Schreber
14 asiaticum L.
15 scordium L.
  (a) scordium
  (b) scordioides (Schreber) Maire &
     Petitmengin 32 Rm Tarnavschi, 1938
16 spinosum L.
17 resupinatum Desf.
18 botrys L. 32 Cz Májovský et al., 1970

19 chamaedrys L. 60 Hu Baksay, 1958
20 webbianum Boiss.
21 krymense Juz.
22 lucidum L.
23 divaricatum Sieber ex Boiss. 62 64 Gr Strid, 1965
24 flavum L.
  (a) flavum
  (b) glaucum (Jordan & Fourr.) Ronniger
  (c) hellenicum Rech. fil.
  (d) gymnocalyx Rech. fil
25 intricatum Lange
26 fragile Boiss.
27 microphyllum Desf.
28 marum L.
29 subspinosum Pourret ex Willd.
30 compactum Clemente ex Lag.
31 pyrenaicum L. 26 Ga Küpfer, 1969
32 rotundifolium Schreber 26 Hs Küpfer, 1969
33 buxifolium Schreber
34 freynii Reverchon ex Willk.
35 cuneifolium Sibth. & Sm.
36 alpestre Sibth. & Sm.
  (a) alpestre 78 Cr Strid, 1965
  (b) gracile (W. Barbey & Major) D. Wood
37 montanum L. 16 Ga Guinochet & Logeois, 1962

                                        26 Ga Küpfer, 1969
                                        30 Hu Baksay, 1956
                                        60 Cz Murín & Váchová, 1967

38 thymifolium Schreber
39 cossonii D. Wood
40 libanitis Schreber
41 carthaginense Lange
42 aragonense Loscos & Pardo
43 pumilum L.
  (a) pumilum
  (b) carolipaui (C. Vicioso ex Pau) D. Wood
44 turredanum Losa & Rivas Goday
45 polium L. 26 Hs Lorenzo-Andreu & Garcia, 1950
                                        78 Ge Strid, 1965
                                        52 Ga Pueck, 1968

```
(a) polium
(b) aureum (Schreber) Arcangeli
(c) capitatum (L.) Arcangeli
(d) pii-fontii Palau
(e) vincentinum (Rouy) D. Wood
46 gnaphalodes L'Hér.
47 eriocephalum Willk.
48 haenseleri Boiss.
49 charidemi Sandwith


3 Scutellaria L.
1 orientalis L.
2 alpina L.
(a) alpina                              22      Ga      Favarger, 1959a
(b) supina (L.) I.B.K. Richardson
3 columnae All.                         34      Hu      Baksay, 1958
(a) columnae
(b) gussonii (Ten.) Rech. fil.
4 altissima L.
5 rubicunda Hornem.
(a) linnaeana (Caruel) Rech. fil.
(b) rubicunda                           34      Gr      Bothmer, 1969
(c) rupestris (Boiss. & Heldr.) I.B.K.
Richardson
(d) geraniana (Tuntas ex Halácsy)
I.B.K. Richardson
6 albida L.
7 velenovskyi Rech. fil.
naxensis Bothmer                        34      Gr      Bothmer, 1969
8 sieberi Bentham
9 hirta Sibth. & Sm.
10 balearica Barc.
11 galericulata L.                    31 32     Ho      Gadella & Kliphuis,
                                                        1968

12 hastifolia L.
13 minor Hudson


4 Prasium L
1 majus L.                              34      Cr Gr   Bothmer, 1970


5 Marrubium L.
1 thessalum Boiss. & Heldr.
2 cylleneum Boiss. & Heldr.
3 incanum Desr.
4 supinum L.
5 velutinum Sibth. & Sm.
6 friwaldskyanum Boiss.
7 leonuroides Desr.
8 alysson L.
9 peregrinum L.                         34      Cz      Murín & Váchová,
                                                        1967
10 vulgare L.                           34      Br      Rutland, 1941
11 pestalozzae Boiss.
12 alternidens Rech. fil.


6 Sideritis L.
1 grandiflora Salzm. ex Bentham
2 foetens Clemente ex Lag.
3 arborescens Salzm. ex Bentham
(a) arborescens
```

(b) luteola (Font Quer) P.W. Ball ex
    Heywood
(c) paulii (Pau) P.W. Ball ex Heywood
4 hirsuta L.
5 endressii Willk.
  (a) endressii
  (b) laxespicata (Degen & Debeaux) Heywood
6 scordioides L.
  (a) scordioides
  (b) cavanillesii (Lag.) P.W. Ball ex
      Heywood

7 glacialis Boiss.	34	Hs	Küpfer & Favarger, 1967
8 hyssopifolia L.	32+6B	Ge He	Favarger, 1965b
	34		
	30	Ga	Favarger & Küpfer, 1968

  (a) hyssopifolia
  (b) guillonii (Timb.-Lagr.) Rouy
9 ovata Cav.
10 linearifolia Lam.
11 angustifolia Lag.
12 reverchonii Willk.
13 javalambrensis Pau

14 spinulosa Barnades ex Asso	22 28 30	Hs	Coutinho & Lorenzo-Andreu, 1948

15 serrata Cav. ex Lag.
16 ilicifolia Willd.
17 leucantha Cav.
18 stachydioides Willk.
19 incana L.
  (a) incana
  (b) virgata (Desf.) Malagarriga
  (c) sericea (Pers.) P.W. Ball ex
      Heywood
  (d) glauca (Cav.) Malagarriga
20 lacaitae Font Quer

21 syriaca L.	24	Rs(K)	Borhidi, 1968
22 clandenstina (Bory & Chaub.) Hayek			
23 scardica Griseb.	32	Bu	Kozuharov & Kuzmanov, 1965b

24 perfoliata L.			
25 montana L.			
(a) montana	16	Cz	Májovský et al., 1970

  (b) remota (D'Urv.) Heywood
  (c) ebracteata (Asso) Murb.
26 lanata L.
27 romana L.

(a) romana	28	Hs	Björkqvist et al., 1969

  (b) purpurea (Talbot ex Bentham)

28 curvidens Stapf	28	Cr Gr	Strid, 1965

7 Melittis L.

1 melissophyllum L.	30	Hu	Baksay, 1958

  (a) melissophyllum
  (b) carpatica (Klokov) P.W. Ball
  (c) albida (Guss.) P.W. Ball

8 Eremostachys Bunge
  1 tuberosa (Pallas) Bunge

9 Phlomis L.
1 tuberosa L.	22	Bu	Markova, 1970

  2 samia L.
  3 herba-venti L.
   (a) herba-venti
   (b) pungens (Willd.) Maire ex
     DeFilipps
  4 purpurea L.
  5 italica L.
  6 floccosa D. Don
  7 cretica C. Presl
  8 ferruginea Ten.

9 lychnitis L.	20	Hs	Lorenzo-Andreu & Garcia, 1950

  10 crinita Cav.

11 lanata Willd.	20	Cr	Strid, 1965
12 fruticosa L.	20	Gr	Strid, 1965

10 Galeopsis L.
1 segetum Necker	16	Rm	Muntzing, 1929
2 pyrenaica Bartl.	16	Ga	Muntzing, 1929
3 ladanum L.	16	Su	Muntzing, 1929
4 angustifolia Ehrh. ex Hoffm.	16	Su	Muntzing, 1929
5 reuteri Reichenb. fil.			
6 speciosa Miller	16	Rm Su	Muntzing, 1929
7 pubescens Besser	16	Au Cz Hu Rm	Muntzing, 1929
8 tetrahit L.	32	Br Be Su	Muntzing, 1929
9 bifida Boenn.	32	Au Rm Su	Muntzing, 1929

11 Lamium L.
  1 orvala L.
2 garganicum L.	18	Co	Contandriopoulos, 1962

   (a) garganicum
   (b) laevigatum Arcangeli
   (c) pictum (Boiss. & Heldr.) P.W. Ball
3 corsicum Gren. & Godron	18	Co	Contandriopoulos, 1962

  4 glaberrimum Taliev
  5 flexuosum Ten.
6 maculatum L.	18	Au	Mattick, 1950
7 album L.	18	Hu	Pólya, 1949
8 moschatum Miller	18	Gr	Strid, 1965

  9 bifidum Cyr.
   (a) bifidum
   (b) balcanicum Velen.
   (c) albimontanum Rech. fil.
10 purpureum L.	18	Fe	Sorsa, 1963
11 hybridum Vill.	36	Is	Löve & Löve, 1956
12 moluccellifolium Fries	36	Is	Löve & Löve, 1956

  13 amplexicaule L.
(a) amplexicaule	18	Hu	Pólya, 1950

   (b) orientale Pacz.

12 Lamiastrum Heister ex Fabr.

1 galeobdolon (L.) Ehrend. & Polatschek
  (a) galeobdolon   18   Ge   Dersch, 1964
  (b) montanum (Pers.) Ehrend. &
      Polatschek   36   Au   Polatschek, 1966
  (c) flavidum (F. Hermann) Ehrend. &
      Polatschek   18   Au   Polatschek, 1966

13 Leonurus L.
   1 cardiaca L.   18   Hu   Pólya, 1949
   2 marrubiastrum L.   24   Rm   Tarnavschi, 1948

14 Moluccella L.
   1 spinosa L.

15 Ballota L.
   1 frutescens (L.) J. Woods
   2 acetabulosa (L.) Bentham   28   Gr   Strid, 1965
   3 pseudodictamnus (L.) Bentham
   4 hirsuta Bentham
   5 rupestris (Biv.) Vis.
   6 macedonica Vandas
   7 nigra L.   22   Br   Rutland, 1941
     (a) nigra
     (b) sericea (Vandas) Patzak
     (c) foetida Hayek
     (d) velutina (Pospichal) Patzak
     (e) uncinata (Fiori & Béguinot) Patzak
     (f) anatolica P.H. Davis

16 Stachys L.
   1 alopecuros (L.) Bentham
   2 monieri (Gouan) P.W. Ball
   3 officinalis (L.) Trevisan   16   Hu   Pólya, 1949
   4 balcanica P.W. Ball
   5 scardica (Griseb.) Hayek
   6 balansae Boiss. & Kotschy
   7 alpina L.
   8 germanica L.   30   Cz   Májovský et al.,
                                                       1970

     (a) germanica
     (b) heldreichii (Boiss.) Hayek
     (c) lusitanica (Hoffmanns. & Link)
         Coutinho
   9 tymphaea Hausskn.
  10 tournefortii Poiret
  11 sericophylla Halácsy
  12 thirkei C. Koch
  13 cassia (Boiss.) Boiss.
  14 cretica L.
     (a) cretica
     (b) salviifolia (Ten.) Rech. fil.
     (c) bulgarica Rech. fil
  15 byzantina C. Koch
  16 heraclea All.
  17 acutifolia Bory & Chaub.
  18 obliqua Waldst. & Kit.
  19 decumbens Pers.
  20 canescens Bory & Chaub.
  21 anisochila Vis. & Pančič
  22 methifolia Vis.
  23 plumosa Griseb.

24 sylvatica L.	66	Hu	Pólya, 1950
25 palustris L.	102	Su	Löve, 1954a
	64	Cz	Májovský et al., 1970
26 recta L.			
(a) recta	34	Ga	Májovský et al., 1970
	32	Hu	Pólya, 1949
(b) labiosa (Bertol.) Briq.	34	It	Favarger, 1959a
(c) subcrenata (Vis.) Briq.			
27 beckeana Dörfler & Hayek			
28 parolinii Vis.			
29 atherocalyx C. Koch			
30 iberica Bieb.			
31 virgata Bory & Chaub.			
32 angustifolia Bieb.			
33 tetragona Boiss. & Heldr.			
34 leucoglossa Griseb.			
35 mucronata Sieber ex Sprengel	30	Cr	Strid, 1965
36 spruneri Boiss.			
37 swainsonii Bentham			
(a) swainsonii	34	Gr	Strid, 1965
(b) argolica (Boiss.) Phitos & Damboldt	34	Gr	Phitos & Damboldt, 1969
(c) scyronica (Boiss.) Phitos & Damboldt			
38 ionica Halácsy			
39 euboica Rech. fil	34	Gr	Strid, 1965
40 circinata L'Hér.			
41 glutinosa L.	32	Co	Contandriopoulos, 1962
42 spinosa L.			
43 iva Griseb.			
44 chrysantha Boiss. & Heldr.			
45 candida Bory & Chaub.			
46 spreitzenhoferi Heldr.			
47 maritima Gouan			
48 pubescens Ten.			
49 arenaria Vahl			
50 corsica Pers.	16	Co	Contandriopoulos, 1962
51 ocymastrum (L.) Briq.	18	Hs	Björkqvist et al., 1969
52 marrubiifolia Viv.	16	Co	Contandriopoulos, 1962
53 annua (L.) L.			
54 spinulosa Sibth. & Sm.			
55 milanii Petrović			
56 arvensis (L.) L.	10	Cr Gr	Strid, 1965
57 brachyclada De Noë ex Cosson			
58 serbica Pančić			

17 Cedronella Moench
  1 canariensis (L.) Webb & Berth.

18 Nepeta L.
  1 multibracteata Desf.
  2 tuberosa L.
    (a) tuberosa

190

(b) reticulata (Desf.) Maire
(c) gienensis (Degen & Hervier) Heywood
3 apuleii Ucria
4 scordotis L.
5 granatensis Boiss.
6 sibthorpii Bentham
7 parnassica Heldr. & Sart.
8 spruneri Boiss.
9 dirphya (Boiss.) Heldr. ex Halácsy
10 camphorata Boiss. & Heldr.
11 heldreichii Halácsy
12 italica L.

13 cataria L.	36	Cz	Májovský et al., 1970
14 nepetella L.	34	He	Contandriopoulos, 1962
15 agrestis Loisel.	18	Co	Contandriopoulos, 1962

16 foliosa Moris
17 grandiflora Bieb.

18 melissifolia Lam.	18	Cr	Strid, 1965

19 latifolia DC.
20 nuda L.
  (a) nuda
  (b) albiflora Gams
21 ucranica L.
22 parviflora Bieb.
23 hispanica Boiss. & Reuter
24 beltranii Pau

19 Glechoma L.

1 hederacea L.	18	Fe	Sorsa, 1963
	18 36	Po	Skalińska et al., 1959

2 hirsuta Waldst. & Kit.

20 Dracocephalum L.
  1 nutans L.

2 thymiflorum L.	c.14	Fe	V. Sorsa, 1962
3 ruyschiana L.	14	Su	Löve & Löve, 1944
4 austriacum L.	14	Hu	Baksay, 1958

5 moldavica L.

21 Prunella L.

1 laciniata (L.) L.	28	Bu	Kozuharov & Kuzmanov, 1969
		Cz	
	32	Cz	Májovský et al., 1970
2 grandiflora (L.) Scholler	28	Cz	Hrubý, 1932
3 vulgaris L.	28	Cz	Hrubý, 1932

4 hyssopifolia L.

22 Cleonia L.
  1 lusitanica (L.) L.

23 Melissa L.
  1 officinalis L.

(a) officinalis	32	Ga	Litardiere, 1945

191

(b) altissima (Sibth. & Sm.) Arcangeli   64   Co   Litardière, 1945

24 Ziziphora L.
   1 capitata L.
   2 hispanica L.
   3 acinoides L.
   4 tenuior L.
   5 taurica Bieb.
   6 persica Bunge

25 Satureja L.
   1 thymbra L.
   2 salzmannii P.W. Ball
   3 spinosa L.
   4 rumelica Velen.
   5 montana L.
     (a) montana
     (b) variegata (Host) P.W. Ball
     (c) illyrica Nyman
     (d) kitaibelii (Wierzb.) P.W. Ball   30   Ju   Lovka et al.,
                                                          1971
     (e) taurica (Velen.) P.W. Ball
   6 parnassica Heldr. & Sart. ex Boiss.
   7 pilosa Velen.
   8 athoa K. Maly
   9 cuneifolia Ten.
  10 obovata Lag.
  11 coerulea Janka
  12 hortensis L.

26 Acinos Miller
   1 corsicus (Pers.) Getliffe
   2 alpinus (L.) Moench.
     (a) alpinus   18   Au   Reese, 1953
     (b) majoranifolius (Miller) P.W. Ball
     (c) meridionalis (Nyman) P.W. Ball   18   Ga   Favarger & Küpfer,
                                                          1968
   3 suaveolens (Sibth. & Sm.) G. Don fil.
   4 arvensis (Lam.) Dandy   18   Ho   Gadella & Kliphuis,
                                                          1966
   5 rotundifolius Pers.

27 Calamintha Miller
   1 grandiflora (L.) Moench   22   He   Wieffering, 1969
   2 sylvatica Bromf.
     (a) sylvatica   24   Hu   Baksay, 1958
     (b) ascendens (Jordan) P.W. Ball   48   Bl   Nilsson & Lassen,
                                                          1971
   3 nepeta (L.) Savi
     (a) nepeta   24   Hs   Bjorkqvist et al.,
                                                          1969
     (b) glandulosa (Req.) P.W. Ball   20   Co   Contandriopoulos,
                                                          1964a
                                          24   Cz   Holub et al.,
                                                          1970
   4 cretica (L.) Lam.
   5 incana (Sibth. & Sm.) Boiss.

28 Clinopodium L.
   1 vulgare L.
     (a) vulgare   20   Ge   Scheerer, 1939

```
(b) arundanum (Boiss.) Nyman              20    Gr    Bothmer, 1967

29 Micromeria Bentham
   1 fruticosa (L.) Druce
     (a) fruticosa
     (b) serpyllifolia (Bieb.) P.H. Davis
   2 thymifolia (Scop.) Fritsch
   3 taygetea P.H. Davis
   4 dalmatica Bentham
   5 pulegium (Rochel) Bentham
   6 frivaldszkyana (Degen) Velen.
   7 filiformis (Aiton) Bentham            30    Bl    Dahlgren et al.,
                                                        1971

   8 microphylla (D'Urv.) Bentham
   9 acropolitana Halácsy
  10 hispida Boiss. & Heldr. ex Bentham
  11 marginata (Sm.) Chater
  12 croatica (Pers.) Schott
  13 nervosa (Desf.) Bentham              30    Gr    Strid, 1965
  14 juliana (L.) Bentham ex Reichenb.    30    Gr    Strid, 1965
  15 myrtifolia Boiss. & Hohen.
  16 cristata (Hampe) Griseb.
  17 cremnophila Boiss. & Heldr.
  18 kerneri Murb.
  19 parviflora (Vis.) Reichenb.
  20 graeca (L.) Bentham ex Reichenb.
     (a) graeca                            20    Hs    Björkqvist et al.,
                                                        1969
     (b) imperica Chater
     (c) garganica (Briq.) Guinea
     (d) longiflora (C. Presl) Nyman
     (e) tenuifolia (Ten.) Nyman
     (f) consentina (Ten.) Guinea
     (g) fruticulosa (Bertol.) Guinea
  21 inodora (Desf.) Bentham

30 Thymbra L.
   1 spicata L.
   2 calostachya (Rech. fil.) Rech. fil.

31 Hyssopus L.
   1 officinalis L.
     (a) canescens (DC.) Briq.
     (b) montanus (Jordan & Fourr.) Briq.
     (c) aristatus (Godron) Briq.
     (d) officinalis

32 Origanum L.
   1 compactum Bentham
   2 heracleoticum L.                      30    Gr    Bothmer, 1970
   3 vulgare L.                            30    Ho    Gadella & Kliphuis,
                                                        1963
                                         ?32    Ge    Scheerer, 1940
   4 virens Hoffmanns. & Link             30    Lu    Fernandes, ined.
   5 majorana L.
   6 onites L.                            30    Gr    Bothmer, 1970
   7 microphyllum (Bentham) Boiss.
```

8 majoricum Camb.
9 dictamnus L.
10 tournefortii Aiton   30   Gr   Bothmer, 1970
11 scabrum Boiss. & Heldr.
   (a) scabrum
   (b) pulchrum (Boiss. & Heldr.)
       P.H. Davis
12 lirium Heldr. ex Halácsy
13 vetteri Briq. & W. Barbey

33 Thymus L.
 1 capitatus (L.) Hoffmanns. & Link   30   Si   Sibiho, 1961
 2 mastichina L.   56   Hs   Kaleva, 1969
 3 tomentosus Willd.
 4 caespititius Brot.
 5 piperella L.
 6 teucrioides Boiss. & Spruner
 7 cephalotos L.
 8 villosus L.
 9 longiflorus Boiss.
10 membranaceus Boiss.
11 antoninae Rouy & Coincy
12 mastigophorus Lacaita
13 dolopicus Form.
14 cherlerioides Vis.   28   Rs(K)   Gogina &
                                                                Suetozacova, 1968
15 parnassicus Halácsy
16 capitellatus Hoffmanns. & Link.
17 camphoratus Hoffmanns. & Link.
18 carnosus Boiss.
19 vulgaris L.   30   Fe   Jalas, 1948
20 hyemalis Lange
21 zygis L.   60   Lu   Jalas & Pohjo,
                                                                1965
22 baeticus Boiss. ex Lacaita
23 hirtus Willd.
24 loscosii Willk.
   (a) loscosii
   (b) fontqueri Jalas   56   Hs   Kaleva, 1969
25 serpylloides Bory
   (a) serpylloides
   (b) gadorensis (Pau) Jalas
26 holosericeus Celak.
27 laconicus Jalas
28 bracteosus Vis. ex Bentham
29 granatensis Boiss.
30 aranjuezii Jalas   28   Hs   Kaleva, 1969
31 bracteatus Lange ex Cutanda   56   · Hs   Kaleva, 1969
32 leptophyllus Lange
33 atticus Celak.
34 plasonii Adamović
35 striatus Vahl   26 54   Ju   Jalas & Kaleva,
                                                                1966
36 spinulosus Ten.
37 aznavourii Velen.
38 kirgisorum Dubjanski
39 zygioides Griseb.
40 comptus Friv.
41 sibthorpii Bentham
42 pannonicus All.   28   Cz   Jalas & Kaleva,
                                                                1967
43 glabrescens Willd.

(a) glabrescens	28 32 56 58	Po	Trela-Sawicha, 1968
(b) decipiens (H. Braun) Domin	52	Ga	Jalas & Kaleva, 1967
(c) urumovii (Velen.) Jalas	56	Rs(K)	Gogina & Svetozocova, 1968

44 longedentatus (Degen & Urum.) Ronniger
45 pallasianus H. Braun
   (a) pallasianus
   (b) brachyodon (Borbas) Jalas

46 herba-barona Loisel	56	Co	Contandriopoulos, 1962
47 nitens Lamotte	28	Ga	Bonnet, 1961

48 willkommii Ronniger
49 richardii Pers.

(a) richardii	28	Ju	Kaleva, 1969

   (b) nitidus (Guss.) Jalas
   (c) ebusitanus (Font Quer) Jalas
50 guberlinensis Iljin
51 binervulatus Klokov & Schost.
52 ocheus Heldr. & Sart. ex Boiss.
53 thracicus Velen.
54 stojanovii Degen

55 longicaulis C. Presl	30	Rm Su	Jalas & Kaleva, 1966
dolomiticus Coste	28	Ga	Bonnet, 1958
embergeri Roussine	48	Ga	Shimoya, 1952

56 praecox Opiz

(a) praecox	c.50 c.54	He Hu	Jalas & Pohjo, 1965
	56	He	
	58	Cz	Jalas & Kaleva, 1967

   (b) skorpilii (Velen.) Jalas

(c) polytrichus (A. Kerner ex	28	Ga	Bonnet, 1966
Borbas) Jalas	c.50 54	It	Jalas & Kaleva, 1966
	55 56	He	Jalas & Pohjo, 1965

   (d) zygiformis (H. Braun) Jalas

(e) arcticus (E. Durand) Jalas	c.50 51 54	Br Ho Hs Ga	Pigott, 1954
57 nervosus Gay ex Willk.	28	Ga	Bonnet, 1966
58 pulcherrimus Schur	56 60	Po	Trela-Sawicka, 1968
59 comosus Heuffel ex Griseb.	28	Rm	Kaleva, 1969

60 bihoriensis Jalas

61 pulegioides L.	28 30	Po	Trela-Sawicka, 1968
62 alpestris Tausch ex A. Kerner	28	Po	Trela-Sawicka, 1968

63 oehmianus Ronniger & Soska
64 alternans Klokov
65 serpyllum L.

(a) serpyllum	24	Po	Trela-Sawicka, 1968
(b) tanaensis (Hyl.) Jalas	24	Fe	Jalas, 1948
66 talijevii Klokov & Schost.			

34 Lycopus L.

1 europaeus L.	22	Su	Ehrenberg, 1945
2 exaltatus L. fil.			

35 Mentha L.

1 requienii Bentham	18	Co	Contandriopoulos, 1962
2 pulegium L.	20	Rm	Tarnavschi, 1948
3 micrantha (Bentham) Schost.			
4 cervina L.			
5 arvensis L.	72	Su	Olsson, 1967
	24 90	Ho	Ouweneel, 1968
x muellerana F.W. Schultz	60	Br	Morton, 1956
	42	Br	Morton, 1956
6 x verticillata L.	78 84 90 120 132	Su	Olsson, 1967
	96	Ho	Ouweneel, 1968
7 x gentilis L.	54 60 84 96 108 120	Br	Morton, 1956
8 x smithiana R.A. Graham	120	Br	Morton, 1956
9 aquatica L.	96	Br	Morton, 1956
10 x piperita L.	66 72	Br	Morton, 1956
x piperita nm. citrata (Ehrh.) Boivin	84 120	Br	Morton, 1956
x maximilianea F.W. Schultz	60 72-78 120	Br	Morton, 1956
x pyramidalis Ten.	c.72	It	Harley, 1977
11 suaveolens Ehrh.	24	Br	Harley, 1977
insularis Req.	24	Bl	Harley, 1977
x rotundifolia (L.) Hudson	24	Ho	Ouweneel, 1968
x villosa Hudson	36	Ho	Ouweneel, 1968
x villosa Hudson nm. alopecuroides (Hull)	36	Br	Morton, 1956
12 longifolia (L.) Hudson	24	Au Ga Gr He Hs	Harley, 1977
13 microphylla C. Koch	48	Gr	Harley, 1977
14 spicata L.	48	Br	Morton, 1956
x villosonervata auct., ?an Opiz	38	Br	Harley, 1977

36 Perilla L.
1 frutescens (L.) Britton

37 Rosmarinus L.

1 officinalis L.	24	Bl	Nilsson & Lassen, 1971
2 eriocalix Jordan & Fourr.			

38 Lavandula L.

1 stoechas L.	30	Gr	Bothmer, 1970
(a) stoechas			
(b) pedunculata (Miller) Samp. ex Rozeira			
(c) lusitanica (Chaytor) Rozeira			
(d) luisieri (Rozeira) Rozeira			
(e) sampaiana Rozeira			

(f) cariensis (Boiss.) Rozeira
2 viridis L'Hér.
3 dentata L.   44   Bl   Nilsson & Lassen,
                                                                1971
4 angustifolia Miller
  (a) angustifolia
  (b) pyrenaica (DC.) Guinea
5 latifolia Medicus
6 lanata Boiss.   50   Hs   Küpfer, 1969
7 multifida L.

39 Horminum L.
1 pyrenaicum L.   12   He   Favarger, 1953

40 Salvia L.
1 officinalis L.
2 lavandulifolia Vahl
3 grandiflora Etlinger   16   Bu   Markova, 1970
4 eichlerana Heldr. ex Halácsy
5 triloba L. fil.
6 candelabrum Boiss.
7 blancoana Webb & Heldr.
8 brachyodon Vandas
9 ringens Sibth. & Sm.
10 pinnata L.
11 scabiosifolia Lam.
12 pomifera L.   14   Gr   Bothmer, 1970
13 phlomoides
14 sclarea L.   22   Bu   Markova, 1970
15 argentea L.   20   Bu   Markova, 1970
16 aethiopis L.   24   Hu   Felföldy, 1947
17 candidissima Vahl
18 glutinosa L.   16   Hu   Pólya, 1949
19 forskaohlei L.
20 bicolor Lam.
21 pratensis L.   18   Cz   Májovský et al.,
                                                                1970
22 dumetorum Andrz. ex Besser
23 teddii Turrill
24 transsylvanica (Schur ex Griseb.) Schur
25 nemorosa L.   12   Cz   Májovský et al.,
                                                                1970
  (a) nemorosa
  (b) tesquicola (Klokov & Pobed.) Soó
26 amplexicaulis Lam.
27 valentina Vahl
28 sclareoides Brot.
29 verbenaca L.   42   Cr   Bothmer, 1970
                                                59 64   Ga Hs   Gadella et al.,
                                                                1966
                                                64   Ga   van Loon et al.,
                                                                1971
                                                60   Bl   Dahlgren et al.,
                                                                1971
30 nutans L.
31 austriaca Jacq.
32 jurisicii Kosanin
33 virgata Jacq.   18   Bu   Markova, 1970
34 viridis L.   16   Cr Gr   Bothmer, 1970
35 verticillata L.   16   Cz   Májovský et al.,
                                                                1970
36 napifolia Jacq.

41 Elsholtzia Willd.
   1 ciliata (Thunb.) Hyl.

CLII SOLANACEAE

1 Nicandra Adanson
  1 physalodes (L.) Gaertner

2 Lycium L.
  1 europaeum L.
  2 intricatum Boiss.   24   Lu   Fernandes &
  3 ruthenicum Murray   Queiros, 1970/71
  4 barbarum L.   24   Ga   Delay, 1967
  5 chinense Miller
  6 afrum L.

3 Atropa L.
  1 bella-donna L.
  2 baetica Willk.

4 Scopolia Jacq.
  1 carniolica Jacq.   48   Hu   Pólya, 1950

5 Hyoscyamus L.
  1 niger L.   34   Au   Griesinger 1937
  2 reticulatus L.
  3 albus L.   68   Bl   Dahlgren et al.,
                                                          1971
  4 aureus L.
  5 pusillus L.

6 Withania Pauquy
  1 somnifera (L.) Dunal
  2 frutescens (L.) Pauquy

7 Physalis L.
  1 alkekengi L.   24   Cz   Májovský et al.,
  2 peruviana L.   1970
  3 pubescens L.
  4 philadelphica Lam.   24   Lu   R. Fernandes, 1970

8 Salpichroa Miers
  1 origanifolia (Lam.) Baillon

9 Capsicum L.
  1 annuum L.

10 Solanum L.
   1 nigrum L.
     (a) nigrum   72   Br   Edmonds, 1977
     (b) schultesii (Opiz) Wessely   72   Ju   Edmonds, 1977
   2 sublobatum Willd. ex Roemer & Schultes
   3 luteum Miller
     (a) luteum   48   Ju   Edmonds, 1977
     (b) alatum (Moench) Dostál   48   Rs(K)   Edmonds, 1977
   4 sarrachoides Sendtner   24   Br   Edmonds, 1977
   5 triflorum Nutt.
   6 pseudocapsicum L.
   7 dulcamara L.   24   Ga   Delay, 1967
   8 tuberosum L.   48   Br   Hawkes, 1956

```
      9 laciniatum Aiton
     10 melongena L.
     11 bonariense L.
     12 sodomeum L.
     13 mauritianum Scop.
     14 cornutum Lam.
     15 elaeagnifolium Cav.

  11 Lycopersicon Miller
      1 esculentum Miller                    24      Br      Haskell, 1964

  12 Mandragora L.
      1 officinarum L.
      2 autumnalis Bertol.                   84      Hs      Tamayo & Grande,
                                                             1967

  13 Triguera Cav.
      1 ambrosiaca Cav.
      2 osbeckii (L.) Willk.

  14 Datura L.
      1 stramonium L.                        24      Hu      Pólya, 1949
      2 ferox L.
      3 innoxia Miller

  15 Cestrum L.
      1 parqui L'Hér.

  16 Nicotiana L.
      1 glauca R.C. Graham
      2 rustica L.
      3 tabacum L.
      4 alata Link & Otto


  CLIII BUDDLEJACEAE

  1 Buddleja L.
      1 davidii Franchet
      2 albiflora Hemsley
      3 japonica Hemsley
      4 lindleyana Fortune ex Lindley


  CLIV SCROPHULARIACEAE

  1 Gratiola L.
      1 officinalis L.                       32      Rm      Tarnavschi, 1948
      2 linifolia Vahl
      3 neglecta Torrey

  2 Lindernia All.
      1 procumbens (Krocker) Philcox
      2 dubia (L.) Pennell

  3 Bacopa Aublet
      1 monnieri (L.) Pennell

  4 Limosella L.
      1 aquatica L.                          40      Is      Löve & Löve, 1956
      2 tenella Quezel & Contandr.           40      Gr      Quézel &
                                                             Contandriopoulos,1965
```
199

```
3 australis R. Br.                        20      Br      Blackburn, 1939

5 Mimulus L.
   1 guttatus DC.                         48      Br      Maude, 1939
   2 luteus L.
   3 moschatus Douglas ex Lindley

6 Dodartia L.
   1 orientalis L.

7 Verbascum L.
   1 spinosum L.
   2 ovalifolium Donn ex Sims
    (a) ovalifolium
    (b) thracicum (Velen.) Murb.
   3 purpureum (Janka) Huber-Morath
   4 adrianopolitanum Podp.
   5 blattaria L.
   6 siculum Tod. ex Lojac.
   7 virgatum Stokes
   8 daenzeri (Fauche & Chaub.) O. Kuntze
   9 rupestre (Davidov) I.K. Ferguson
  10 roripifolium (Halácsy) I.K.
      Ferguson
  11 boissieri (Heldr. & Sart. ex Boiss.)
      O. Kuntze
  12 barnadesii Vahl.
  13 laciniatum (Poiret) O. Kuntze
  14 creticum (L.) Cav.
  15 arcturus L.
  16 levanticum I.K. Ferguson
  17 bugulifolium Lam.
  18 spectabile Bieb.
  19 hervieri Degen
  20 phoeniceum L.                        32      Cz      Májovský et al.,
                                                          1970
                                          36      Hu      Baksay, 1961
  21 xanthophoeniceum Griseb.
  22 cylleneum (Boiss. & Heldr.) O. Kuntze
  23 acaule (Bory & Chaub.) O. Kuntze
  24 pyramidatum Bieb.
  25 zuccarinii (Boiss.) I.K. Ferguson
  26 orientale (L.) All.
  27 phlomoides L.
  28 densiflorum Bertol.
  29 samniticum Ten.
  30 guicciardii Heldr. ex Boiss.
  31 vandasii (Rohlena) Rohlena
  32 niveum Ten.
    (a) niveum
    (b) visianianum (Reichenb.) Murb.
    (c) garganicum (Ten.) Murb.
  33 macrurum Ten.
  34 thapsus L.                           32      Lu      Fernandes, 1950
                                          36      Ho      Gadella & Kliphuis,
                                                          1966

    (a) thapsus
    (b) crassifolium (Lam.) Murb.
    (c) giganteum (Willk.) Nyman
  35 litigiosum Samp.
  36 longifolium Ten.
```

37 boerhavii L.
38 argenteum Ten.
39 nevadense Boiss.
40 lagurus Fischer & C.A. Meyer
41 georgicum Bentham
42 nicolai Rohlena
43 anisophyllum Murb.
44 eriophorum Godron
45 baldaccii Degen
46 epixanthinum Boiss. & Heldr.
47 euboicum Murb. & Rech. fil.
48 pelium Halácsy
49 dieckianum Borbás & Degen
50 macedonicum Kosanin & Murb.
51 botuliforme Murb.
52 dentifolium Delile
53 speciosum Schrader
  (a) speciosum
  (b) megaphlomos (Boiss. & Heldr.)
     Nyman
54 adeliae Heldr. ex Boiss.
55 lasianthum Boiss. ex Bentham
56 undulatum Lam.
57 pentelicum Murb.
58 sinuatum L.   30   It   Mori, 1957
59 halacsyanum Sint. & Bornm. ex Halácsy
60 pinnatifidum Vahl
61 leucophyllum Griseb.
62 cylindrocarpum Griseb.
63 pseudonobile Stoj. & Stefanov
64 humile Janka
65 adenanthum Bornm.
66 nobile Velen.
67 graecum Heldr. & Sart. ex Boiss.
68 herzogii Bornm.
69 mallophorum Boiss. & Heldr.
70 decorum Velen.
71 mucronatum Lam.
72 pulverulentum Vill.
73 gnaphalodes Bieb.
74 glandulosum Delile
75 davidoffii Murb.
76 jankaeanum Pančić
77 durmitoreum Rohlena
78 rotundifolium Ten.
  (a) rotundifolium
  (b) haenseleri (Boiss.) Murb.
  (c) conocarpum (Moris) I.K. Ferguson
79 reiseri Halácsy
80 delphicum Boiss. & Heldr.
81 banaticum Schrader
82 lychnitis L.
83 chaixii Vill.
  (a) chaixii
  (b) austriacum (Schott ex Roemer &
     Schultes) Hayek
  (c) orientale Hayek
84 bithynicum Boiss.
85 nigrum L.
  (a) nigrum
  (b) abietinum (Borbás) I.K. Ferguson

86 lanatum Schrader
87 glabratum Friv.
  (a) glabratum
  (b) brandzae (Franchet ex Brandza) Murb.
  (c) bosnense (K. Maly) Murb.

8 Scrophularia L.

1 vernalis L.	40	Ge Po	Vaarama & Leikas, 1970
2 bosniaca G. Beck			
3 aestivalis Griseb.	44	Bu	Grau, 1976
4 divaricata Ledeb.			
5 peregrina L.	36	Ga	Vaarama & Leikas, 1970
6 arguta Aiton			
7 pyrenaica Bentham	58	Ga	Grau, 1976
8 scorodonia L.	58	Br Hs Lu	Grau, 1976
	60	Ga Lu	Vaarama & Leikas, 1970
9 grandiflora DC.			
(a) grandiflora	60	Lu	Vaarama & Leikas, 1970
(b) reuteri (Daveau) I.B.K. Richardson	58	Hs	Grau, 1976
10 sambucifolia L.	58	Hs	Grau, 1976
11 trifoliata L.	c.84	Co	Contandriopoulos, 1962
12 sciophila Willk.	58	Hs	Grau, 1976
13 sublyrata Brot.			
schousboei Lange	60	Lu	Grau, 1976
14 laevigata Vahl			
15 scopolii Hoppe	26	Bu	Vaarama & Leikas, 1970
16 alpestris Gay ex Bentham	68	Hs	Grau, 1976
	72	Ga	Vaarama & Leikas, 1970
17 herminii Hoffmanns. & Link	c.52	Lu	Vaarama & Hiirsalmi, 1967
18 nodosa L.	36	It	Mori, 1957
19 auriculata L.	78 84 86	Ho	Gadella & Kliphuis, 1966
	80	Ga It Lu	Vaarama & Ieikas, 1970
lyrata Willd.	58	Hs	Grau, 1976
20 umbrosa Dumor .	52	Cz	Májovský et al., 1970
	26	Po Rm	Grau, 1976
21 lucida L.	24 26	Ga Gr	Grau, ined.
22 myriophylla Boiss. & Heldr.			
23 heterophylla Willd.			
(a) heterophylla	24	Gr	Grau, ined.
(b) laciniata (Waldst. & Kit.) Maire			
& Petitmengin	26		
24 taygetea Boiss.			
25 frutescens L.	26	Lu	Rodrigues, 1953
26 ramosissima Loisel.			
27 spinulescens Degen & Hausskn.			
28 canina L.			
(a) canina	24 26	Ga It	Vaarama & Leikas, 1970

```
  (b) crithmifolia Boiss.                   24      Hs    Grau, ined.
  (c) hoppii (Koch) P. Fourn.            24 26      He    Vaarama & Hiirschmi,
                                                           1967
29 cretacea Fischer ex Sprengel
30 rupestris Bieb. ex Willd.

 9 Anarrhinum Desf.
   1 bellidifolium (L.) Willd.             18      Co    Contandriopoulos,
                                                           1962

   2 laxiflorum Boiss.
   3 longipedicellatum R. Fernandes
   4 duriminium (Brot.) Pers.
   5 corsicum Jordan & Fourr.              18      Co    Contandriopoulos,
                                                           1962

10 Antirrhinum L.
   1 valentinum Font Quer
   2 sempervirens Lapeyr.
   3 pulverulentum Laz.-Ibiza
   4 pertegasii Rothm.
   5 microphyllum Rothm.
   6 charidemi Lange
   7 molle L.
   8 grosii Font Quer
   9 hispanicum Chav.
     (a) hispanicum
     (b) mollissimum (Roth.) D.A. Webb
  10 meonanthum Hoffmanns. & Link
  11 braun-blanquetii Rothm.
  12 siculum Miller
  13 barrelieri Boreau
  14 graniticum Rothm.
  15 australe Rothm.
  16 latifolium Miller                     16      It    Pogliani, 1964
  17 majus L.
     (a) majus
     (b) linkianum (Boiss. & Reuter) Rothm.  16    Hs    Björkqvist et al., 1969
     (c) cirrhigerum (Ficalho) Franco
     (d) tortuosum (Bosc) Rouy

11 Asarina Miller
   1 procumbens Miller

12 Misopates Rafin.
   1 orontium (L.) Rafin.                   16      Ga    van Loon et al.,
                                                           1971

   2 calycinum Rothm.

13 Chaenorhinum (DC.) Reichenb.
   1 origanifolium (L.) Fourr.             14      Ga    van Loon et al.,
                                                           1971

     (a) origanifolium
     (b) cadevallii (O. Bolós & Vigo) Laínz
     (c) crassifolium (Cav.) Rivas Goday &
         Borja
     (d) segoviense (Willk.) R. Fernandes
   2 glareosum (Boiss.) Willk.             14      Hs    Küpfer, 1968
   3 villosum (L.) Lange
   4 tenellum (Cav.) Lange
   5 macropodum (Boiss. & Reuter) Lange
     (a) macropodum
     (b) degenii (Hervier) R. Fernandes
```

6 serpyllifolium (Lange) Lange
 (a) serpyllifolium
 (b) lusitanicum R. Fernandes
7 robustum Loscos
8 grandiflorum (Cosson) Willk.
9 rubrifolium (Robill. & Cast. ex DC.)
  Fourr.
 (a) rubrifolium
 (b) formenterae (Gand.) R. Fernandes
 (c) raveyi (Boiss.) R. Fernandes
10 minus (L.) Lange
 (a) minus
 (b) litorale (Willd.) Hayek   14   Ju   Nilsson & Lassen,
                                                            1971
 (c) idaeum (Rech. fil.) R. Fernandes

14 Linaria Miller
 1 reflexa (L.) Desf.
 2 pedunculata (L.) Chaz.
 3 flava (Poiret) Desf.
 4 pseudolaxiflora Lojac.
 5 cretacea Fischer ex Sprengel
 6 sabulosa Czern. ex Klokov.
 7 tonzigii Lona
 8 cavanillesii Chav.
 9 oligantha Lange
10 huteri Lange
11 bipunctata (L.) Dum.-Courset
12 ficalhoana Rouy
13 hirta (L.) Moench
14 triphylla (L.) Miller
15 genistifolia (L.) Miller   12   Cz   Damboldt, 1968
 (a) genistifolia   12   Bu   Markova & Ivanova,
                                                          1970

 (b) sofiana (Velen.) Chater & D.A. Webb
 (c) dalmatica (L.) Maire & Petitmengin
16 peloponnesiaca Boiss. & Heldr.
17 spartea (L.) Willd.
18 viscosa (L.) Dum.-Courset   12   Hs   Valdés, 1970
19 incarnata (Vent.) Sprengel
20 hellenica Turrill
21 algarviana Chav.
22 elegans Cav.
23 nigricans Lange
24 clementei Haenseler ex Boiss.
25 chalepensis (L.) Miller
26 canadensis (L.) Dum.-Courset
27 purpurea (L.) Miller   12   Si   Valdés, 1970
28 capraria Moris & De Not.
29 nivea Boiss. & Reuter
30 repens (L.) Miller
31 microsepala A. Kerner
32 triornithophora (L.) Willd.
33 pelisseriana (L.) Miller   24   Bl   Dahlgren et al.,
                                                          1971
34 vulgaris Miller   12   Ho   Gadella & Kliphuis,
                                                          1966
35 angustissima (Loisel.) Borbás
36 biebersteinii Besser
37 macroura (Bieb.) Bieb.

38 debilis Kuprian.
39 odora (Bieb.) Fischer
40 loeselii Schweigger
41 latifolia Desf.
42 platycalyx Boiss.
43 anticaria Boiss. & Reuter   12   Hs   Valdés, 1970
44 lilacina Lange
45 verticillata Boiss.
46 thymifolia (Vahl) DC.
47 lamarckii Rouy
48 tristis (L.) Miller   12   Hs   Valdés, 1970
49 aeruginea (Gouan) Cav.   12   Hs   Küpfer, 1969
  (a) aeruginea
  (b) pruinosa (Sennen & Pau) Chater &
      Valdés
50 amoi Campo ex Amo   12   Hs   Valdés, 1970
51 caesia (Pers.) DC. ex Chav.   12   Hs   Valdés, 1970
52 supina (L.) Chaz.   12   Ga   van Loon, et al.,
                                               1971
53 oblongifolia (Boiss.) Boiss. & Reuter 12   Hs   Valdés, 1970
  (a) oblongifolia
  (b) haenseleri (Boiss. & Reuter) Valdés
54 badalii Willk.
55 propinqua Boiss. & Reuter
56 glauca (L.) Chaz.
  (a) glauca
  (b) olcadium Valdés & D.A. Webb
  (c) aragonensis (Lange) Valdés
  (d) bubanii (Font Quer) Valdés
57 diffusa Hoffmanns. & Link
58 coutinhoi Valdés
59 satureioides Boiss.
60 depauperata Leresche ex Lange
61 alpina (L.) Miller   12   Ga   Favarger & Huynh,
                                             1964
62 faucicola Leresche & Levier
63 glacialis Boiss.   12   Hs   Küpfer, 1968
64 ricardoi Coutinho
65 amethystea (Lam.) Hoffmanns. & Link
  (a) amethystea
  (b) multipunctata (Brot.) Chater & D.A.
      Webb
66 saxatilis (L.) Chaz.   12   Hs   Valdés, 1970
67 arenaria DC.
68 arvensis (L.) Desf.
69 simplex (Willd.) DC.
70 micrantha (Cav.) Hoffmanns. & Link

15 Cymbalaria Hill.
  1 muralis P. Gaertner, B. Meyer &
    Scherb.
    (a) muralis
    (b) visianii D.A. Webb
    (c) pubescens (C. Presl)
  2 longipes (Boiss. & Heldr.) A. Cheval.
  3 aequitriloba (Viv.) A. Cheval.
    (a) aequitriloba   56   Bl   Dahlgren et al., 1971
    (b) fragilis (Rodr.) D.A. Webb
  4 muelleri (Moris) A. Cheval.
  5 pallida (Ten.) Wettst.
  6 hepaticifolia (Poiret) Wettst.   56   Co   Contandriopoulos,
                                            1962

```
 7 pilosa (Jacq.) L.H. Bailey
 8 microcalyx (Boiss.) Wettst.
   (a) ebelii (Cuf.) Cuf.
   (b) minor (Cuf.) W. Greuter              28      Gr      Damboldt, 1971
   (c) dodekanesi W. Greuter
   (d) microcalyx

16 Kickxia Dumort.
   1 cirrhosa (L.) Fritsch
   2 commutata (Bernh. ex Reichenb.) Fritsch
   (a) commutata
   (b) graeca (Bory & Chaub.) R. Fernandes
   3 elatine (L.) Dumort.
   (a) elatine                              36      Ge      Wulff, 1939
   (b) crinita (Mabille) W. Greuter
   4 spuria (L.) Dumort
   (a) spuria
   (b) integrifolia (Brot.) R. Fernandes
   5 lanigera (Desf.) Hand.-Mazz.

17 Digitalis L.
   1 obscura L.                             56      Hs      Carpio, 1957
   (a) obscura
   (b) laciniata (Lindley) Maire
   2 purpurea L.
   (a) purpurea                            56      Ga      Lungeanu, 1967
   (b) mariana (Boiss.) Rivas Goday
   (c) heywoodii P. & M. Silva             56      Lu      Lungeanu, 1967
   3 thapsi L.
   4 dubia Rodr.
   5 grandiflora Miller                    56      Po      Skalińska et al.,
                                                           1959
   6 lutea L.
   (a) lutea                               56      Hs      Carpio, 1957
   (b) australis (Ten.) Arcangeli          56      It      Larsen, 1955b
   7 viridiflora Lindley                   56      Bu      Lungeanu, 1967
   8 parviflora Jacq.                      56      Hs      Lungeanu, 1967
   9 laevigata Waldst. & Kit.              56      Ju      Lungeanu, 1967
   (a) laevigata
   (b) graeca (Ivanina) Werner
  10 ferruginea L.                         56      Bu      Markova, 1970
  11 lanata Ehrh.                          56      Bu      Markova, 1970
  12 leucophaea Sibth. & Sm.

18 Erinus L.
   1 alpinus L.                            14      He      Favarger, 1953

19 Lagotis Gaertner
   1 minor (Willd.) Standley
   2 uralensis Schischkin

20 Wulfenia Jacq.
   1 carinthiaca Jacq.                     18      Au      Favarger, 1965
   2 baldaccii Degen

21 Veronica L.
   1 gentianoides Vahl
   2 bellidioides L.
   (a) bellidioides                        36      Au      Fischer, 1969
   (b) lilacina (Townsend) Nyman           18      Ga      Küpfer, 1968
   3 serpyllifolia L.
```

(a) serpyllifolia	14	Is	Löve & Löve, 1956
(b) humifusa (Dickson) Syme	14	Br	Rutland, 1941
4 repens Clarion ex DC.	14	Co	Contandriopoulos, 1962
5 alpina L.	18	Au	Fischer, 1969
	18 36	No	Knaben & Engelskjon, 1967
6 fruticans Jacq.	16	Au	Fischer, 1969
7 fruticulosa L.			
8 nummularia Gouan	16	Ca	Favarger & Kupfer, 1968
9 saturejoides Vis.			
10 erinoides Boiss. & Spruner	16	Gr	Quezel & Contandrio-poulos, 1965
contandriopouli Quezel	32	Gr	Quezel, 1967
11 ponae Gouan			
12 urticifolia Jacq.	18	He	Brandt, 1961
13 austriaca L.			
(a) teucrium (L.) D.A. Webb	64	He	Brandt, 1952
(b) vahlii (Gaudin) D.A. Webb	16	Ga	Küpfer, 1969
(c) dentata (F.W. Schmidt) Watzl	64	He	Brandt, 1952
	48	Cz	Májovský et al., 1970
(d) austriaca	48	Au	Fischer, 1969
14 multifida L.			
15 orientalis Miller			
16 tenuifolia Asso			
17 prostrata L.			
(a) prostrata	16	Ju	Fischer, 1969
(b) scheereri J.P. Brandt	32	Ge He	Brandt, 1952
18 turrilliana Stoj. & Stefanov			
19 rhodopaea (Velen.) Degen ex Stoj. & Stefanov			
20 pectinata L.			
21 rosea Desf.			
22 aragonensis Stroh			
23 thymifolia Sibth. & Sm.			
24 thessalica Bentham	16	Gr	Quezel & Contandrio-poulos, 1965
25 aphylla L.	18	Po	Skalińska et al., 1964
26 baumgartenii Roemer & Schultes			
27 officinalis L.	18	Da Su	Böcher, 1944
	36	Ho	Gadella & Kliphuis, 1968
28 allionii Vill.	18	Ga It	Bocquet et al., 1967
29 debneyi Hochst.			
30 chamaedrys L.			
(a) chamaedrys	32	Ho	Gadella & Kliphuis, 1963
(b) vindobonensis M. Fischer	16	Au	Fischer, 1970
31 melissifolia Poiret			
32 micrantha Hoffmanns. & Link			
33 peduncularis Bieb.			
34 montana L.	18	Ho	Gadella & Kliphuis, 1963

35 scutellata L.	18	Ho	Gadella & Kliphuis, 1963
36 beccabunga L.	18	Ho	Gadella & Kliphuis, 1963
	18 36	Po	Skalińska et al., 1964
37 scardica Griseb.	18	Hu Ju Rm	Marchant, ined.
38 anagalloides Guss.	18 + 0-2 B	Hs	Bjorkqvist et al., 1969
39 anagallis-aquatica L.	36	Ho	Gadella & Kliphuis, 1968
40 catenata Pennell	36	Ho	Gadella & Kliphuis, 1966
41 acinifolia L.	14	Europe	Fischer, ined.
	16	Rm	Tarnavschi, 1948
42 praecox All.	18	Au	Fischer, 1969
43 triphyllos L.	14	Au	Fischer, 1969
44 glauca Sibth. & Sm.			
45 aznavourii Dörfler			
46 arvensis L.	16	Au	Fischer, 1969
	?14	Is	Löve & Löve, 1956
47 verna L.	16	Co	Contandriopoulos, 1962
48 dillenii Crantz	16	Ge	Hofelich, 1935
49 peregrina L.			
50 chamaepithyoides Lam.			
51 grisebachii Walters			
52 agrostis L.	28	Ge	Wulff, 1937
53 polita Fries	14	Au	Fischer, 1969
	18	Rs(K)	Borhidi, 1968
54 opaca Fries			
55 persica Poiret	28	It	Gadella & Kliphuis, 1970
56 filiformis Sm.	14	Ge	Beatus, 1936
57 hederifolia L.			
(a) triloba (Opiz) Celak.	18	Au	Fischer, 1967
(b) sibthorpioides (Debeaux, Degen & Hervier) Walters			
(c) lucorum (Klett & Richter) Hartl	36	Au He Po	Fischer, 1967
(d) hederifolia	54	Au Ga Ge Gr It Si Su	Fischer, 1967
58 cymbalaria Bodard	18	Bl	Nilsson & Lassen, 1971
	36	Sa Si	Fischer, 1969
	54	Sa Tu	Fischer, 1969
59 longifolia L.	34	Rm	Borhidi, 1968
	68	Fe	Raitanen, 1967
60 paniculata L.	34	Hu	Graze, 1933
61 bachofenii Heuffel			
62 spicata L.			
(a) spicata	34	He	Graze, 1933
		Fe	Raitanen, 1967
(b) incana (L.) Walters	68	Hu	Graze, 1933

```
(c) barrelieri (Schott ex Roemer &
      Schultes) Murb.                    34      Ju      Graze, 1933
(d) orchidea (Crantz) Hayek             34      Au      Fischer, 1969

(e) crassifolia (Nyman) Hayek
```

22 Hebe Commerson
  1 salicifolia (G. Forster) Pennell
  2 speciosa (R. Cunn. ex A. Cunn.) Andersen
  3 elliptica (G. Forster) Pennell

```
23 Paederota L.
   1 bonarota (L.) L.                    36      It      Favarger, 1965b
   2 lutea Scop.                         36      Au      Favarger, 1965b
                                       c.54      Au      Fischer, 1969


24 Sibthorpia L.
   1 europaea L.                         18      Br      I. & O. Hedburg, 1961
   2 africana L.
   3 peregrina L.
```

25 Lafuentea Lag.
  1 rotundifolia Lag.

```
26 Castilleja Mutis ex L. fil.
   1 pallida (L.) Sprengel
   2 schrenkii Rebr.
   3 lapponica (Gand.) Rebr.             48      Rs(N)   Löve & Löve, 1961b


27 Melampyrum L.
   1 cristatum L.                        18      Br      Hambler, 1954
   2 ciliatum Boiss. & Heldr.
   3 arvense L.                          18      Au      van Witsch, 1932
   4 barbatum Waldst. & Kit. ex Willd.
    (a) barbatum
    (b) carstiense Ronniger
   5 variegatum Huter, Porta & Rigo
   6 fimbriatum Vandas
   7 nemorosum L.
    (a) nemorosum                        18      Ge      Reese, 1953
    (b) debreceniense (Rapaics) Soó
   8 polonicum (Beauverd) Soó
   9 velebiticum Borbás                  18      He      Favarger, 1969b
  10 vaudense (Ronniger) Soó
  11 catalaunicum Freyn
  12 italicum Soó
  13 bohemicum A. Kerner
  14 subalpinum (Juratzka) A. Kerner
  15 bihariense A. Kerner
  16 hoermannianum K. Maly
  17 doerfleri Ronniger
  18 scardicum Wettst.
  19 trichocalycinum Vandas
  20 heracleoticum Boiss. & Orph.
  21 sylvaticum L.                       18      Is      Löve & Löve, 1956
  22 herbichii Woloszczak
  23 saxosum Baumg.
  24 pratense L.                         18      Fe      Sorsa, 1962
```

28 Tozzia L.
  1 alpina L.

```
    (a) alpina                              20        Au      Mattick, 1950

    (b) carpathica (Woloszczak) Dostál

29 Euphrasia L.
   1 grandiflora Hochst.
   2 azorica H.C. Watson
   3 rostkoviana Hayne
     (a) montana (Jordan) Wettst.          22        Au      van Witsch, 1932
     (b) rostkoviana                       22        Au      van Witsch, 1932
     (c) campestris (Jordan) P. Fourn.
   4 rivularis Pugsley                      22        Br      Yeo, 1954
   5 anglica Pugsley                        22        Br      Yeo, 1954
   6 vigursii Davey                         22        Br      Yeo, 1956
   7 hirtella Jordan ex Reuter              22        He      Yeo, 1970
   8 drosocalyx Freyn
   9 picta Wimmer
     (a) picta                             22        Au      Yeo, 1970
     (b) kerneri (Wettst.) Yeo
  10 marchesettii Wettst. ex Marchesetti
  11 arctica Lange ex Rostrup
     (a) tenuis (Brenner) Yeo              44        Is      Löve & Löve, 1956
     (b) arctica
     (c) minor Yeo
     (d) borealis (Townsend) Yeo           44        Br      Yeo, 1954
     (e) slovaca Yeo
  12 tetraquetra (Breb.) Arrondeau         44        Br      Yeo, 1954
  13 nemorosa (Pers.) Wallr.               44        Br      Yeo, 1954
  14 coerulea Hoppe & Fürnrohr
  15 pseudokerneri Pugsley                 44        Br      Yeo, 1954
  16 confusa Pugsley                       44        Br      Yeo, 1954
  17 stricta D. Wolff ex J.F. Lehm.        44        Br      Yeo, 1954
  18 hyperborea Joerg.
  19 pectinata Ten.                        44        He      Favarger, 1969
  20 liburnica Wettst.
  21 alpina Lam.                           22        Ga      Favarger, 1969
  22 christii Favrat
  23 cisalpina Pugsley
  24 minima Jacq. ex DC.
     (a) minima                           44        Au      van Witsch, 1932
     (b) tatrae (Wettst.) Hayek
  25 frigida Pugsley                       44        Is      Löve & Löve, 1956
  26 foulaensis Townsend ex Wettst.        44        Br      Yeo, 1954
  27 cambrica Pugsley
  28 ostenfeldii (Pugsley) Yeo
  29 marshallii Pugsley                    44        Br      Yeo, 1954
  30 rotundifolia Pugsley
  31 dunensis Wiinst.
  32 campbelliae Pugsley
  33 micrantha Reichenb.                   44        Br      Yeo, 1954
  34 scottica Wettst.                      44        Br      Yeo, 1954
  35 calida Yeo
  36 saamica Juz.
  37 atropurpurea (Rostrup) Ostenf.
  38 bottnica Kihlman
  39 willkommii Freyn
  40 taurica Ganeschin
  41 salisburgensis Funck                  44        Hb      Yeo, 1966
  42 portae Wettst.
  43 illyrica Wettst.
  44 dinarica (G. Beck) Murb.
```

45 cuspidata Host
46 tricuspidata L.

30 Odontites Ludwig
  1 tenuifolia (Pers.) G. Don fil.
  2 longiflora (Vahl) Webb
  3 glutinosa (Bieb.) Bentham
  4 viscosa (L.) Clairv.
   (a) viscosa
   (b) hispanica (Boiss. & Reuter) Rothm.
  5 granatensis Boiss.
  6 purpurea (Desf.) G. Don fil.
  7 rigidifolia (Biv.) Bentham

8 lutea (L.) Clairv.	20	Ge	Schneider, 1964

  9 lanceolata (Gaudin) Reichenb.
 10 jaubertiana (Boreau) D. Dietr. ex
    Walpers
   (a) jaubertiana
   (b) chrysantha (Boreau) P. Fourn.
   (c) cebennensis (Coste & Soulie)
     P. Fourn.
 11 kaliformis (Pourret) Pau
 12 bocconei (Guss.) Walpers
 13 linkii Heldr. & Sart. ex Boiss.
 14 corsica (Loisel.) G. Don fil.
 15 verna (Bellardi) Dumort.

(a) verna	40	Ge	Schneider, 1964
(b) litoralis (Fries) Nyman	20	Ge	Schneider, 1964
(c) serotina (Dumort.) Corb.	20	Ge	Rottgardt, 1956

31 Bartsia L.

1 alpina L.	12	Au Ga He	Favarger, 1953
	36	Ga	Favarger & Küpfer, 1968
	24	Is	Löve & Löve, 1956

  2 spicata Ramond
  3 aspera (Brot.) Lange

32 Parentucellia Viv.

1 viscosa (L.) Caruel	48	Br	Hambler, 1955

  2 latifolia (L.) Caruel

33 Bellardia All.

1 trixago (L.) All.	24	Ga	Delay, 1969

34 Pedicularis L.
  1 acaulis Scop.

2 sceptrum-carolinum L.	32	No	Knaben & Engelskjon, 1967
3 foliosa L.	16	Au	Witsch, 1932

  4 hoermanniana K. Maly
  5 hacquetii Graf
  6 exaltata Besser

7 recutita L.	16	Au	Mattick, 1950

  8 limnogena A. Kerner

9 rosea Wulfen	16	It	Favarger & Huynh, 1964

   (a) rosea
   (b) allionii (Reichenb. fil.) E. Mayer
 10 orthantha Griseb.
 11 dasyantha Hadač

12 hirsuta L.	16	No	Knaben & Engelskjon, 1967
13 oederi Vahl	16	No	Knaben, 1950
14 flammea L.	16	Is	Löve & Löve, 1956
15 verticillata L.	12	Ga	Küpfer, 1968
16 amoena Adams ex Steven			
17 arguteserrata Vved.			
18 palustris L.	16	Ge	Reese, 1953
(a) palustris			
(b) opsiantha (E.L. Ekman) Almquist			
(c) borealis (J.W. Zett.) Hyl.			
19 labradorica Wirsing			
20 sylvatica L.			
(a) sylvatica	16	Ho	Gadella & Kliphuis, 1968
(b) lusitanica (Hoffmanns. & Link) Coutinho			
(c) hibernica D.A. Webb			
21 sudetica Willd.	16	Rs(N)	Sokolovskaya & Strelkova, 1960
22 comosa L.			
(a) comosa			
(b) campestris (Griseb.) Soó			
23 kaufmannii Pinzger			
24 sibthorpii Boiss.			
25 uralensis Vved.			
26 asparagoides Lapeyr.			
27 schizocalyx (Lange) Steininger			
28 brachyodonta Schlosser & Vuk.			
(a) brachyodonta			
(b) moesiaca (Stadlm.) Hayek			
(c) montenegrina (Janka ex Nyman) D.A. Webb			
(d) grisebachii (Wettst.) Hayek			
29 heterodonta Pančić			
30 leucodon Griseb.			
(a) leucodon			
(b) occulta (Janka) E. Mayer			
31 physocalyx Bunge			
32 dasystachys Schrenk			
33 graeca Bunge			
34 friderici-augusti Tommasini			
35 petiolaris Ten.			
36 ferdinandi Bornm.			
37 compacta Stephan ex Willd.			
38 lapponica L.	16	Su	Löve & Löve, 1944
39 tuberosa L.	16	Au	Van Witsch, 1932
40 elongata A. Kerner	16	It	Favarger, 1965
41 julica E. Mayer			
42 ascendens Schleicher ex Gaudin	16	He	Favarger, 1953
43 baumgartenii Simonkai			
44 rostratospicata Crantz	16		
45 rostratocapitata Crantz	16	Au	Mattick, 1950
(a) rostratocapitata			
(b) glabra H. Kunz			
46 kerneri Dalla Torre	16	He	Favarger, 1959a
47 pyrenaica Gay	16	Ga	Küpfer & Favarger, 1968
48 mixta Gren.			
49 cenisia Gaudin			
50 gyroflexa Vill.			

```
    (a) gyroflexa                            16      He      Favarger, 1963
    (b) praetutiana (Levier ex
        Steininger) H. Kunz & E. Mayer
 51 elegans Ten.
 52 portenschlagii Sauter ex Reichenb.
 53 asplenifolia Floerke ex Willd.          16      Au      Mattick, 1950
 54 resupinata L.

 35 Rhinanthus L.
  1 groenlandicus (Ostenf.) Chab.           22      Is      Löve & Löve, 1956
  2 minor L.                                 22      Au      Tschermak-Woess, 1967
  3 asperulus (Murb.) Soó
  4 wettsteinii (Steneck) Soó
  5 pubescens (Sterneck) Boiss. & Heldr.
    ex Soó
  6 pindicus (Sterneck) Soó
  7 antiquus (Sterneck) Schinz & Thell.
  8 dinaricus Murb.
  9 aristatus Celak.
 10 pampaninii Chab.
 11 alpinus Baumg.
 12 carinthiacus Widder
 13 burnatii (Chab.) Soó
 14 rumelicus Velen.
 15 wagneri Degen
 16 mediterraneus (Sterneck) Adamović       22      Ga      Campion-Bourget, 1967
 17 melampyroides (Borbás & Degen) Soó
 18 songeonii Chab.
 19 ovifugus Chab.
 20 subulatus (Chab.) Soó
 21 borbasii (Dörfler) Soó
    (a) borbasii
    (b) songaricus (Sterneck) Soó
 22 halophilus U. Schneider
 23 angustifolius C.C. Gmelin               22      Au      Tschermak-Woess, 1967
    (a) angustifolius
    (b) cretaceus (Vassilcz.) Soó
    (c) grandiflorus (Wallr.) D.A. Webb
 24 freynii (A. Kerner ex Sterneck) Fiori
 25 alectorolophus (Scop.) Pollich          22      Au      Tschermak-Woess, 1967

 36 Rhynchocorys Griseb.
  1 elephas (L.) Griseb.

 37 Siphonostegia Bentham
  1 syriaca (Boiss. & Reuter) Boiss.

 38 Cymbaria L.
  1 borysthenica Pallas ex Schlecht.

 39 Lathraea L.
  1 squamaria L.                            36      Au      van Witsch, 1932
  2 rhodopea Dingler
  3 clandestina L.


CLV GLOBULARIACEAE

  1 Globularia L.
  1 incanescens Viv.                        16      It      Schwarz, 1964
  2 trichosantha Fischer & C.A. Meyer
  3 punctata Lapeyr.                        16      Au Ga Hu   Larsen, 1957b
```

213

4 vulgaris L.	32	Gs Su	Larsen, 1957b
5 valentina Willk.			
6 cambessedesii Willk.	64	Bl	Schwarz, 1964
7 spinosa L.			
8 alypum L.	16	Ga	Schwarz, 1964
9 cordifolia L.	32	Ga He	Larsen, 1957
10 meridionalis (Podp.) O. Schwarz	16	Al Bu	Schwarz, 1964
11 repens Lam.	16	Ga Hs	Schwarz, 1964
12 neapolitana O. Schwarz			
13 stygia Orph. ex Boiss.			
14 nudicaulis L.	16 24	Ge	Schwarz, 1964
x fuxeensis Giraud.	16	Ga	Contandriopoulos & Cauwet, 1968
15 gracilis Rouy & Richter			

CLVI ACANTHACEAE

1 Acanthus L.

1 mollis L.	56	Ga	van Loon et al., 1971

2 balcanicus Heywood & I.B.K. Richardson
3 spinosus L.

2 Justicia L.
1 adhatoda L.

CLVII PEDALIACEAE

1 Sesamum L.
1 indicum L.

CLVIII MARTYNIACEAE

1 Proboscidea Schmidel
1 louisianica (Miller) Thell.

CLIX GESNERIACEAE

1 Haberlea Friv.

1 rhodopensis Friv.	38	Bu	Borhidi, 1968
	48	Gr	Contandriopoulos, 1967

2 Ramonda L.C.M. Richard

1 myconi (L.) Reichenb.	48	Ga	Contandriopoulos, 1967
2 nathaliae Pančić & Petrović	48	Ju Gr	Contandriopoulos, 1967
3 serbica Pančić	c.96	Gr	Contandriopoulos, 1967

3 Jankaea Boiss.

1 heldreichii (Boiss.) Boiss.	56	Gr	Contandriopoulos, 1967

CLX OROBANCHACEAE

1 Cistanche Hoffmanns. & Link

```
 1 phelypaea (L.) Coutinho                     40      Lu        Gardé, 1952
 2 salsa (C.A. Meyer) G. Beck

2 Orobanche L.
   1 ramosa L.                                  24      Lu        Gardé, 1952
     (a) ramosa
     (b) Reuter) Coutinho
     (c) mutelii (F.W. Schultz) Coutinho
   2 aegyptiaca Pers.                           24      Rs(E)     Zhukov, 1939
   3 lavandulacea Reichenb.
   4 trichocalyx (Webb & Berth.) G. Beck
   5 oxyloba (Reuter) G. Beck
   6 schultzii Mutel
   7 caesia Reichenb.
   8 arenaria Borkh.
   9 purpurea Jacq.                             24      Br        Hambler, 1958
  10 coerulescens Stephan
  11 cernua Loefl.                              24      Rs(E)     Zhukov, 1939
  12 amoena C.A. Meyer
  13 crenata Forskål                            38      Lu        Gardé, 1952
  14 alba Stephan ex Willd.
  15 reticulata Wallr.                          38      Br        Hambler, 1958
  16 haenseleri Reuter
  17 pancicii G. Beck
  18 serbica G. Beck & Petrović
  19 pubescens D'Urv.
  20 densiflora Salzm. ex Reuter
  21 amethystea Thuill.
     (a) amethystea
     (b) castellana (Reuter) Rouy
  22 canescens C. Presl
  23 esulae Pančić
  24 calendulae Pomel
  25 grisebachii Reuter
  26 loricata Reichenb.                        c.38     Br        Hambler, 1958
  27 minor Sm.                                  38      Europe    Löve, 1954a
  28 clausonis Pomel
  29 hederae Duby                               38      Br        Hambler, 1958
  30 latisquama (F.W. Schultz) Batt.
  31 caryophyllacea Sm.                         38      Ga        Lévêque &
                                                                  Gorenflot, 1969
  32 teucrii Holandre
  33 lutea Baumg.                               38      Rs(E)     Zhukov, 1939
  34 elatior Su-ton                             38      Br        Hambler, 1958
  35 alsatica Kirschleger
  36 chironii Lojac.
  37 laserpitii-sileris Reuter ex Jordan
  38 flava C.F.P. Mart. ex F.W. Schultz         38      Po        Skalińska et al.,
                                                                  1966
  39 lucorum A. Braun
  40 salviae F.W. Schultz ex Koch
  41 rapum-genistae Thuill.                     38      Br        Hambler, 1958
     (a) rapum-genistae
     (b) benthamii (Timb.-Lagr.) P. Fourn.
     (c) rigens (Loisel.) P. Fourn.
  42 gracilis Sm.
  43 variegata Wallr.
  44 foetida Poiret
  45 sanguinea C. Presl

3 Boschniakia C.A. Meyer
  1 rossica (Cham. & Schlecht.) B. Fedtsch.
```

4 Phelypaea L.
  1 coccinea Poiret
  2 boissieri (Reuter) Stapf

CLXI LENTIBULARIACEAE

1 Pinguicula L.
  1 lusitanica L.   12   Ga   Contandriopoulos, 1962
  2 hirtiflora Ten.   16   It   Honsell, 1959
  3 alpina L.   32   Rs(N)   Sokolovskaya &
                                                                   Strelkova, 1960
  4 villosa L.   16   No   Knaben, 1950
  5 corsica Bernard & Gren.   16   Co   Favarger &
                                                                   Contandriopoulos, 1961
  6 leptoceras Reichenb.   32   Au   Casper, 1962
  7 balcanica Casper   32   Bu   Casper, 1966
  8 nevadensis (Lindb.) Casper
  9 vallisneriifolia Webb
 10 longifolia Ramond ex DC.
    (a) longifolia
    (b) caussensis Casper
    (c) reichenbachiana (Schindler) Casper   32   Ga   Casper, 1962
 11 grandiflora Lam.   32   Ga   Contandriopoulos, 1962
 12 vulgaris L.   64   Da   Larsen, 1965

2 Utricularia L.
  1 subulata L.
  2 gibba L.
  3 minor L.   c.40   Ge   Reese, 1952
  4 intermedia Hayne
    ochroleuca R. Hartman   c.40   Ge   Reese, 1952
  5 vulgaris L.   c.40   Ge   Reese, 1952
  6 australis R. Br.   c.40   Ge   Reese, 1952

CLXII MYOPORACEAE

1 Myoporum Solander ex G. Forster
  1 tenuifolium G. Forster
  2 tetrandrum (Labill.) Domin

CLXIII PLANTAGINACEAE

1 Plantago L.
  1 major L.
    (a) major   12   Cz   Májovský et al.,
                                                                   1970
    (b) winteri (Wirtgen.) W. Ludwig
    (c) intermedia (DC.) Arcangeli   12   Cz   Májovský et al.,
                                                                   1970
  2 tenuiflora Waldst. & Kit.   24   Cz   Májovský et al.,
                                                                   1970
  3 cornuti Gouan   12   Rm   Rahn, 1966
  4 coronopus L.
    (a) coronopus   10 + 0 - 1B   Lu   Fernandes &
                                                                   Franca, 1972
    (b) commutata (Guss.) Pilger   20   Co   Gorenflot, 1960
    (c) cupanii (Guss.) Nyman

(d) purpurascens Pilger

5 macrorhiza Poiret	10	Lu	Böcher et al., 1955
6 serraria L.	10 + 0 - 3B 20 + 2B	Lu	Fernandes & Franca, 1972
7 crassifolia Forskål	20	Bl	Dahlgren et al., 1971
8 maritima L.	12	Ho	Gadella & Kliphuis, 1967a
	18	Po	Skalińska et al., 1961
	24	Fe	V. Sorsa, 1962
9 subulata L.	12	It	Rahn, 1966
holosteum Scop.	12	Lu	Fernandes & Franca, 1973
insularis (Gren. & Godron) Nyman	24	Co	Contandriopoulos, 1962
10 alpina L.	12	Ga	Cartier & Lennoir, 1968
	24	Ga	Cartier & Lennoir, 1968
11 schwarzenbergiana Schur	12	Rm	Rahn, 1966
12 media L.	12	Ju	Gadella & Kliphuis, 1972
	24	Ho	Gadella & Kliphuis, 1968
13 maxima Juss. ex Jacq.			
14 reniformis G. Beck	12	Ju	Sušnik & Lovka, 1973
15 atrata Hoppe	12	It	Rahn, 1966
	24 + 0 - 2B	Cz	Holub et al., 1971
	36	Ga	Cartier, 1965
16 monosperma Pourret	12	Ga	Cartier, 1965
17 nivalis Boiss.	12	Hs	Rahn, 1966
18 gentianoides Sibth. & Sm.	12	Bu	Kozuharov & Petrova, 1974
19 amplexicaulis Cav.	10	Gr	Runemark, 1967b
20 lanceolata L.	12 + 0 - 1B	Lu	Fernandes & Franca, 1972
21 altissima L.	72	Cz	Májovský et al., 1970
22 argentea Chaix	12 12 + 2B	Rm	Rahn, 1966
23 lagopus L.	12 + 0 - 1B	Lu	Fernandes & Franca, 1972
24 albicans L.	20	Hs	Lorenzo-Andreu, 1950
	30	Gr	Runemark, 1967b
25 ovata Forskål	8	Hs	Rahn, 1957
26 minuta Pallas			
27 loeflingii L.			
28 notata Lag.			
29 bellardii All.			
(a) bellardii	10	Lu	Rodriguez, 1953
(b) deflexa (Pilger) Rech. fil.	10	Gr	Runemark, 1967b
30 cretica L.	10	Gr	Runemark, 1967b
31 squarrosa Murray			
32 arenaria Waldst. & Kit.	12	Bu	Kozuharov & Petrova, 1974
33 afra L.	12	Bl	Dahlgren et al., 1971

```
34 sempervirens Crantz                      12    Ga    Rahn, 1966
35 asperrima (Gand.) Hervier


2 Littorella Bergius
  1 uniflora (L.) Ascherson                  24    Ge    Wulff, 1939a


CLXIV CAPRIFOLIACEAE

1 Sambucus L.
  1 ebulus L.
  2 nigra L.                                 36    Hu    Pólya, 1950
  3 racemosa L.                              36    Fe    V. Sorsa, 1962


2 Viburnum L.
  1 opulus L.                                18    Su    Löve, 1954 a
  2 lantana L.                               18    Br    Egolf, 1962
  3 tinus L.
    (a) tinus
    (b) subcordatum (Trelease) P. Silva


3 Symphoricarpos Duh.
  1 albus (L.) S.F. Blake


4 Linnaea L.
  1 borealis L.                              32    Fe    V. Sorsa, 1962


5 Leycesteria Wall.
  1 formosa Wall.


6 Lonicera L.
  1 caerulea L.
    (a) caerulea                             18    Su    Löve, 1954a
    (b) pallasii (Ledeb.) Browicz
  2 pyrenaica L.
    (a) pyrenaica
    (b) majoricensis (Gand.) Browicz
  3 alpigena L.
    (a) alpigena                             18    Au    Löve, ined.
    (b) formanekiana (Halácsy) Hayek
  4 glutinosa Vis.
  5 hellenica Orph. ex Boiss.
  6 nigra L.                                 18    Cz    Májovský et al.,
                                                         1970
  7 tatarica L.
  8 xylosteum L.                             18    Fe    Sorsa, 1963
  9 arborea Boiss.
 10 nummariifolia Jaub. & Spach
 11 biflora Desf.
 12 japonica Thunb.                          18    Ga    Löve, ined.
 13 implexa Aiton
 14 splendida Boiss.
 15 caprifolium L.                           18    It    Löve, ined.
 16 etrusca G. Santi
 17 periclymenum L.
    (a) periclymenum                         18    Br    Rutland, 1941
                                             36    Br    Janaki-Ammal &
                                                         Saunders, 1952
                                             54    Ho    Gadella & Kliphuis,
                                                         1963
    (b) hispanica (Boiss. & Reuter) Nyman
```

CLXV ADOXACEAE

1 Adoxa L.
  1 moschatellina L.

CLXVI VALERIANACEAE

1 Patrinia Juss.
  1 sibirica (L.) Juss.

2 Valerianella Miller
  1 coronata (L.) DC.   14   Bl   Dahlgren et al.,
                                                           1971
  2 divaricata Lange
  3 lasiocarpa (Steven) Betcke
  4 pumila (L.) DC.   14   Ga Rm   Ernet, 1972
  5 hirsutissima Link
  6 kotschyi Boiss.
  7 discoidea (L.) Loisel.   14   Ga Lu   Ernet, 1972
  8 obtusiloba Boiss.
  9 vesicaria (L.) Moench
 10 locusta (L.) Laterrade   16   Ge   Ernet, 1972
 11 martinii Loscos
 12 carinata Loisel.   16   Au   Ernet, 1972
 13 turgida (Steven)Betcke
 14 costata (Steven) Betcke
 15 echinata (L.) DC.   16   It   Ernet, 1972
 16 dentata (L.) Pollich   16   Rm   Ernet, 1972
 17 rimosa Bast.   16   Ga Ge Su Ernet, 1972
 18 eriocarpa Desv.   16   Si   Ernet, 1972
    muricata (Steven ex Bieb.) J.W.
    Loudon   16   Hs It Lu Ernet, 1972
 19 microcarpa Loisel.   16   It   Ernet, 1972
 20 puberula (Bertol. ex Guss.) DC.
 21 pontica Lipsky
 22 uncinata (Bieb.) Dufresne

3 Fedia Gaertner
  1 cornucopiae (L.) Gaertner

4 Valeriana L.
  1 officinalis L.   14   Au   Titz, 1964
                                            28   Cz   Májovský et al.,
                                                           1970
                                            49   Po   Skalinska, 1958
                                            56   Ho   Gadella & Kliphuis,
                                                           1963
  (a) officinalis   14   Po   Skalińska, 1950b
  (b) collina (Wallr.) Nyman   28   Po   Skalińska, 1950b
  (c) sambucifolia (Mikan fil.) Čelak.   56   Po   Skalińska, 1950b
     salina Pleijel   56   Su   Jonsell in litt.,
                                                           1973
     repens Host   56   Ge   Walther, 1949
  2 dioscoridis Sibth. & Sm.   16   Gr   Seitz, 1969
  3 tuberosa L.   16   Ga   Guinochet &
                                                           Logeois, 1962
  4 asarifolia Dufresne
  5 alliariifolia Adams
  6 pyrenaica L.
  7 globulariifolia Ramond ex DC.   16   Ga Hs   Küpfer &
                                                           Favarger, 1968

```
 8 olenaea Boiss. & Heldr.
 9 dioica L.
   (a) dioica                              16      Ju      Lovka et al., 1971
                                           32      Ge      Engel, 1972
   (b) simplicifolia (Reichenb.) Nyman     16      Po      Skalińska, 1950
10 tripteris L.                            16      He      Larsen, 1954
11 montana L.                              32      Au      Tischler, 1950
   phitosiana                              32      Cr      Quézel & Contand-
                                                           riopoulos, 1965
12 bertiscea Pančić
13 crinii Orph. ex Boiss.
14 capitata Link                           56      Rs(N)   Sokolovskaya &
                                                           Strelkova, 1960
15 saxatilis L.
   (a) saxatilis                           24      Au      Favarger, 1965
   (b) pancicii (Halácsy & Bald.)
       Ockendon                            24      Ju      Engel, 1972
16 celtica L.                             c.48     Ga      Favarger, 1965
                                          c.72     Au      Polatschek, 1966
                                          c.96     Au      Favarger, 1965
17 elongata Jacq.                          24      It      Favarger, 1965
18 supina Ard.                             16      He      Favarger, 1965
19 saliunca All.                           16      Ga      Favarger, 1965
20 longiflora Willk.


 5 Centranthus DC.
   1 ruber (L.) DC.
     (a) ruber                             32      It      Larsen, 1958b
     (b) sibthorpii (Heldr. & Sart. ex
         Boiss.) Hayek
   2 angustifolius (Miller) DC.
   3 lecoqii Jordan
   4 longiflorus Steven
     (a) junceus (Boiss. & Heldr.) I.B.K.
         Richardson
     (b) kellereri (Stoj. Stefanov &
         Georgiev) I.B.K. Richardson
   5 nevadensis Boiss.
     (a) nevadensis
     (b) sieberi (Heldr.) I.B.K.
         Richardson
   6 trinervis (Viv.) Béguinot            28      Co      Contandriopoulos,
                                                           1962
   7 calcitrapae (L.) Dufresne            32      Bl      Dahlgren et al., 1971
     (a) calcitrapae
     (b) trichocarpus I.B.K. Richardson
   8 macrosiphon Boiss.                   32      Hs      Murray, 1975


CLXVII DIPSACACEAE

1 Morina L.
  1 persica L.

2 Cephalaria Schrader
  1 squamiflora (Sieber) W. Greuter
    (a) balearica (Willk.) W. Greuter
    (b) squamiflora
  2 leucantha (L.) Roemer & Schultes      18      Ga      Ritter, 1973
  3 radiata Griseb. & Schenk
  4 laevigata (Waldst. & Kit.) Schrader
```

```
  5 coriacea (Willd.) Roemer & Schultes ex
    Steudel
  6 uralensis (Murray) Roemer & Schultes      18      Rm      Lungeanu, 1972
  7 syriaca (L.) Roemer & Schultes
  8 transylvanica (L.) Roemer & Schultes      18      Cz      Májovský et al.,
                                                              1970
  9 joppica (Sprengel) Beguinot
 10 ambrosioides (Sibth. & Sm.) Roemer
    & Schultes
 11 flava (Sibth. & Sm.) Szabo
 12 pastricensis Dörfler & Hayek
 13 alpina (L.) Roemer & Schultes
 14 litvinovii Bobrov

3 Dipsacus L.
  1 gmelinii Bieb.
  2 sativus (L.) Honckeny                      18      Ga      Poucques, 1948
  3 fullonum L.                                18      Po      Skalińska et al.,
                                                              1964
  4 laciniatus L.                              16    Bot. Gard.   Poucques, 1948
                                               18    Bot. Gard.   Kachidze, 1929
  5 comosus Hoffmanns. & Link
  6 ferox Loisel.
  7 pilosus L.                                 18    Bot. Gard.   Kachidze, 1929
  8 strigosus Willd.

4 Succisa Haller
  1 pratensis Moench                           20      Fe      Sorsa, 1963
  2 pinnatifida Lange

5 Succisella G. Beck
  1 inflexa (Kluk) G. Beck                     20      Hu      Baksay, 1955
  2 petteri (J. Kerner & Murb.) G. Beck
  3 carvalhoana (Mariz) Baksay
  4 microcephala (Willk.) G. Beck

6 Knautia L.
  1 drymeia Heuffel
    (a) nympharum (Boiss. & Heldr.)
        Ehrend.                                20      Ju      Ehrendorfer, 1962
    (b) tergestina (G. Beck) Ehrend.           20    It Ju     Ehrendorfer, 1962
    (c) centrifrons (Borbas) Ehrend.           20      It      Ehrendorfer, 1962
    (d) drymeia (Jacq.) Wettst.                40   Au Hu Ju   Ehrendorfer, 1962
                                            38 42-44   Ju      Ehrendorfer, 1962
    (e) intermedia (Pernh. & Wettst.)
        Ehrend.                                20      Au      Ehrendorfer, ined.
                                               40      Au      Ehrendorfer, 1962
  2 gussonei Szabo
  3 arvernensis (Briq.) Szabo                  40      Ga      Ehrendorfer, 1962
  4 sarajevensis (G. Beck) Szabo               40      Ju      Ehrendorfer, 1962
  5 dipsacifolia Kreutzer
    (a) lancifolia (Heuffel) Ehrend.
    (b) pocutica (Szabo) Ehrend.               40      Hu      Baksay, 1956
    (c) turocensis (Borbas) Jav. ex Kiss       40      Cz      Májovský et al.,
                                                              1978
    (d) gracilis (Szabo) Ehrend.               40      Ge      Breton-Sintes, 1971
    (e) sixtina (Briq.) Ehrend.                60              Ehrendorfer, ined.
    (f) dipsacifolia                           60   Au Ga He   Ehrendorfer, 1962
  6 ressmannii (Pacher) Briq.                  60      It      Ehrendorfer, 1962
  7 nevadensis (M. Winkler ex Szabo)
    Szabo                                      64      Hs      Ehrendorfer, 1962
```

8 basaltica Chassagne & Szabo	20	Ga	Breton-Sintes, 1971
9 foreziensis Chassagne & Szabo	40	Ga	Breton-Sintes, 1971
10 godetii Reuter	20	Ga	Sintes & Cauderon, 1965
11 salvadoris Sennen ex Szabo	20		Ehrendorfer, ined.
12 longifolia (Waldst. & Kit.) Koch	20	Au It	Ehrendorfer, 1962
13 pancicii Szabo			
14 midzorensis Form.	20	Ju	Ehrendorfer, 1962
15 magnifica Boiss. & Orph.			
16 dinarica (Murb.) Borbás			
(a) dinarica	20	Ju	Ehrendorfer, 1962
	40	Ju	Ehrendorfer, 1962
(b) silana (Grande) Ehrend.	40		Ehrendorfer, ined.
17 lucana Lacaita & Szabo			
18 subcanescens Jordan	40	Ga	Ehrendorfer, 1962
19 baldensis A. Kerner ex Borbás	40	It	Ehrendorfer, 1962
20 velutina Briq.	20	It	Ehrendorfer, 1962
21 transalpina (Christ) Briq.	40	It	Ehrendorfer, 1962
22 persicina A. Kerner	40	It	Ehrendorfer, 1962
23 carinthiaca Ehrend.	20	Au	Ehrendorfer, 1962
24 x norica Ehrend.	40	Au	Ehrendorfer, 1962
25 albanica Briq.	20	Ju	Ehrendorfer, 1962
26 velebitica Szabo	20	Ju	Ehrendorfer, 1962
27 mollis Jordan	20	Ga	Ehrendorfer, ined
28 calycina (C. Presl) Guss.	20	It	Ehrendorfer, 1962
29 arvensis (L.) Coulter	20	Au Hu Ju	Ehrendorfer, 1962
	40	Ga He It	Ehrendorfer, 1962
	43	Au	Ehrendorfer, 1962
	46	Au	Ehrendorfer, 1962
30 kitaibelii (Schultes) Borbás			
(a) kitaibelii	40	Au Po	Ehrendorfer, 1962
(b) tomentella (Szabo) Baksay	40	Hu	Baksay, 1956
31 ambigua Boiss. & Orph.	20	Gr Ju	Ehrendorfer, 1962
32 macedonica Griseb.	20	Ju	Ehrendorfer, 1962
33 visianii Szabo	20	Ju	Ehrendorfer, 1962
34 purpurea (Vill.) Borbás	20	He It	Ehrendorfer, 1962
35 rupicola (Willk.) Szabo			
36 subscaposa Boiss. & Reuter	20	Hs	Ehrendorfer, ined.
37 illyrica G. Beck	40	Ju	Ehrendorfer, 1962
38 pectinata Ehrend.	20	Ju	Ehrendorfer, 1962
39 clementii (G. Beck) Ehrend.	40	Ju	Ehrendorfer, 1962
40 adriatica Ehrend.	40	Ju	Ehrendorfer, ined.
41 dalmatica G. Beck	20	Ju	Ehrendorfer, ined.
42 travnicensis (G. Beck) Szabo	60	Ju	Ehrendorfer, 1962
43 fleischmannii (Hladnik ex Reichenb.) Pacher	40	Ju	Ehrendorfer, 1962
44 tatarica (L.) Szabo	20	Rs(C)	Ehrendorfer, ined.
45 byzantina Fritsch			
46 integrifolia (L.) Bertol.	20	Gr Ju	Ehrendorfer, 1962
47 degenii Borbás			
48 orientalis L.	16	Gr	Ehrendorfer, ined.

7 Pterocephalus Adanson
  1 papposus (L.) Coulter
  2 brevis Coulter
  3 diandrus (Lag.) Lag.
  4 intermedius (Lag.) Coutinho
  5 spathulatus (Lag.) Coulter
  6 perennis Coulter
   (a) perennis
   (b) bellidifolius (Boiss.) Vierh.

8 Scabiosa L.
  1 limonifolia Vahl
  2 saxatilis Cav.
   (a) saxatilis
   (b) grosii Font Quer
  3 cretica L.
  4 albocincta W. Greuter
  5 minoana (P.H. Davis) W. Greuter
  6 variifolia Boiss.
  7 hymettia Boiss. & Spruner
  8 epirota Halácsy & Bald.
  9 sphaciotica Roemer & Schultes
 10 pulsatilloides Boiss.
   (a) pulsatilloides
   (b) macropoda (Costa ex Willk.) Nyman
 11 isetensis L.
 12 crenata Cyr.
   (a) crenata
   (b) dallaportae (Heldr. ex Boiss.)
      Hayek
   (c) breviscapa (Boiss. & Heldr.) Hayek
 13 graminifolia L.   16   Ga   Ritter, 1973
                                             18   Ju   Lovka et al., 1971
 14 rhodopensis Stoj. & Stefanov   18   Bu   Frey, 1970
 15 argentea L.   16   Ju   Gadella & Kliphuis,
                                                        1972

 16 stellata L.
   (a) stellata
   (b) simplex (Desf.) Coutinho
 17 monspeliensis Jacq.
 18 rotata Bieb.   18   Bu   Frey, 1970
 19 micrantha Desf.
 20 sicula L.
 21 hispidula Boiss.
 22 cosmoides Boiss.
 23 atropurpurea L.   16   Ga   van Loon et al.,
                                                        1971

 24 semipapposa Salzm. ex DC.
 25 silenifolia Waldst. & Kit.
 26 vestina Facch. ex Koch   16   It   Damboldt, 1966
 27 canescens Waldst. & Kit.   16   Ge Po Frey, 1969a
 28 parviflora Desf.
 29 tenuis Spruner ex Boiss.
 30 cinerea Lapeyr. ex Lam.   16   Ju   Sušnik & Lovka, 1973
   (a) cinerea
   (b) hladnikiana (Host) Jasiewicz
 31 turolensis Pau ex Willk.
 32 holosericea Bertol.
 33 taygetea Boiss. & Heldr.
 34 triandra L.   16   Co   Contandriopoulos,
                                                        1964a

 35 achaeta Vis. & Pančić
 36 lucida Vill.   16   Hu   Baksay, 1956
 37 nitens Roemer & Schultes
 38 columbaria L.   16   Bu Ge Frey, 1969a
   (a) columbaria
   (b) pseudobanatica (Schur) Jáv. &
      Csapody
   (c) portae (A. Kerner ex Huter) Hayek
 39 balcanica Velen.
 40 ochroleuca L.   16   Ge Po Frey, 1969a

                          223

```
41 webbiana D. Don
42 fumarioides Vis. & Pančić
43 triniifolia Friv.                          16      Bu      Frey, 1969

9 Tremastelma Rafin.
  1 palaestinum (L.) Janchen                  16      Al      Strid, 1971

10 Pycnocomon Hoffmanns. & Link
   1 rutifolium (Vahl) Hoffmanns. & Link      18      Co      Contandriopoulos,
                                                              1962


CLXVIII CAMPANULACEAE

1 Campanula L.
  1 fastigiata Dufour ex A. DC.
  2 uniflora L.                                34     No      Knaben & Engelskjon,
                                                              1967
  3 cenisia L.                                 34     He      Podlech & Damboldt,
                                                              1964
                                         34 + 3B     He      Podlech & Damboldt,
                                                              1964
  4 zoysii Wulfen                              34     Ju      Damboldt & Podlech,
                                                              1964
  5 carpatica Jacq.                            34     Rm      Borhidi, 1968
  6 arvatica Lag.                              28     Hs      Damboldt & Podlech,
                                                              1964
    (a) arvatica
    (b) adsurgens (Leresche & Levier)
        Damboldt
  7 raineri Perpenti                           32     It      Damboldt & Podlech,
                                                              1964
  8 aizoon Boiss. & Spruner                    16     Gr      Contandriopoulos,
                                                              1964b
    (a) aizoon
    (b) aizoides (Zaffran) Fedorov
  9 primulifolia Brot.                         36     Lu      Gadella, 1964
 10 lusitanica L.                              18     Hs      Damboldt & Podlech,
                                                              1964
    (a) lusitanica
    (b) transtagana (R. Fernandes) Fedorov
 11 phrygia Jaub. & Spach                      16     Ju      Damboldt & Podlech,
                                                              1964
 12 sparsa Friv.
    (a) sparsa
    (b) frivaldskyi (Steudel) Hayek           20     Gr      Damboldt & Podlech,
                                                              1964
    (c) sphaerothrix (Griseb.) Hayek          20     Gr      Contandriopoulos,
                                                              1966
 13 ramosissima Sibth. & Sm.                   20     Gr      Damboldt & Podlech,
                                                              1964
 14 spatulata Sibth. & Sm.
    (a) spatulata                              20     Gr      Contandriopoulos,
                                                              1966
    (b) sprunerana (Hampe) Hayek              20     Gr      Contandriopoulos,
                                                              1966
    (c) filicaulis (Halácsy) Phitos
 15 patula L.
    (a) patula                                 20     Ju      Gadella & Kliphuis,
                                                              1972
                                               40     Ju      Gadella & Kliphuis,
                                                              1972
```

(b) costae (Willk.) Fedorov
(c) epigaea (Janka) Hayek   20   Bu   Ancev, 1975
(d) abietina (Griseb.) Simonkai   40   Po   Gadella, 1966
16 decumbens A. DC.
17 hemschinica C. Koch
18 stevenii Bieb.
(a) wolgensis (P. Smirnov) Fedorov
(b) altaica (Ledeb.) Fedorov
19 rapunculus L.   20   Ga It   Larsen, 1956a
20 persicifolia L.   16   Au Bu Ga Gadella, 1966
21 andrewsii A. DC.   34   Gr   Phitos, 1965
(a) andrewsii
(b) hirsutula Phitos
22 topaliana Beauverd   34   Gr   Phitos, 1965
23 lavrensis (Tocl & Rohlena) Phitos   34   Gr   Phitos, 1965
24 goulimyi Turrill   34   Gr   Phitos, 1965
25 celsii A. DC.   34   Gr   Phitos, 1965
(a) celsii
(b) parnesia Phitos
(c) spathulifolia (Turrill) Phitos
(d) carystea Phitos
26 anchusiflora Sibth. & Sm.   34   Gr   Phitos, 1965
27 euboica Phitos   34   Gr   Phitos, 1965
28 reiseri Halácsy   34   Gr   Phitos, 1963
29 rechingeri Phitos   34   Gr   Phitos, 1964
30 merxmuelleri Phitos   34   Gr   Phitos, 1965
31 medium L.   34   Ga   Gadella, 1964
32 pelviformis Lam.   34   Gr   Phitos, 1964
33 tubulosa Lam.   34   Cr   Phitos, 1965
34 carpatha Halácsy   34   Cr   Phitos, 1965
35 lyrata Lam.   34   Gr   Phitos, 1963
36 saxatilis L.   34   Gr   Phitos, 1963
3 (a) saxatilis
(b) cytherea Rech. fil. & Phitos
37 laciniata L.   34   Gr   Phitos, 1964
38 rupestris Sibth. & Sm.   34   Gr   Phitos, 1964
39 cymaea Phitos   34   Gr   Phitos, 1964
40 constantini Beauverd & Top.   34   Gr   Phitos, 1964
41 scopelia Phitos   34   Gr   Phitos, 1964
42 sciathia Phitos   34   Gr   Phitos, 1964
43 thessala Maire   34   Gr   Phitos, 1964
44 barbata L.   34   Au   Mattick, 1950

45 alpina Jacq.   34   Po   Skalińska, 1959
(a) alpina
(b) orbelica (Pančić) Urum.
46 speciosa Pourret   34   Ga   Gadella, 1964
47 affinis Schultes
(a) affinis
(b) bolsoii (Vayr.) Fedorov
48 formanekiana Degen & Dörfler   24   Gr   Contandriopoulos,
                                                                     1966
49 lingulata Waldst. & Kit.   34   Gr Ju   Contandriopoulos,
                                                                     1966
50 sibirica L.   34 Au Cz Hu Rm   Gadella, 1964
(a) sibirica
(b) taurica (Juz.) Fedorov
(c) talievii (Juz.) Fedorov
(d) elatior (Fomin) Fedorov
(e) divergentiformis (Jáv.) Domin
(f) charkeviczii (Fedorov) Fedorov

```
51 incurva Aucher ex A. DC.              32    Gr         Contandriopoulos,
                                                          1964
52 grossekii Heuffel
53 lanata Friv.
54 dichotoma L.
55 alpestris All.                        34    Ga         Gadella, 1966
56 oreadum Boiss. & Heldr.               34    Gr         Contandriopoulos,
                                                          1966
57 calaminthifolia Lam.
58 hierapetrae Rech. fil.
59 amorgina Rech. fil.
60 heterophylla L.
61 mollis L.                             26    Hs         Podlech & Damboldt,
                                                          1964
62 papillosa Halácsy
63 orphanidea Boiss.
64 rupicola Boiss. & Spruner             32    Gr         Contandriopoulos,
                                                          1964
65 petraea L.                            34    Ga         Guinochet & Logeois,
                                                          1962
66 tymphaea Hausskn.                     34    Gr         Contandriopoulos,
                                                          1966
67 stenosiphon Boiss. & Heldr.           34    Gr         Contandriopoulos,
                                                          1964
68 transsilvanica Schur ex Andrae
69 moesiaca Velen.
70 glomerata L.                          30    Au Cz Ga   Gadella, 1964
                                               Da It Su
   (a) glomerata
   (b) serotina (Wettst.) O. Schwarz
   (c) farinosa (Rochel) Kirschleger     30    Cz         Májovský et al., 1974
   (d) elliptica (Kit. ex Schultes) O. Schwarz
   (e) subcapitata (M. Popov) Fedorov
   (f) cervicarioides (Schultes) P. Fourn.
   (g) hispida (Witasek) Hayek
71 foliosa Ten.                          34    Gr         Contandriopoulos,
                                                          1966
72 cervicaria L.                         34    Bu Rm      Gadella, 1964
73 macrostachya Waldst. & Kit. ex Willd. 18    Hu         Baksay, 1958
74 spicata L.                            34    Au Ga It   Gadella, 1964
75 thyrsoides L.                         34    Ga He      Gadella, 1964
   (a) thyrsoides
   (b) carniolica (Sünd.) Podl.
76 pyramidalis                           34    Ju         Podlech & Damboldt,
                                                          1964
77 versicolor Andrews                    34    Gr         Damboldt, 1971
78 morettiana Reichenb.                  34    It         Podlech & Damboldt,
                                                          1964
79 radicosa Bory & Chaub.                34    Gr         Contandriopoulos,
                                                          1964
80 secundiflora Vis. & Pančić
81 hawkinsiana Hausskn. & Heldr.         22    Gr         Contandriopoulos,
                                                          1964
82 sartorii Boiss. & Heldr.              34    Gr         Damboldt & Podlech,
                                                          1964
83 herminii Hoffmanns. & Link            32    Hs         Damboldt & Podlech,
                                                          1964
84 waldsteiniana Schultes                34    Ju         Merxmüller &
                                                          Damboldt, 1962
85 tommasiniana Koch                     34    Ju         Damboldt, 1965
86 isophylla Moretti                     32    It         Merxmüller &
                                                          Damboldt, 1962
```

87 fragilis Cyr.   32   It   Damboldt, 1965
   (a) fragilis
   (b) cavolinii (Ten.) Damboldt
88 elatinoides Moretti   34   It   Damboldt, 1965
89 elatines L.   34   It   Damboldt, 1965
90 portenschlagiana Schultes   34   Ju   Merxmüller &
                                                                Damboldt, 1962
91 poscharskyana Degen   34   Ju   Merxmüller &
                                                                Damboldt, 1962
92 garganica Ten.   34   It   Merxmüller &
                                                                Damboldt, 1962
   (a) garganica
   (b) cephallenica (Feer) Hayek
   (c) acarnanica (Damboldt) Damboldt
93 fenestrellata Feer
   (a) fenestrellata
   (b) istriaca (Feer) Fedorov
   (c) debarensis (Rech. fil.) Damboldt   34   Ju   Merxmüller &
                                                                Damboldt, 1962
94 specularioides Cosson
95 scutellata Griseb.   14   Ju   Contandriopoulos,
                                                                1966
96 drabifolia Sibth. & Sm.   28   Gr   Contandriopoulos,
                                                                1964
   (a) drabifolia
   (b) creutzburgii (W. Greuter) Fedorov
   (c) pinatzii (W. Greuter & Phitos)
       Fedorov
97 delicatula Boiss.
98 erinus L.   28   Lu   Gadella, 1964
99 latifolia L.   34   No   Laane, 1971
                                         34 + 5B   Cz Su   Gadella, 1964
100 trachelium L.   34   He   Gadella, 1966
    (a) trachelium
    (b) athoa (Boiss. & Heldr.) Hayek
101 rapunculoides L.   68   Rm   Gadella, 1964
                                               102   Au Cz Ga Gadella, 1964
102 bononiensis L.   34 Bu Cz Ge Hu   Gadella, 1964
103 aparinoides Pursh
104 trichocalycina Ten.   32   Gr   Contandriopoulos,
                                                                1966
105 macrorhiza Gay ex A. DC.   34   Ga   Podlech, 1962
    gracillima Podl.   34   Ga   Podlech, 1965
106 sabatia De Not.   34   It   Podlech, 1962
107 carnica Schiede ex Mert. & Koch   34   It   Gadella, 1964
108 tanfanii Podl.   34   It   Podlech, 1965
109 xylocarpa Kovanda   34   Cz   Kovanda, 1966
110 crassipes Heuffel   34   Ju   Lovka et al., 1972
111 praesignis G. Beck   34   Au   Podlech, 1962
112 forsythii (Arcangeli) Podl.   34   Sa   Podlech, 1962
113 hispanica Willk.
    (a) hispanica   34   Hs   Podlech & Damboldt,
                                                                1964

    ruscinonensis   34   Ga   Podlech & Damboldt,
                                                                1964

    (b) catalanica Podl.
114 justiniana Witasek   34   Ju   Podlech, 1962
115 hercegovina Degen & Fiala   34   Ju   Podlech & Damboldt,
                                                                1964

116 albanica Witasek
    (a) albanica   34   Gr   Podlech, 1962

(b) sancta (Hayek) Podl.	34	Gr	Podlech & Damboldt, 1964
117 romanica Savul.			
118 moravica (Spitzner) Kovanda			
(a) moravica	68	Cz	Kovanda, 1966
(b) xylorrhiza (O. Schwarz) Kovanda	102	Cz	Kovanda, 1970
119 apennina (Podl.) Podl.	34	It	Podlech, 1970
120 willkommii Witasek	68	He	Podlech & Damboldt, 1964
121 fritschii Witasek	68	Ga	Podlech, 1962
122 longisepala Podl.	34	Ga	Podlech, 1965
	68	Ga	Podlech, 1965
123 marchesettii Witasek	68	Ju	Podlech, 1962
124 velebitica Borbás			
125 bertolae Colla	102	It	Podlech & Damboldt, 1964
126 pseudostenocodon Lacaita	102	It	Podlech, 1962
127 rhomboidalis L.	34	He	Favarger, 1949a
128 cantabrica Feer	34	Hs	Podlech & Damboldt, 1964
129 serrata (Kit.) Hendrych	34	Cz	Kovanda, 1970a
	34 + 2B	Rm	Podlech & Damboldt, 1964
130 recta Dulac	34	Ga	Podlech & Damboldt, 1964
131 precatoria Timb.-Lagr.			
132 witasekiana Vierh.	34	Ju	Podlech, 1962
133 cochlearifolia Lam.	34	Au Cz Ge	Gadella, 1964
134 cespitosa Scop.	34	Ju	Podlech, 1962
135 jaubertiana Timb.-Lagr.			
(a) jaubertiana			
(b) andorrana (Br.-Bl.) P. Monts.			
136 excisa Schleicher ex Murith	34	He	Podlech, 1962
137 stenocodon Boiss. & Reuter	34	It	Podlech, 1962
138 pulla L.	34	Au	Gadella, 1964
139 scheuchzeri Vill.	68	Au Ge	Kovanda, 1970a
	102	It	Podlech & Damboldt, 1964
bohemica Hruby	68	Ga Hs	Podlech & Damboldt, 1964
gelida Kovanda	68	Cz	Kovanda, 1970a
140 ficarioides Timb.-Lagr.	102	Ga	Podlech & Damboldt, 1963
141 rotundifolia L.	34	Ga	Guinochet, 1942
	68	Ga	Guinochet, 1942
	102	Ga	Kovanda, 1970b
142 giesekiana Vest	34	Sb	Flovik, 1940
143 baumgartenii J. Becker	68	Ge	Podlech, 1962
144 beckiana Hayek	68	Au	Gutermann, 1961
2 Azorina Feer			
1 vidalii (H.C. Watson) Feer	56	Lu	Rodrigues, 1954
3 Symphyandra A. DC.			
1 cretica A. DC.			
(a) cretica	36	Cr	Phitos, 1966
(b) samothracica (Degen) Hayek	36	Gr	Phitos, 1966
(c) sporadum (Halácsy) Hayek	36	Gr	Phitos, 1966
2 wanneri (Rochel) Heuffel			
3 hofmannii Pant.			

4 Adenophora Fischer
  1 lilifolia (L.) Ledeb. ex A. DC.   34   Ju   Lovka et al., 1971
  2 taurica (Suk.) Juz.

5 Legousia Durande
  1 falcata (Ten.) Fritsch
  2 castellana (Lange) Samp.
  3 hybrida (L.) Delarbre   20   Br   Rutland, 1941
  4 speculum-veneris (L.) Chaix   20   Ga   Gadella, 1966
  5 pentagonia (L.) Druce

6 Trachelium L.
  1 caeruleum L.
    (a) caeruleum   32   Lu   Damboldt, ined.
    (b) lanceolatum (Guss.) Arcangeli
  2 jacquinii (Sieber) Boiss.
    (a) jacquinii
    (b) rumelianum (Hampe) Tutin   32 34   Gr   Damboldt, 1968
  3 asperuloides Boiss. & Orph.   34   Gr   Damboldt, 1968

7 Petromarula Vent. ex Hedwig fil.
  1 pinnata (L.) A. DC.   30   Cr   Podlech & Damboldt,
                                                           1964

8 Asyneuma Griseb. & Schenk
  1 anthericoides (Janka) Bornm.
  2 limonifolium (L.) Janchen   24   Gr   Damboldt, 1968
  3 comosiforme Hayek & Janchen
  4 canescens (Waldst. & Kit.) Griseb. &
    Schenk
    (a) canescens   32   Bu   Ančev, 1975
    (b) cordifolium (Bornm.) Damboldt

9 Phyteuma L.
  1 spicatum L.
    (a) spicatum   22   Au   Polatschek, 1966
                                           22 + 1 - 4B Po   Ochlewska, 1966
    (b) coeruleum R. Schulz   22   Au   Polatschek, 1966
  2 vagneri A. Kerner
  3 pyrenaicum R. Schulz
  4 ovatum Honckeny   22   Au   Polatschek, 1966
                                           26   He   Contandriopoulos,
                                                               1962
  5 nigrum F.W. Schmidt   22   Cz   Kovanda, 1970
  6 gallicum R. Schulz
  7 tetramerum Schur
  8 michelii All.
  9 scorzonerifolium Vill.
 10 zahlbruckneri Vest   24   Au   Polatschek, 1966
 11 betonicifolium Vill.   24   He   Favarger, 1953
 12 cordatum Balbis   24   Ga   Favarger, 1953
 13 orbiculare L.   22 + 0 - 1B Au   Polatschek, 1966
 14 pseudorbiculare Pant.
 15 sieberi Sprengel   20   Au   Polatschek, 1966
 16 scheuchzeri All.   26   He It   Contandriopoulos,
                                                               1962
    (a) scheuchzeri
    (b) columnae (Gaudin) Becherer
 17 charmelii Vill.   26   Ga   Contandriopoulos,
                                                               1962

    villarsii R. Schulz   26   Ga   Contandriopoulos,
                                                               1962

229

18 serratum Viv.	28	Co	Contandriopoulos, 1962
19 hemisphaericum L.	28	Ga	Contandriopoulos, 1962
20 hedraianthifolium R. Schulz	28	He	Contandriopoulos, 1962
21 humile Schleicher ex Gaudin	28	He	Contandriopoulos, 1962

22 globulariifolium Sternb. & Hoppe

(a) globulariifolium	28	Au	Polatschek, 1966
(b) pedemontanum (R. Schulz) Becherer	28	Ga	Contandriopoulos, 1962

23 rupicola Br.-Bl.

24 confusum A. Kerner	28	Au	Polatschek, 1966
	28 + 1B		

10 Physoplexis (Endl.) Schur

1 comosa (L.) Schur	34	It	Damboldt, 1965

11 Wahlenbergia Schrader ex Roth
  1 hederacea (L.) Reichenb.
  2 nutabunda (Guss.) A. DC.

12 Edraianthus A. DC.

1 parnassicus (Boiss. & Spruner) Halácsy	32	Gr	Contandriopoulos, 1964b
2 graminifolius (L.) A. DC.	32	Gr	Contandriopoulos, 1964b

  (a) graminifolius
  (b) niveus (G. Beck) Janchen

3 tenuifolius (Waldst. & Kit.) A. DC.	32	Gr	Contandriopoulos, 1966

4 dalmaticus (A. DC.) A. DC.

5 serbicus Petrović	32	Bu	Kuzmanov, 1973

6 serpyllifolius (Vis.) A. DC.
7 pumilio (Portenschl.) A. DC.
8 dinaricus (A. Kerner) Wettst.
9 wettsteinii Halácsy & Bald.

13 Jasione L.

1 montana L.	12	Au Cz Lu	Kovanda,
blepharodon Boiss. & Reuter	12	Hs	Björkqvist et al., 1969

2 penicillata Boiss.
3 corymbosa Poiret ex Schultes
4 lusitanica A. DC.
5 crispa (Pourret) Samp.

(a) amethystina (Lag. & Rodr.) Tutin	36	Hs	Küpfer & Favarger, 1967

  (b) centralis (Rivas Martínez) Rivas
     Martínez
  (c) arvernensis Tutin

(d) crispa	36	Ga Hs	Küpfer & Favarger, 1967

  (e) serpentinica P. Silva
  (f) mariana (Willk.) Rivas Martínez
  (g) maritima (Duby) Tutin
  (h) sessiliflora (Boiss. & Reuter)
     Rivas Martínez
  (i) tomentosa (A. DC.) Rivas Martínez
  (j) cavanillesii (C. Vicioso)

```
6 laevis Lam.                                12    Ju       Gadella & Kliphuis,
                                                            1972
                                             24    Hs       Küpfer, 1971
  (a) laevis
  (b) carpetana (Boiss. & Reuter) Rivas
      Martínez
  (c) orbiculata (Griseb. ex Velen.)
      Tutin
  (d) rosularis (Boiss. & Reuter) Tutin
7 bulgarica Stoj. & Stefanov
8 heldreichii Boiss. & Orph.                 12    Gr       Contandriopoulos,
                                                            1966
9 foliosa Cav.
  (a) foliosa
  (b) minuta (Agardh ex Roemer & Schultes)
      Font Quer

14 Lobelia L.
  1 urens L.
  2 dortmanna L.                             14    Fe Ge Po  Reese, 1952

15 Laurentia Adanson
  1 gasparrinii (Tineo) Strobl


CLXIX COMPOSITAE

Eupatorieae Cass.

1 Eupatorium L.
  1 cannabinum L.
    (a) cannabinum                           20    Hu       Pólya, 1949
                                             40    Po       Skalińska et al.,
                                                            1971
    (b) corsicum (Req. ex Loisel.) P. Fourn.
  2 adenophorum Sprengel                     51    Lu       Fernandes &
                                                            Queirós, 1971

Astereae Cass.

2 Grindelia Willd.
  1 squarrosa (Pursh) Dunal

3 Solidago L.
  1 virgaurea L.                             18    It       Garbari &
                                                            Tornadore, 1972
    ssp. minuta (L.) Arcangeli              18    Ga       Ritter, 1973
  2 sempervirens L.
  3 canadensis L.
  4 gigantea Aiton                           36    Ga       Beaudry, 1970
  5 graminifolia (L.) Salisb.

4 Dichrocephala L'Hér. ex DC.
  1 integrifolia (L. fil.) O. Kuntze         18    non-Europe

5 Bellis L.
  1 annua L.
    (a) annua                                18    Lu       Fernandes &
                                                            Queirós, 1971
    (b) vandasii (Velen.) D.A. Webb
  2 perennis L.                              18    Hu       Felföldy, 1947
  3 bernardii Boiss. & Reuter                18    Co       Contandriopoulos,
                                                            1957
```

4 azorica Hochst.
5 longifolia Boiss. & Heldr.
6 sylvestris Cyr.   36   Lu   Fernandes &
                                                          Queirós, 1971
7 rotundifolia (Desf.) Boiss. & Reuter

6 Bellium L.
  1 bellidioides L.   18   Bl   Dahlgren et al.,
                                                          1971
  2 minutum (L.) L.
  3 crassifolium Moris

7 Aster L.
  1 macrophyllus L.
  2 schreberi Nees
  3 divaricatus L.
  4 novae-angliae L.
  5 puniceus L.
  6 laevis L.
  7 novi-belgii L.
  8 x salignus Willd.
  9 x versicolor Willd.
 10 lanceolatus Willd.
 11 pilosus Willd
 12 squamatus (Sprengel) Hieron.   20   Lu   Fernandes &
                                                          Queirós, 1971
 13 sibiricus L.
 14 pyrenaeus Desf. ex DC.   18   Ga   Delay, 1969
 15 amellus L.   18   Cz   Holub et al., 1970
                                         36
                                         54   Cz   Májovský et al.,
                                                          1970
 16 willkommii Schultz Bip.
 17 alpinus L.   18   Ga   Favarger &
                                                          Küpfer, 1968
                                         36   Cz   Favarger, 1959a
 18 bellidiastrum (L.) Scop.   18   He   Larsen, 1954
 19 tripolium L.
   (a) tripolium   18   Su   I. & O. Hedberg, 1964
   (b) pannonicus (Jacq.) Soó   18   Cz   Májovský et al.,
                                                          1970
 20 sedifolius L.
   (a) sedifolius   36   It   Garbari &
                                                          Tornadore, 1972
   (b) dracunculoides (Lam.) Merxm.
   (c) canus (Waldst. & Kit.) Merxm.
   (d) illyricus (Murb.) Merxm.
   (e) trinervis (Pers.) Thell.
   (f) angustissimus (Tausch) Merxm.
 21 albanicus Degen
 22 kirghisorum (Fischer ex Bieb.) Korsh.
 23 aragonensis Asso   20   Lu   Fernandes &
                                                          Queirós, 1971
 24 creticus (Gand.) Rech. fil.
 25 linosyris (L.) Bernh.   18   Cz   Májovský et al.,
                                                          1970
                                         36   Hu   Baksay, 1958
 26 oleifolius (Lam.) Wagenitz
 27 tarbagatensis (C. Koch) Merxm.

8 Erigeron L.

1 annuus (L.) Pers.   26   Po   Bijok et al., 1972
                               27 36 N. Amer. Montgomery & Yang, 1960

  (a) annuus
  (b) septentrionalis (Fernald & Wieg.)
      Wagenitz
  (c) strigosus (Muhl. ex Willd.)
      Wagenitz

2 karvinskianus DC.   36   Lu   Fernandes & Queirós, 1971

3 acer L.
  (a) acer   18   Hs   Gadella & Kliphuis, 1968

  (b) angulosus (Gaudin) Vacc.
  (c) droebachiensis (O.F. Mueller)
      Arcangeli
  (d) macrophyllus (Herbich) Guterm.
  (e) politus (Fries) H. Lindb. fil. 18   No   Knaben & Engelskjon, 1967

4 orientalis Boiss.
5 atticus Vill.   18   He   Favarger, 1954
6 gaudinii Brügger   18   He   Favarger, 1954
7 alpinus L.   18   Po   Skalińska et al., 1966

8 epiroticus (Vierh.) Halácsy
9 neglectus A. Kerner
10 nanus Schur   18   Po   Skalińska et al., 1966

11 borealis (Vierh.) Simmons 18   Is   Löve & Löve, 1956
12 glabratus Hoppe & Hornsch. ex Bluff
   & Fingerh.   18   Au   Reese, 1953
13 major (Boiss.) Vierh.
14 uniflorus L.   18   Is   Löve & Löve, 1956
  subsp. eriocephalus (J. Vahl) Cronq. 18   Is   Löve & Löve, 1956
  aragonensis Vierh.   18   Hs Ga Küpfer & Favarger, 1967

15 humilis R.C. Graham   36   Is   Löve & Löve, 1956
16 frigidus Boiss. ex DC.   18   Hs   Küpfer & Favarger, 1967

17 silenifolius (Turcz. ex DC.) Botsch.

9 Conyza Less.
  1 canadensis (L.) Cronq.   18   Po   Frey, 1971
  2 bonariensis (L.) Cronq.   36   Lu   Fernandes & Queirós, 1971
                        54   Al   Strid, 1971

10 Nolletia Cass.
  1 chrysocomoides (Desf.) Cass. ex Less.

11 Baccharis L.
  1 halimifolia L.

Inuleae Cass.

12 Karelinia Less.
  1 caspia (Pallas) Less.

13 Filago L.
  1 vulgaris Lam.
  2 eriocephala Guss.   28   Gr   Runemark, 1970

3 aegaea Wagenitz
  (a) aegaea   28   Gr   Runemark, 1970
  (b) aristata Wagenitz
4 cretensis Gand.
  (a) cretensis   28   Gr   Runemark, 1970
  (b) cycladum Wagenitz   28   Gr   Runemark, 1970
5 lutescens Jordan   28   Lu   Fernandes &
                                                        Queirós, 1971
  (a) lutescens
  (b) atlantica Wagenitz
6 fuscescens Pomel
7 pyramidata L.   28   Bl   Dahlgren et al.,
                                                   1971
8 desertorum Pomel
9 ramosissima Lange
10 congesta Guss. ex DC.
11 duriaei Cosson ex Lange
12 micropodioides Lange
13 mareotica Delile
14 filaginoides (Kar. & Kir.) Wagenitz
15 eriosphaera (Boiss. & Heldr.) Chrtek
   & J. Holub
16 hispanica (Degen & Hervier) Chrtek &
   J. Holub

14 Ifloga Cass.
  1 spicata (Forskål) Schultz Bip.

15 Logfia Cass.
  1 heterantha (Rafin.) J. Holub
  2 arvensis (L.) J. Holub   28   Cz   Holub et al., ined.
  3 minima (Sm.) Dumort.   28   Ge   Reese, 1952
  4 clementei (Willk.) J. Holub
  5 gallica (L.) Cosson & Germ.   28   Lu   Fernandes &
                                                        Queirós, 1971
  6 neglecta (Soyer-Willemet) J. Holub

16 Evax Gaertner
  1 pygmaea (L.) Brot.
  (a) pygmaea
  (b) ramosissima (Mariz) R. Fernandes &
     Nogueira
  2 contracta Boiss.
  3 asterisciflora (Lam.) Pers.
  4 carpetana Lange
  5 lusitanica Samp.
  6 rotundata Moris   26   Co   Contandriopoulos,
                                                   1962
  7 perpusilla Boiss. & Heldr.
  8 nevadensis Boiss.

17 Bombycilaena (DC.) Smolj.
  1 erecta (L.) Smolj.
  2 discolor (Pers.) Lainz

18 Micropus L.
  1 supinus L.

19 Evacidium Pomel
  1 discolor (DC.) Maire

20 Omalotheca Cass.
  1 sylvatica (L.) Schultz Bip. & F.W.
    Schultz
  2 norvegica (Gunn.) Schultz Bip. &

F.W. Schultz	56	Is	Löve & Löve, 1956

  3 hoppeana (Koch) Schultz Bip. & F.W.
    Schultz
  4 roeseri (Boiss. & Heldr.) J. Holub
  5 pichleri (Murb.) J. Holub

6 supina (L.) DC.	28	Cz	Májovský et al., 1970

21 Gamochaeta Weddell
  1 subfalcata (Cabrera) Cabrera

2 purpurea (L.) Cabrera	28	Lu	Fernandes & Queirós, 1971

22 Filaginella Opiz

1 uliginosa (L.) Opiz	14	Ho	Gadella & Kliphuis, 1966

  (a) uliginosa
  (b) sibirica (Kirp.) J. Holub
  (c) kasachstanica (Kirp.) J. Holub
  (d) rossica (Kirp.) J. Holub

23 Gnaphalium L.

1 luteo-album L.	14	Lu	Fernandes & Queirós, 1971

  2 undulatum L.

24 Helichrysum Miller
  1 amorginum Boiss. & Orph.
  2 sibthorpii Rouy
  3 doerfleri Rech. fil.

4 frigidum (Labill.) Willd.	28	Co	Contandriopoulos, 1962
5 stoechas (L.) Moench	28	Lu	Fernandes & Queirós, 1971

  (a) stoechas
  (b) barrelieri (Ten.) Nyman
  6 rupestre (Rafin.) DC.
  7 heldreichii Boiss.

8 ambiguum (Pers.) C. Presl	28	Bl	Guinochet & Lefranc, 1972

  9 saxatile Moris
  (a) saxatile
  (b) errerae (Tineo) Nyman

10 italicum (Roth) G. Don fil.	28	It	Mori, 1957

  (a) italicum
  (b) microphyllum (Willd.) Nyman

(c) serotinum (Boiss.) P. Fourn.	28	Lu	Rodrigues, 1953

11 orientale (L.) Gaertner
12 plicatum DC.

13 arenarium (L.) Moench	14		
	28	Po	Skalińska et al., 1968

  (a) arenarium
  (b) ponticum (Velen.) Clapham
14 graveolens (Bieb.) Sweet

15 foetidum (L.) Cass.	14	Lu	Fernandes & Queirós, 1971

16 bracteatum (Vent.) Andrews

25 Lasiopogon Cass.
  1 muscoides (Desf.) DC.

26 Antennaria Gaertner

1 dioica (L.) Gaertner	28	Ga He It	Urbanska-Worytkiewicz, 1968
2 nordhageniana Rune & Rönning	28	No	Rune & Rönning, 1956
3 alpina (L.) Gaertner	70	No	Urbanska-Worytkiewicz, 1967
	84	Is	Löve & Löve, 1956
	85	No	Urbanska-Worytkiewicz, 1967
boecherana A.E. Porsild	56	Is	Löve, ined.
4 porsildii Elis. Ekman	63	No	Urbanska-Worytkiewicz, 1967
	70	No	Urbanska-Worytkiewicz, 1967
5 carpatica (Wahlenb.) Bluff & Fingerh.	56	Au He Po	Urbanska-Worytkiewicz, 1967
6 villifera Boriss.	28	No	Urbanska-Worytkiewicz, 1970
	42	No	Urbanska-Worytkiewicz, 1970

27 Leontopodium (Pers.) R. Br.

1 alpinum Cass.	52	He It	Urbanska-Worytkiewicz, 1968b

  (a) alpinum
  (b) nivale (Ten.) Tutin

28 Anaphalis DC.

1 margaritacea (L.) Bentham	28	Br	Maude, 1940

29 Phagnalon Cass.
  1 sordidum (L.) Reichenb.

2 rupestre (L.) DC.	18	Hs	Gadella et al., 1966

  3 metlesicsii Pignatti
  4 graecum Boiss. & Heldr.

5 saxatile (L.) Cass.	18	Lu	Fernandes & Queirós, 1971

30 Leysera L.
  1 leyseroides (Desf.) Maire

31 Inula L.

1 helenium L.	20	Br	Rutland, 1941
2 helvetica Weber			
3 germanica L.	16	Hu	Baksay, 1958
4 salicina L.			
(a) salicina	16	Hu	Baksay, 1958
(b) aspera (Poiret) Hayek			
5 spiraeifolia L.	16	Ju	Nilsson & Lassen, 1971
6 hirta L.	16	Cz	Májovský et al., 1970
7 ensifolia L.	16	Cz	Májovský et al., 1970
8 britannica L.	32	Po	Skalińska, 1959

```
 9 caspica Blume
10 oculus-christi L.                          32      Hu     Baksay, 1958
11 helenioides DC.
12 montana L.                                 16      Ga     Guinochet &
                                                             Logeois, 1962
13 candida (L.) Cass.
   (a) candida
   (b) decalvans (Halácsy) P.W. Ball ex
       Tutin
   (c) limonella (Heldr.) Rech. fil.
14 verbascifolia (Willd.) Hausskn.
   (a) verbascifolia                          16      Ju     Larsen, 1956a
   (b) aschersoniana (Janka) Tutin
   (c) parnassica (Boiss. & Heldr.) Tutin
   (d) methanea (Hausskn.) Tutin
   (e) heterolepis (Boiss.) Tutin
15 subfloccosa Rech. fil.
16 conyza DC.                                 32      Cz     Májovský et al.,
                                                             1970
17 thapsoides Sprengel
18 bifrons (L.) L.
19 crithmoides L.                             18      Ga     Delay, 1968

32 Dittrichia W. Greuter
 1 viscosa (L.) W. Greuter                    18      Lu     Fernandes &
                                                             Queirós, 1971
                                              34      Ga     Gadella et al.,
   (a) viscosa                                               1966
   (b) revoluta (Hoffmanns. & Link) P.
       Silva & Tutin
 2 graveolens (L.) W. Greuter                 18      Lu     Fernandes &
                                                             Queirós, 1971
                                    20 + 0 - 2B  Bl     Nilsson & Lassen,
                                                             1971

33 Pulicaria Gaertner
 1 odora (L.) Reichenb.                       18      Lu     Fernandes &
                                                             Queirós, 1971
                                    18 + 0 - 6B  Lu     Queirós, 1973
 2 dysenterica (L.) Bernh.                    18      Al     Strid, 1971
                                              20      Cz     Májovský et al.,
                                                             1970
 3 vulgaris Gaertner                          18      Ge     Wulff, 1937b
 4 paludosa Link                              18      Bu     Kuzmanov & Kozuharov,
                                                             1970
                                    15 + 1B     Lu     Fernandes &
                                                             Queirós, 1971
 5 sicula (L.) Moris

34 Carpesium L.
 1 cernuum L.
 2 abrotanoides L.                            40      Hu     Baksay, 1956

35 Jasonia Cass.
 1 glutinosa (L.) DC.
 2 tuberosa (L.) DC.                          18      Lu     Fernandes &
                                                             Queirós, 1971

36 Buphthalmum L.
 1 salicifolium L.                            20      Cz     Májovský et al.,
                                                             1970
 2 inuloides Moris
```

37 Telekia Baumg.
  1 speciosa (Schreber) Baumg.   20   Cz   Májovský et al.,
                                                                  1970
  2 speciosissima (L.) Less.   20   It   Favarger, 1959a

38 Pallenis (Cass.) Cass.
  1 spinosa (L.) Cass.
    (a) spinosa   10   Lu   Fernandes &
                                                                  Queirós, 1971
    (b) microcephala (Halácsy) Rech. fil.

39 Asteriscus Miller
  1 aquaticus (L.) Less.   14   Bl   Dahlgren et al.,
                                                                  1971
  2 maritimus (L.) Less.   12   Lu   Fernandes &
                                                                  Queirós, 1971

Heliantheae Cass.

40 Guizotia Cass.
  1 abyssinica (L. fil.) Cass.

41 Bidens L.
  1 tripartita L.   48   Hu   Pólya, 1950
  2 connata Muhl. ex Willd.
  3 radiata Thuill.   48   Da   Hagerup, 1944
  4 aurea (Aiton) Sherff   72   Lu   Fernandes &
                                                                  Queirós, 1971
  5 cernua L.   24   Po   Skalińska et al.,
                                                                  1959
  6 frondosa L.   48   Lu   Fernandes &
                                                                  Queirós, 1971
  7 pilosa L.   72   Lu   Fernandes &
                                                                  Queirós, 1971
  8 bipinnata L.
  9 subalternans DC.

42 Sigesbeckia L.
  1 orientalis L.   60   Rm   Solbrig et al.,
                                                                  1972
  2 jorullensis Kunth

43 Eclipta L.
  1 prostrata (L.) L.   22   Lu   Fernandes &
                                                                  Queirós, 1971

44 Rudbeckia L.
  1 hirta L.
  2 laciniata L.   76   Cz   Holub et al.,
                                                                  1972

45 Helianthus L.
  1 annuus L.
  2 tuberosus L.
  3 x laetiflorus Pers.

46 Verbesina L.
  1 encelioides (Cav.) Bentham & Hooker
    (a) encelioides
    (b) exauriculata (Robinson & Greenman)
        J.R. Coleman

47 Silphium L.
   1 perfoliatum L.

48 Iva L.
   1 xanthifolia Nutt.   36   Cz   Feráková, 1968

49 Ambrosia L.
   1 maritima L.
   2 artemisiifolia L.   36   non-Europe
   3 tenuifolia Sprengel
   4 coronopifolia Torrey & A. Gray   72   non-Europe
   5 trifida L.   24   non-Europe

50 Xanthium L.
   1 strumarium L.
     (a) strumarium   36   Cz   Májovský et al.,
                                                                       1970
     (b) italicum (Moretti) D. Love   36   It   Fagidi & Fabbri,
                                                                       1971
   2 spinosum L.   36   Cz   Májovský et al.,
                                                                       1970

51 Heliopsis Pers.
   1 helianthoides (L.) Sweet
     (a) scabra (Dunal) Fisher
     (b) occidentalis Fisher

52 Galinsoga Ruiz & Pavón
   1 parviflora Cav.   16   Lu   Fernandes & Queirós,
                                                                       1971
   2 ciliata (Rafin.) S.F. Blake   32   Po   Skalińska et al.,
                                                                       1964
                                                          36   Ge   Reese, 1952

53 Schkuhria Roth
   1 pinnata (Lam.) O. Kuntze

54 Gaillardia Foug.
   1 aristata Pursh

55 Tagetes L.
   1 minuta L.

Anthemideae Cass.

56 Santolina L.
   1 oblongifolia Boiss.
   2 elegans Boiss. ex DC.
   3 viscosa Lag.
   4 rosmarinifolia L.   18   Lu   Fernandes & Queirós,
                                                                       1971
                                                          36   Lu   Fernandes & Queirós,
                                                                       1971
   (a) rosmarinifolia
   (b) canescens (Lag.) Nyman
   5 chamaecyparissus L.
   (a) chamaecyparissus
   (b) squarrosa (DC.) Nyman
   (c) tomentosa (Pers.) Arcangeli
   (d) insularis (Genn. ex Fiori) Yeo

57 Anthemis L.
 1 trotzkiana Claus ex Bunge
 2 carpatica Willd.
  (a) carpatica   36   Hs   Küpfer &
                                                   Favarger, 1967
                                       54   Hs   Küpfer &
                                                   Favarger, 1967
  (b) pyrethriformis (Schur) Beldie
  (c) petraea (Ten.) R. Fernandes
 3 sibthorpii Griseb.
 4 aetnensis Schouw   36   Si   Larsen &
                                                   Laegaard, 1971
 5 punctata Vahl
 6 cretica L.
  (a) cretica
  (b) calabrica (Arcangeli) R. Fernandes
  (c) saxatilis (DC.) R. Fernandes
  (d) alpina (L.) R. Fernandes
 7 sterilis Steven
 8 abrotanifolia (Willd.) Guss.
 9 panachaica Halácsy
10 argyrophylla (Halácsy & Georgiev)
   Velen.
11 gerardiana Jordan
12 alpestris (Hoffmanns. & Link) R.
   Fernandes
13 maritima L.   36   Lu   Fernandes & Queirós,
                                                   1971
14 virescens Velen.
15 orbelica Pančić
16 hydruntina Groves
17 tuberculata Boiss.
  (a) tuberculata
  (b) turolensis (Pau ex A. Caballero)
      R. Fernandes & Borja
18 ismelia Lojac.
19 tenuiloba (DC.) R. Fernandes
  (a) tenuiloba
  (b) cronia (Boiss. & Heldr.) R. Fernandes
20 anatolica Boiss.
21 spruneri Boiss. & Heldr.
22 orientalis (L.) Degen
23 pindicola Heldr. ex Halácsy
24 meteorica Hausskn.
25 rumelica (Velen.) Stoj. & Acht.
26 stribrnyi Velen.
27 arvensis L.
  (a) arvensis   18   Lu   Rodriguez, 1953
  (b) incrassata (Loisel.) Nyman
  (c) cyllenea (Halácsy) R. Fernandes
  (d) sphacelata (C. Presl) R. Fernandes
28 ruthenica Bieb.   18   Cz   Májovský et al.,
                                                   1970
29 auriculata Boiss.
30 werneri Stoj. & Acht.
31 flexicaulis Rech. fil.
32 scopulorum Rech. fil
33 tomentosa L.
  (a) tomentosa   18   It   Carpineri, 1971
  (b) heracleotica (Boiss. & Heldr.)
      R. Fernandes

240

34 rigida (Sibth. & Sm.) Boiss. & Heldr.
35 chrysantha Gay
36 secundiramea Biv.
  (a) secundiramea
  (b) intermedia (Guss.) R. Fernandes
37 muricata (DC.) Guss.   18   Si   Carpineri, 1971
38 macedonica Boiss. & Orph.
39 cotula L.   18   Cz   Májovský et al.,
                                                          1970
40 lithuanica (DC.) Besser ex Trautv.
41 chia L.
42 tinctoria L.
  (a) tinctoria   18   Cz   Májovský et al.,
                                                          1970
  (b) subtinctoria (Dobrocz.) Soó
  (c) australis R. Fernandes
  (d) fussii (Griseb.) Beldie
43 gaudium-solis Velen.
44 sancti-johannis Turrill
45 cretacea Zefirov
46 monantha Willd.   18   Rs(K)   Chouksanova et al.,
47 parnassica (Boiss. & Heldr.) R.   1968
   Fernandes
48 dubia Steven
49 triumfetti (L.) DC.
50 dumetorum D. Sossn.
51 macrantha Heuffel
52 jailensis Zefirov   18   Rs(K)   Chouksanova et al.,
                                                          1968
53 altissima L.   18   Ga   Delay, 1971
54 coelopoda Boiss.
55 syriaca Bornm.
56 segetalis Ten.   18   It   Carpineri, 1971
57 austriaca Jacq.   18   Bu   Kuzmanov &
                                                          Kozuharov, 1970
58 brachmannii Boiss. & Heldr.
59 filicaulis (Boiss. & Heldr.) W. Greuter
60 tomentella W. Greuter
61 ammanthus W. Greuter
  (a) ammanthus
  (b) paleacea W. Greuter
62 glaberrima (Rech. fil.) W. Greuter

58 Achillea L.
  1 ageratifolia (Sibth. & Sm.) Boiss.   18   Gr   Contandriopoulos
                                                          & Martin, 1967
  2 servica Nyman   18   Ju   Lovka et al., 1971
  3 oxyloba (DC.) Schulz Bip.   18   It   Favarger & Huynh,
    (a) oxyloba   1964
    (b) mucronulata (Bertol.) I.B.K.
        Richardson
    (c) schurii (Schultz Bip.) Heimerl
  4 barrelieri (Ten.) Schultz Bip.
  5 erba-rotta All.   18   He   Favarger & Huynh,
    (a) olympica (Heimerl) I.B.K. Richardson   1964
    (b) moschata (Wulfen) I.B.K. Richardson
    (c) ambigua (Heimerl) I.B.K. Richardson
    (d) erba-rotta   18   Ga   Favarger & Huynh,
                                                          1964
    (e) calcarea (Porta) I.B.K. Richardson
    (f) rupestris (Porta) I.B.K. Richardson

6 nana L.	18	He	Favarger, 1953
7 umbellata Sibth. & Sm.	18	Gr	Contandriopoulos & Martin, 1967
	18 + 1 - 3B	Gr	Contandriopoulos & Martin, 1967
9 atrata L.	18	Au	Reese, 1952
10 clusiana Tausch	18	Au	Favarger & Huynh, 1964
11 clavennae L.	18	Gr	Contandriopoulos & Martin, 1967
12 ambrosiaca (Boiss. & Heldr.) Boiss.	18	Gr	Contandriopoulos & Martin, 1967
13 abrotanoides (Vis.) Vis.	18	Gr	Contandriopoulos & Martin, 1967
14 fraasii Schultz Bip.	18	Gr	Contandriopoulos & Martin, 1967
15 macrophylla L.			
16 lingulata Waldst. & Kit.			
17 ptarmica L.	18	Ho	Gadella & Kliphuis, 1968
18 cartilaginea Ledeb. ex Reichenb.	18	Po	Dabrowska, 1971
19 pyrenaica Sibth. ex Godron	18	Hs	Küpfer, 1968
20 impatiens L.			
21 grandifolia Friv.	18	Gr	Contandriopoulos & Martin, 1967
22 distans			
(a) distans	54	Cz	Májovský et al., 1970
(b) tanacetifolia Janchen			
23 stricta (Koch) Schleicher ex Gremli	54	Europe	Ehrendorfer, 1952
24 millefolium L.			
(a) millefolium	54	Ho	Gadella & Kliphuis, 1966
(b) sudetica (Opiz) Weiss	54	Cz	Májovský et al., 1970
monticola	72	Hs	Ehrendorfer, 1959
25 pannonica Scheele	72	Cz	Májovský et al., 1970
26 setacea Waldst. & Kit.	18	Cz	Májovský et al., 1970
27 asplenifolia Vent.	18	Hu	Pólya, 1949
28 roseo-alba Ehrend.	18	It	Ehrendorfer, 1959
	36	Au	Ehrendorfer, 1959
29 collina J. Becker ex Reichenb.	36	Hu	Pólya, 1948
30 crithmifolia Waldst. & Kit.	18	Hu	Baksay, 1958
31 odorata L.	18	Au	Ehrendorfer, 1959
32 nobilis L.			
(a) nobilis	18	Po	Skalińska et al., 1970
(b) neilreichii (A. Kerner) Velen.	45	Cz	Májovský et al., 1970
33 virescens (Fenzl) Heimerl	36		
34 ligustica All.	18	It	Caldo, 1971
	c.54	Si	Larsen & Laeggaard, 1971
35 chamaemelifolia Pourret			
36 tomentosa L.			
37 chrysocoma Friv.			
38 holosericea Sibth. & Sm.	18	Gr	Contandriopoulos & Martin, 1967
39 absinthoides Halácsy			

40 leptophylla Bieb.
41 glaberrima Klokov
42 ageratum L.   18   Lu   Fernandes & Queirós,
                                                1971

43 ochroleuca Ehrh.	18	Hu	Baksay, 1958
44 depressa Janka			
45 aegyptiaca L.	18	Gr	Contandriopoulos & Martin, 1967
46 clypeolata Sibth. & Sm.	18	Bu	Kuzmanov & Kozuharov, 1970
	18 + 1B	Gr	Contandriopoulos & Martin, 1967
47 thracica Velen.			
48 coarctata Poiret	18	Rm	Lungeanu, 1973
	36	Bu	Kuzmanov & Kozuharov, 1970
49 micrantha Willd.			
50 biebersteinii C. Afan.			
51 santolinoides Lag.			
52 cretica L.	18	Cr	Contandriopoulos & Martin, 1967

59 Chamaemelum Miller
  1 nobile (L.) All.
  2 mixtum (L.) All.
  3 fuscatum (Brot.) Vasc.

60 Matricaria L.
  1 trichophylla (Boiss.) Boiss.
  2 conoclinia (Boiss. & Balansa) Nyman
  3 caucasica (Willd.) Poiret
  4 maritima L.

(a) maritima	18	Da	Larsen, 1965
	18 + 1 - 4B	Br	Kay, 1969
(b) subpolaris (Pobed.) Rauschert			
(c) phaeocephala (Rupr.) Rauschert	18	Fe	Hamet-Ahti
5 perforata Merat	18	Ge	Rottgardt, 1956
	18 + 1B	Br	Kay, 1969
	36	Ge	Fritsch, 1973

  6 rosella (Boiss. & Orph.) Nyman
  7 tempskyana (Freyn & Sint.) Rauschert
  8 parviflora (Willd.) Poiret

61 Chamomilla S.F. Gray

1 recutita (L.) Rauschert	18	Lu	Fernandes & Queirós, 1971
2 tzvelevii (Pobed.) Rauschert			
3 suaveolens (Pursh) Rybd.	18	Lu	Fernandes & Queirós, 1971

  4 aurea (Loefl.) Gay ex Cosson & Kralik

62 Cladanthus Cass.
  1 arabicus (L.) Cass

63 Anacyclus L.

1 clavatus (Desf.) Pers.	18	Ga	Van Loon et al., 1971
2 radiatus Loisel.	18	Lu	Fernandes & Queirós, 1971

  3 valentinus L.
  4 pyrethrum (L.) Link

64 Lonas Adanson
   1 annua (L.) Vines & Druce

65 Otanthus Hoffmanns. & Link
   1 maritimus (L.) Hoffmanns. & Link   18   Lu   Fernandes & Queirós, 1971

66 Chrysanthemum L.
   1 segetum L.   18   Ho   Gadella & Kliphuis, 1968

   2 coronarium L.   18   Lu   Fernandes & Queirós, 1971

67 Heteranthemis Schott
   1 viscidehirta Schott

68 Dendranthema (DC.) Desmoulins
   1 zawadskii (Herbich) Tzvelev   54   Po   Skalińska et al., 1959
   2 arcticum (L.) Tzvelev

69 Tanacetum L.
   1 vulgare L.   18   Cz   Májovský et al., 1970
   2 microphyllum DC.
   3 annuum L.   18   Lu   Queirós, 1975
   4 achilleifolium (Bieb.) Schultz Bip.
   5 paczoskii (Zefirov) Tzvelev
   6 millefolium (L.) Tzvelev
   7 bipinnatum (L.) Schultz Bip.   54   Rs(N)   Sokolovskaya & Strelkova, 1960

   8 santolina Winkler
   9 corymbosum (L.) Schultz Bip.
     (a) corymbosum   36   Lu   Fernandes & Queirós, 1971

     (b) clusii (Fischer ex Reichenb.)
         Heywood   18   Cz   Májovský et al., 1970
   10 mucronulatum (Hoffmanns. & Link)
      Heywood
   11 parthenium (L.) Schultz Bip.   18   Ju   Lovka et al., 1972
   12 parthenifolium (Willd.) Schultz Bip.
   13 macrophyllum (Waldst. & Kit.) Schultz
      Bip.
   14 cinerariifolium (Trev.) Schultz Bip.

70 Leucanthemella Tzvelev
   1 serotina (L.) Tzvelev   18   Hu   Baksay, 1958

71 Balsamita Miller
   1 major Desf.

72 Phalacrocarpum Willk.
   1 oppositifolium (Brot.) Willk.   18   Lu   Fernandes & Queirós, 1971

   2 hoffmannseggii (Samp.) Lainz   18   Lu   Fernandes & Queirós, 1971

73 Otospermum Willk.
   1 glabrum (Lag.) Willk.   18   Lu   Fernandes & Queirós, 1971

74 Leucanthemopsis (Giroux) Heywood
   1 alpina (L.) Heywood

(a) alpina	18	Cz	Májovský et al., 1970
	36	Ga	Tombal, 1969
	54	Ga Hs	Küpfer & Favarger, 1967
(b) tomentosa (Loisel.) Heywood	18	Co	Contandriopoulos, 1962
(c) cuneata (Pau) Heywood	54	Hs	Küpfer, 1971
2 pallida (Miller) Heywood	36	Hs	Küpfer, 1971
(a) pallida			
(b) virescens (Pau) Heywood			
(c) spathulifolia (Gay) Heywood			
3 flaveola (Hoffmanns. & Link	36	Lu	Küpfer, 1971
4 radicans (Cav.) Heywood	18	Hs	Favarger & Küpfer, 1967

5 pulverulenta (Lag.) Heywood
  (a) pulverulenta
  (b) pseudopulverulenta (Heywood) Heywood

75 Prolongoa Boiss.
  1 pectinata (L.) Boiss.

76 Lepidophorum Cass.			
1 repandum (L.) DC.	18	Lu	Fernandes & Queirós, 1971

77 Daveaua Willk. ex Mariz
  1 anthemoides Mariz

78 Glossopappus G. Kunze
  1 macrotus (Durieu) Briq.

79 Hymenostemma (G. Kunze) Willk.
  1 pseudanthemis (G. Kunze) Willk.

80 Coleostephus Cass.			
1 myconis (L.) Reichenb. fil.	18	Lu	Fernandes & Queirós, 1971
2 clausonis Pomel			

81 Leucanthemum Miller			
1 vulgare Lam.	36 54	Po	Skalińska et al., 1961
praecox (Horvatić) Horvatić	18	Po	Frey, 1969b
leucolepis (Briq. & Cavillier) Horvatic	18	Hs	Favarger, 1975
	36	Ju	Mirković, 1966
	54	Bu Ju	Polatschek, 1966
adustum (Koch) Gremli	54	Po	Przywara, 1970
heterophyllum (Willd.) DC.	45	Ju	Papeš, 1975
	54	Au	Polatschek, 1966
	72	It	Favarger & Villard, 1965
cuneifolium Le Grand ex Coste	72	Ga	Villard, 1970
maximum (Ramond) DC.	90	It	Favarger & Villard, 1965
	108	Ga	Favarger & Villard, 1965
pallens (Gay) DC.	54	Ga	Favarger & Villard, 1965
subglaucum De Laramb.	54	Ge	Favarger, 1975b

		90	Ga It	Favarger & Villard, 1965
		72 + 1B	Ga	Favarger, 1975
sylvaticum (Hoffmanns. & Link) Nyman		36	Lu	Fernandes & Queirós, 1971
		54	Lu	Fernandes & Queirós, 1971
2 graminifolium (L.) Lam.		18	Ga	Favarger & Villard, 1965
3 gracilicaule (Dufour) Alavi & Heywood				
4 burnatii Briq. & Cavillier		18	Ga	Favarger & Villard, 1965
5 monspeliense (L.) Coste		36	Ga	Favarger & Villard, 1965
6 atratum (Jacq.) DC.				
(a) atratum		54	It	Favarger & Villard, 1965
(b) halleri (Suter) Heywood		18	Au	Polatschek, 1966
(c) coronopifolium (Vill.) Horvatić		54	It	Marchi, 1971
(d) ceratophylloides (All.) Horvatić		54	He	Favarger & Villard, 1965
(e) tridactylites (A. Kerner & Huter ex Rigo) Heywood				
(f) platylepis (Borbás) Heywood		27	Ju	Papeš, 1972
		36	Ju	Polatschek, 1966
		45	Ju	Papeš, 1975
		54	Ju	Favarger & Villard, 1965
		108	Ju	Mirkovic, 1966
(g) lithopolitanicum (E. Mayer) Horvatić		18	Ju	Lovka et al., 1972
		72	Ju	Mirković, 1966
7 chloroticum A. Kerner & Murb.				
8 corsicum (Less.) DC.		36	Co	Contandriopoulos, 1964a
9 waldsteinii (Schultz Bip.) Pouzar		18	Cz	Villard, 1970
10 discoideum (All.) Coste		18	Ga	Forissier, 1975
11 arundanum (Boiss.) Cuatrec.				
12 paludosum (Poiret) Bonnet & Barratte				
82 Plagius L'Hér. ex DC.				
1 flosculus (L.) Alavi & Heywood		18	Co	Contandriopoulos, 1962
83 Cotula L.				
1 coronopifolia L.		20	Lu	Fernandes & Queirós, 1971
2 australis (Sieber ex Sprengel) Hooker fil.				
84 Chlamydophora Ehrenb. ex Less.				
1 tridentata (Delile) Ehrenb. ex Less.				
85 Nananthea DC.				
1 perpusilla (Loisel.) DC.		18	Co	Contandriopoulos, 1962
86 Soliva Ruiz & Pavón				
1 pterosperma (Juss.) Less.		c.110	Lu	Fernandes & Queirós, 1971

246

87 Gymnostyles Juss.
  1 stolonifera (Brot.) Tutin

88 Artemisia L.
  1 vulgaris L.   16   Ho   Gadella & Kliphuis,
                                                                      1963
  2 verlotiorum Lamotte   54   It   Vignoli, 1945
  3 tilesii Ledeb.
  4 absinthium L.   18   Hu   Pólya, 1950
  5 siversiana Ehrh. ex Willd.
  6 arborescens L.   18   It   Martinoli, 1943
  7 stellerana Besser
  8 alba Turra   36   It   Cesca, 1972
  9 maritima L.
    (a) maritima   36   Ge   Löve & Löve, in
                                                                      litt. 1972
                                                   54   Ho   Gadella & Kliphuis,
                                                                      1968
                                              (50-56) Br Da Ge   Persson, 1974
                                                            Su
    (b) humifusa (Fries ex Hartman)
        K. Persson
 10 caerulescens L.
    (a) caerulescens   18   It   Fagioli & Fabbri,
                                                                      1971
    (b) gallica (Willd.) K. Persson   18   It   Löve & Löve, unpub.
 11 vallesiaca All.   36   It   Löve & Löve, unpub.
 12 lerchiana Weber
 13 nitrosa Weber
 14 taurica Willd.
 15 santonicum L.
    (a) santonicum
    (b) patens (Neilr.) K. Persson   18   Hu   Pólya, 1945
 16 pauciflora Weber
 17 gracilescens Krasch. & Iljin
 18 lessingiana Besser
 19 laciniata Willd.   18   Au   Ehrendorfer, 1964
 20 armeniaca Lam.
 21 latifolia Ledeb.
 22 pancicii (Janka) Ronniger   54   Au Ju   Ehrendorfer, 1964
 23 insipida Vill.
 24 oelandica (Besser) Komarov.   54   Su   Ehrendorfer, 1964
 25 atrata Lam.   18   Ga   Ehrendorfer, 1964
 26 norvegica Fries   18   Ho   Ehrendorfer, 1964
 27 abrotanum L.   18   Po   Skalińska et al.,
                                                                      1959
 28 molinieri Quezel, Barbero & R. Loisel
 29 santolinifolia Turcz. ex Krasch.
 30 pontica L.   18   Hu   Pólya, 1948
 31 austriaca Jacq.   16   Po   Skalińska et al.,
                                                                      1959
 32 chamaemelifolia Vill
    (a) chamaemelifolia   18   It   Löve & Löve, in
                                                                      litt. 1972
    (b) cantabrica Laínz
 33 macrantha Ledeb.
 34 umbelliformis Lam.   34   He   Favarger & Huynh,
                                                                      1964
    gabriellae Br.-Bl.   (not 34) 36   Ga   Favarger & Küpfer,
                                                                      1968

247

35 nitida Bertol.	54	It	Löve & Löve, in litt. 1972
36 eriantha Ten.	18	Po	Skalińska et al., 1959
37 genipi Weber	18	He	Favarger, 1953
38 glacialis L.	16	It	Favarger & Huynh, 1964
39 granatensis Boiss.			
40 pedemontana Balbis	16	It	Marchetti, 1962
41 frigida Willd.			
42 sericea Weber			
43 reptans C. Sm. ex Link			
44 barrelieri Besser			
45 herba-alba Asso			
46 hololeuca Bieb. ex Besser			
47 rupestris L.	18	Su	Ehrendorfer, 1964
48 annua L.	18	Hu	Pólya, 1948
49 dracunculus L.	18	Su	Rousi, 1969
	36	Ga	Rousi, 1969
	90	Da Fe	Rousi, 1969
50 glauca Pallas ex Willd.			
51 trautvetterana Besser			
52 salsoloides Willd.			
53 tschernieviana Besser			
54 commutata Besser			
55 bargusinensis Sprengel			
56 campestris L.			
(a) campestris	36	Cz	Májovský et al., 1970
(b) glutinosa (Gay ex Besser) Batt.			
(c) maritima Arcangeli	54	Lu	Fernandes & Queirós, 1971
(d) alpina (DC.) Arcangeli	36		
(e) bottnica A.N. Lundstrom ex Kinb.			
(f) borealis (Pallas) H.M. Hall & Clements	18	He	Damboldt, 1964
	36	Non-Europe	
57 scoparia Waldst. & Kit.			

Senecioneae Cass.

89 Tussilago L.			
1 farfara L.	60	Ho	Gadella & Kliphuis, 1966
90 Petasites Miller			
1 albus (L.) Gaertner	60	Ge	Scheerer, 1939
2 hybridus (L.) P. Gaertner, B. Meyer & Scherb.	60	Ho	Gadella & Kliphuis, 1966
(a) hybridus			
(b) ochroleucus (Boiss. & Huet) Sourek			
3 paradoxus (Retz.) Baumg.	60	Au	Sørensen & Christiansen, 1964
4 kablikianus Tausch ex Berchtold	60	Cz	Löve & Löve, 1961
5 radiatus (J.F. Gmelin) J. Toman			
6 spurius (Retz) Reichenb.	60	Da	Sørensen & Christiansen, 1964
7 frigidus (L.) Fries	60	Sb	Flovik, 1940
8 fragrans (Vill.) C. Presl	58 59 60 61	Br It	Sørensen & Christiansen, 1964

```
 9 japonicus (Siebold & Zucc.) Maxim.     84-87    Da    Sørensen &
                                                         Christiansen, 1964
10 doerfleri Hayek
11 sibiricus (J.F. Gmelin) Dingwall

91 Homogyne Cass.
 1 alpina (L.) Cass.                       120      Au    Favarger, 1971
                                           160      Cz    Májovský et al.,
                                                            1970
 2 discolor (Jacq.) Cass.                   60      Au    Favarger, 1971
 3 sylvestris Cass.                         58      Au    Favarger, 1969

92 Adenostyles Cass.
 1 alliariae (Gouan) A. Kerner
   (a) alliariae
   (b) hybrida (Vill.) DC.
 2 alpina (L.) Bluff & Fingerh.            38      He    Contandriopoulos,
                                                            1964a
   (a) alpina
   (b) briquetii (Gamisans) Tutin
 3 leucophylla (Willd.) Reichenb.          38      Ga    Delay, 1970

93 Arnica L.
 1 angustifolia Vahl
   (a) alpina (L.) I.K. Ferguson           76      Su    Löve, 1954a
   (b) iljinii (Maguire) I.K. Ferguson     56      Sb    Flovik, 1940
 2 montana L.                              38      Lu    Fernandes & Queirós,
                                                            1971
   (a) montana
   (b) atlantica A. Bolos

94 Doronicum L.
 1 corsicum (Loisel.) Poiret              60      Co    Contandriopoulos,
                                                            1962
 2 austriacum Jacq.                       60      Hu    Baksay, 1956
 3 columnae Ten.
 4 orientale Hoffm.                        60      Hu    Baksay, 1956
 5 carpetanum Boiss. & Reuter ex Willk.   60      Hs    Küpfer, 1969
                                          120      Lu    Fernandes & Queirós,
                                                            1971
 6 plantagineum L.                        120      Lu    Fernandes & Queirós,
                                                            1971
 7 hungaricum Reichenb. fil.              60      Hu    Baksay, 1956
 8 pardalianches L.                       60      Ga    Delay, 1970
 9 cataractarum Widder
10 grandiflorum Lam.                      60      Ga    Küpfer &
                                                           Favarger, 1968
11 glaciale (Wulfen) Nyman
   (a) glaciale
   (b) calcarium (Vierh.) Hegi            60      Au    Polatschek, 1966
12 clusii (All.) Tausch                   60      Ju    Lovka et al.,
                                                            1972
                                          120      Po    Skalińska, 1950

95 Erechtites Rafin.
 1 hieracifolia (L.) Rafin. ex DC.
```

96 Senecio L.
    1 mikanioides Otto ex Walpers   20   Lu   Fernandes & Queirós,
                                                            1971
    2 gnaphalodes Sieber
    3 ambiguus (Biv.) DC.
      (a) ambiguus
      (b) gibbosus (Guss.) Chater
    4 bicolor (Willd.) Tod.
      (a) bicolor
      (b) nebrodensis (Guss.) Chater
      (c) cineraria (DC.) Chater   40   Lu   Fernandes & Queirós,
                                                            1971
    5 thapsoides DC.
      (a) thapsoides
      (b) visianianus (Papaf. ex Vis.) Vandas
    6 incanus L.
      (a) incanus   40   He   Favarger, in litt.
                                                            1970
      (b) carniolicus (Willd.) Br.-Bl.   c.120   It   Favarger, in litt.
                                                            1974
                                          c.140   He   Favarger, in litt.
                                                            1974
    7 persoonii De Not.   40   It   Küpfer, 1971
    8 leucophyllus DC.   40   Ga   Küpfer &
                                                            Favarger, 1967
    9 halleri Dandy
   10 boissieri DC.   40   Hs   Küpfer, 1968
   11 linifolius L.   40   Bl   Dahlgren et al.,
                                                            1971
   12 inaequidens DC.
   13 nevadensis Boiss. & Reuter
   14 quinqueradiatus Boiss. ex DC.
   15 othonnae Bieb.
   16 fluviatilis Wallr.   40   Cz   Májovský et al.,
                                                            1970
   17 nemorensis L.
      (a) nemorensis   40   Cz   Májovský et al.,
                                                            1970
      (b) fuchsii (C.C. Gmelin) Čelak.   40   Cz   Májovský et al.,
                                                            1970
   18 cacaliaster Lam.
   19 doria L.
      (a) doria
      (b) umbrosus (Waldst. & Kit.) Soó   40   Cz   Májovský et al.,
                                                            1970
      (c) kirghisicus (DC.) Chater
      (d) legionensis (Lange) Chater
   20 lopezii Boiss.
   21 paludosus L.   40   Ga   Zickler, 1968
   22 eubaeus Boiss. & Heldr.
   23 macedonicus Griseb.
   24 auricula Bourgeau ex Cosson
   25 pyrenaicus L.   40   Hs   Küpfer & Favarger,
                                                            1967
   26 doronicum (L.) L.
      (a) doronicum   40   Lu   Fernandes & Queirós,
                                                            1971
                                           80   Ga   Guinochet, 1967
      (b) ruthenensis (Mazuc & Timb.-Lagr.)
      (c) gerardii (Godron & Gren.) Nyman   40   Ga   Guinochet, 1967
   27 scopolii Hoppe & Hornsch. ex Bluff &
      Fingerh.

28 lagascanus DC.
29 eriopus Willk.
30 smithii DC.
31 malvifolius (L'Hér.) DC.
32 petasitis (Sims) DC.
33 integrifolius (L.) Clairv.
   (a) integrifolius

	48	Br	Rutland, 1941
	c.90	Rs(N)	Sokolovskaya & Strelkova, 1960

   (b) maritimus (Syme) Chater
   (c) serpentini (Gayer) Jáv.
   (d) capitatus (Wahlenb.) Cuf.

	64	Cz	Májovský,et al., 1970
	96	Ga	Zickler, 1968
96 + 1 - 4B		Ga	Favarger, 1965

   (e) aurantiacus (Hoppe ex Willd.)
      Briq. & Cavillier
   (f) czernjaevii (Minder) Chater
   (g) tundricola (Tolm.) Chater
   (h) atropurpureus (Ledeb.) Cuf.
34 balbisianus DC.
35 elodes Boiss. ex DC.
36 helenitis (L.) Schinz & Thell.
   (a) helenitis   48   Ga   Delay, 1971
   (b) candidus (Corb.) Brunerye
   (c) macrochaetus (Willk.) Brunerye
   (d) salisburgensis Cuf.
37 rivularis (Waldst. & Kit.) DC.
   (a) rivularis
   (b) pseudocrispus (Fiori) E. Mayer
38 ovirensis (Koch) DC.
   (a) ovirensis
   (b) gaudinii (Gremli) Cuf.
39 papposus (Reichenb.) Less.   40   Po   Frey, 1969b
   (a) papposus
   (b) wagneri (Degen) Cuf.
   (c) kitaibelii (Jáv.) Cuf.
40 congestus (R. Br.) DC.
41 cordatus Koch   40   It   Zickler, 1968
42 subalpinus Koch   40   Po   Skalińska et al., 1966

43 pancicii Degen   100   Bu   Kozuharov & Kuzmanov, 1964

44 jacobaea L.

	40	Ho	Gadella & Kliphuis, 1973
	80	Cz	Májovský et al., 1970

45 aquaticus Hill
   (a) aquaticus
   (b) barbareifolius (Wimmer & Grab.)
      Walters
46 erucifolius L.
47 carpetanus Boiss. & Reuter
48 squalidus L.

	20	Ga	Zickler, 1968

49 siculus All.   20   Si   Larsen & Laegaard, 1971

50 cambrensis Rosser   60   Br   Rosser, 1955
51 aethnensis Jan ex DC.
52 nebrodensis L.
53 abrotanifolius L.

(a) abrotanifolius L.	c.40	He	Favarger, 1965
(b) carpathicus (Herbich) Nyman	40	Cz	Májovský et al., 1970
54 adonidifolius Loisel.	40	Ga	Delay, 1970
55 resedifolius Less.			
56 delphinifolius Vahl			
57 minutus (Cav.) DC.			
58 gallicus Chaix	20	Lu	Fernandes & Queirós, 1971
59 leucanthemifolius Poiret	20	Lu	Fernandes & Queirós, 1971
rodriguezii Willk. ex Rodr.	20	Bl	Dahlgren et al., 1971
60 vernalis Waldst. & Kit.	40	Cz	Májovský et al., 1970
61 petraeus Boiss. & Reuter			
62 sylvaticus L.	40	Lu	Fernandes & Queirós, 1971
63 lividus L.	40	Lu	Fernandes & Queirós, 1971
64 viscosus L.	40	Ga	Zickler, 1968
65 vulgaris L.	40	Ho	Gadella & Kliphuis, 1963
66 elegans L.			
67 flavus (Decne) Schultz Bip.			

97 Ligularia Cass.

1 sibirica (L.) Cass.	60	Cz	Májovský et al., 1970
2 glauca (L.) O. Hoffm.	60	Rm	Tarnavschi, 1948
3 dentata (A. Gray) Hara			

98 Kleinia Miller
1 mandraliscae Tineo
2 repens (L.) Haw.

99 Cacalia L.
1 hastata L.

Calenduleae Cass.

100 Calendula L.
1 suffruticosa Vahl
(a) suffruticosa

(b) lusitanica (Boiss.) Ohle	32	Lu	Fernandes & Queirós, 1971
(c) fulgida (Rafin.) Ohle	32	Si	Meusel & Ohle, 1966
(d) tomentosa Murb.			
(e) maritima (Guss.) Meikle			
(f) algarbiensis (Boiss.) Nyman	32	Lu	Rodriguez, 1953
	32 + 2B	Lu	Fernandes & Queirós, 1971
2 officinalis L.			
3 stellata Cav.			
4 arvensis L.	44	Lu	Fernandes & Queirós, 1971
5 tripterocarpa Rupr.			

101 Chrysanthemoides Fabr.
1 monilifera (L.) T. Norlindh

Arctotideae Cass.

102 Arctotis L.
  1 stoechadifolia Bergius

103 Arctotheca Wendl.
  1 calendula (L.) Levyns   18   Lu   Fernandes & Queirós,
                                                    1971

104 Gazania Gaertner
  1 rigens (L.) Gaertner

Cardueae Cass.

105 Amphoricarpos Vis.
  1 neumayeri Vis
    (a) neumayeri   24   Ju   Sušnik & Lovka
                                                  (Löve, in litt.)
    (b) murbeckii Bošnjak

106 Carlina L.
  1 diae (Rech. fil.) Meusel &
    Kästner
  2 tragacanthifolia Klatt
  3 corymbosa L.   18   Lu   Fernandes & Queirós,
                                                  1971
                            20   Si   Arata, 1944
    (a) corymbosa
    (b) graeca (Boiss.) Nyman
    (c) curetum (Heldr. ex Halácsy) Rech. fil.
  4 sicula Ten.
  5 barnebiana B.L. Burtt & P.H. Davis
  6 vulgaris L.   20   Cz   Májovský et al.,
    (a) vulgaris   1970
    (b) intermedia (Schur) Hayek
    (c) longifolia Nyman
  7 frigida Boiss. & Heldr.
  8 fiumensis Simonkai
  9 macrocephala Moris
    (a) macrocephala   20   Co   Contandriopoulos,
                                                1962
    (b) nebrodensis (Guss. ex DC.) D.A. Webb
  10 acaulis L.
    (a) acaulis
    (b) simplex (Waldst. & Kit.) Nyman   20   It   Arata, 1944
  11 acanthifolia All.   20   Ga   Delay, 1970
    (a) acanthifolia
    (b) cynara (Pourret ex Duby) Rouy
  12 lanata L.   20   Si   Arata, 1944
  13 racemosa L.   20   Lu   Fernandes & Queirós,
                                                    1971

107 Atractylis L.
  1 gummifera L.   20   Lu   Fernandes & Queirós,
                                                      1971
  2 tutinii Franco
  3 humilis L.
  4 cancellata L.

108 Xeranthemum L.
  1 annuum L.   12   Ju   Gadella & Kliphuis,
                                                    1972

```
  2 inapertum (L.) Miller
  3 cylindraceum Sibth. & Sm.              20     Bu     Kuzmanov &
                                                         Kozuharov, 1970

109 Cardopatum Pers.
  1 corymbosum (L.) Pers.

110 Echinops L.
  1 spinosus L.
  2 spinosissimus Turra
   (a) spinosissimus
   (b) bithynicus (Boiss.) Kozuharov
   (c) neumayeri (Vis.) Kozuharov
  3 orientalis Trautv.
  4 strigosus L.                           32     Lu     Fernandes & Queiros,
                                                         1971
  5 graecus Miller
  6 fontqueri Pau
  7 sphaerocephalus L.                     32     Cz     Májovský et al.,
                                                         1970
   (a) sphaerocephalus
   (b) albidus (Boiss. & Spruner) Kozuharov
   (c) taygeteus (Boiss. & Heldr.)
       kozuharov
  8 exaltatus Schrader
  9 bannaticus Rochel ex Schrader
 10 ritro L.                               32     Bu     Kuzmanov &
                                                         Kozuharov, 1970
   (a) ritro
   (b) thracicus (Velen.) Kozuharov
   (c) sartorianus (Boiss. & Heldr.)
       Kozuharov
   (d) meyeri (DC.) Kozuharov
   (e) ruthenicus (Bieb.) Nyman
 11 microcephalus Sibth. & Sm.
 12 oxyodontus Bornm. & Diels.

111 Berardia Vill.
  1 subacaulis Vill.                       36     He     Favarger, 1959a

112 Arctium L.
  1 tomentosum Miller                      36     Cz     Májovský et al.,
                                                         1970
  2 lappa L.                               36     Cz     Májovský et al.,
                                                         1970
  3 pubens Bab.
  4 minus Bernh.                           36     Cz     Májovský et al.,
                                                         1970
  5 nemorosum Lej.                         36     Cz     Májovský et al.,
                                                         1970

113 Cousinia Cass.
  1 astracanica (Sprengel) Tamamsch.

114 Saussurea DC.
  1 amara (L.) DC.
  2 salsa (Pallas) Sprengel
  3 turgaiensis B. Fedtsch.
  4 pygmaea (Jacq.) Sprengel              52     Ju     Lovka et al., 1972
  5 parviflora (Poiret) DC.
  6 porcii Degen
```

```
7 alpina (L.) DC.
  (a) alpina                              52      No      Knaben &
                                                          Engelskjon, 1967
                                          54      Su      Löve & Löve, 1944
  (b) macrophylla (Sauter) Nyman
  (c) depressa (Gren.) Nyman           50-54      Ga      Favarger, 1965
  (d) esthonica (Baer ex Rupr.) Kupffer
8 discolor (Willd.) DC.                  26      Ju      Lovka et al.,
                                                          1972
9 controversa DC.

115 Staehelina L.
  1 fruticosa (L.) L.
  2 uniflosculosa Sibth. & Sm.
  3 arborea Schreber
  4 baetica DC.
  5 dubia L.                             30      Lu      Fernandes & Queirós,
                                                          1971

116 Jurinea Cass.
  1 linearifolia DC.
  2 stoechadifolia (Bieb.) DC.
  3 tzar-ferdinandii Davidov
  4 pinnata (Lag.) DC.
  5 tanaitica Klokov
  6 albicaulis Bunge
    (a) kilaea (Aznav.) Kozuharov        30      Bu      Kuzmanov & Ancev, 1973
    (b) laxa (Fischer ex Iljin)
        Kozuharov
  7 kirghisorum Janisch.
  8 cyanoides (L.) Reichenb.
    (a) cyanoides
    (b) tenuiloba (Bunge) Nyman
  9 ewersmanii Bunge
 10 mollis (L.) Reichenb.                30      Bu      Kuzmanov &
                                                          Kozuharov, 1967
                                         34      Cz      Holub et al., 1971
                                         35      Cz      Holub et al., 1971
                                         36      Rm      Lungeanu, 1973
    (a) mollis
    (b) moschata (DC.) Nyman
    (c) transylvanica (Sprengel) Hayek
    (d) anatolica (Boiss.) Stoj. &
        Stefanov
 11 polyclonos (L.) DC.
 12 ledebourii Bunge                     36      Rm      Lungeanu, 1973
 13 consanguinea DC.
    (a) consanguinea
    (b) neicevii Kozuharov
    (c) arachnoidea (Bunge) Kozuharov    36      Asia    Griff, 1965
    (d) bulgarica (Velen.) Kozuharov
 14 glycacantha (Sibth. & Sm.) DC.       30      Hu      Baksay, 1956
 15 humilis (Desf.) DC.                  34      Hs      Küpfer, 1969
 16 taygetea Halácsy
 17 fontqueri Cuatrec.

117 Carduus L.
  1 macrocephalus Desf.
    (a) macrocephalus
    (b) sporadum (Halácsy) Franco
    (c) siculus Franco
```

2 granatensis Willk.
3 taygeteus Boiss. & Heldr
  (a) taygeteus
  (b) insularis Franco
4 thoermeri Weinm.

5 nutans L.	16	Cz	Májovský et al., 1970

  (a) natans
  (b) alpicola (Gillot) Chassagne &
    J. Arenes
  (c) platylepis (Reichenb. & Sauter)
    Nyman
6 micropterus (Borbás) Teyber
  (a) micropterus
  (b) perspinosus (Fiori) Kazmi

7 broteroi Welw. ex Coutinho	20	Lu	Fernandes & Queirós, 1971

8 sandwithii Kazmi
9 platypus Lange
10 chrysacanthus Ten.
  (a) chrysacanthus
  (b) hispanicus Franco
11 aurosicus Vill.

12 acanthoides L.	16	Ju	Podlech, 1964
	20	Ge	Fritsch, 1973
	22	Hu	Baksay, 1958

13 ramosissimus Pančić
14 tmoleus Boiss.
  (a) tmoleus
  (b) armatus (Boiss. & Heldr.) Franco
15 thessalus Boiss. & Heldr.
16 cronius Boiss. & Heldr.
  (a) cronius
  (b) baldaccii Kazmi

17 personata (L.) Jacq.	18	He	Larsen, 1954

  (a) personata
  (b) albidus (Adamović) Kazmi
18 crispus L.

(a) crispus	16	Fe	Hamet-Ahti & Virrankoski, 1970

  (b) multiflorus (Gaudin) Franco
19 litigiosus Nocca & Balbis
  (a) litigiosus
  (b) horridissimus (Briq. & Cavillier)
    Franco
20 euboicus Franco
21 vivariensis Jordan
  (a) vivariensis
  (b) australis Nyman
  (c) assoi (Willk.) Kazmi
22 nigrescens Vill.
23 hamulosus Ehrh.
  (a) hamulosus
  (b) hystrix (C.A. Meyer) Kazmi
24 uncinatus Bieb.
25 defloratus L.

(a) defloratus	18	Ga It	Favarger & Küpfer, 1970
(b) glaucus Nyman	22	Hu	Baksay, 1958

26 argemone Pourret ex Lam.

(a) argemone	22	Hs	Favarger & Küpfer, 1970

(b) obtusisquamus Franco	22	He	Favarger & Küpfer, 1970
27 carlinifolius Lam.	22	Ga Hs	Favarger & Küpfer, 1970

28 candicans Waldst. & Kit.
  (a) candicans
  (b) globifer (Velen.) Kazmi

29 collinus Waldst. & Kit.	16	Hu	Baksay, 1958
	32	Ga	Kliphuis & Wieffering, 1972

  (a) collinus
  (b) cylindricus (Borbás) Soó
  (c) glabrescens (Sagorski) Kazmi
30 carduelis (L.) Gren.
31 kerneri Simonkai
  (a) lobulatiformis (Csuros & E.I.
    Nyarady) Soó
  (b) kerneri
  (c) scardicus (Griseb.) Kazmi
  (d) austro-orientalis Franco
32 affinis Guss.
  (a) affinis
  (b) brutius (Porta) Kazmi
33 adpressus C.A. Meyer

34 argyroa Biv.	26	Si	Podlech, 1964

35 bourgeanus Boiss. & Reuter
  (a) bourgeanus
  (b) valentinus (Boiss. & Reuter) Franco
36 myriacanthus Salzm. ex DC.
37 asturicus Franco

38 carpetanus Boiss. & Reuter	16	Lu	Fernandes & Queirós, 1971
39 carlinoides Gouan	18	Hs	Küpfer, 1968

  (a) carlinoides
  (b) hispanicus (Kazmi) Franco
40 corymbosus Ten.
41 acicularis Bertol.
42 argentatus L.

43 meonanthus Hoffmanns. & Link	16	Lu	Queirós, 1973
44 tenuiflorus Curtis	54	Ga	van Loon et al., 1971
45 pycnocephalus L.	62-64	Bl	Dahlgren et al., 1971

  (a) pycnocephalus
  (b) albidus (Bieb.) Kazmi
46 australis L. fil.

47 cephalanthus Viv.	22	Co	Contandriopoulos, 1962
48 fasciculiflorus Viv.	22	Co	Contandriopoulos, 1962

118 Cirsium Miller
  1 ferox (L.) DC.
  2 heldreichii Halácsy
   (a) heldreichii
   (b) euboicum Petrak
  3 bulgaricum DC.
  4 polycephalum DC.
  5 morinifolium Boiss. & Heldr.
  6 hypopsilum Boiss. & Heldr.
  7 vallis-demonis Lojac.

```
 8 lacaitae Petrak
 9 tenoreanum Petrak
10 lobelii Ten.
11 morisianum Reichenb. fil.
12 richteranum Gillot
13 costae (Sennen & Pau) Petrak
14 giraudiasii Sennen & Pau
15 eriophorum (L.) Scop.                    34      Cz    Májovský et al.,
                                                         1970

16 odontolepis Boiss. ex DC.
17 spathulatum (Moretti) Gaudin
18 ligulare Boiss.
19 grecescui Rouy
20 decussatum Janka
21 boujartii (Piller & Mitterp.) Schultz Bip.
   (a) boujartii
   (b) wettsteinii Petrak
22 furiens Griseb. & Schenk
23 ciliatum Moench
24 serrulatum (Bieb.) Fischer
25 laniflorum (Bieb.) Fischer
26 scabrum (Poiret) Bonnet & Barratte
27 echinatum (Desf.) DC.
28 vulgare (Savi) Ten.                      68      Lu    Fernandes & Queirós,
                                                          1971

                                           102      Lu    Fernandes & Queirós,
                                                          1971
29 italicum (Savi) DC.
30 dissectum (L.) Hill
31 filipendulum Lange                       34      Lu    Fernandes & Queirós,
                                                          1971
32 tuberosum (L.) All.                      34      He    Favarger, 1969
33 rivulare (Jacq.) All.                    34      Ga    Delay, 1970
34 montanum (Waldst. & Kit. ex Willd.)
     Sprengel
35 erisithales (Jacq.) Scop.                34      Hu    Baksay, 1956
36 appendiculatum Griseb.                   68      Bu    Kuzmanov & Ancev,
                                                          1973
37 waldsteinii Rouy                         68      Po    Mizianty & Frey,
                                                          1973
38 hypoleucum DC.
39 carniolicum Scop.
   (a) carniolicum                          16      Ju    Lovka & Sušnik, 1973
   (b) rufescens (Ramond ex DC.) P. Fourn.
40 oleraceum (L.) Scop.                     34      Cz    Májovský et al.,
                                                          1970
41 spinosissimum (L.) Scop.
   (a) spinosissimum                        34      He    Larsen, 1954
   (b) bertolonii (Sprengel) Werner         34      It    Miceli & Garbari,
                                                          1976
42 glabrum DC.                              34      Ga    Küpfer, 1971
43 candelabrum Griseb.
44 acaule Scop.
   (a) acaule                               34      Po    Skalińska et al.,
                                                          1959
   (b) gregarium (Boiss. ex DC.) Werner 34 + 2B     Hs    Küpfer, 1968
   (c) esculentum (Sievers) Werner
45 mairei Halácsy
46 valentinum Porta & Rigo
47 helenioides (L.) Hill                    c.34    Fe    V. Sorsa, 1962
```

```
 48 pannonicum (L. fil.) Link            34    Hu    Baksay, 1956
 49 heterotrichum Pančić
 50 canum (L.) All.                       34    Ju    Nilsson & Lassen,
                                                      1971
 51 tymphaeum Hausskn.
 52 monspessulanum (L.) Hill
 53 welwitschii Cosson                    34    Lu    Fernandes & Queirós,
                                                      1971
 54 alatum (S.G. Gmelin) Bobrov
 55 brachycephalum Juratzka
 56 bourgaeanum Willk.
 57 palustre (L.) Scop.                   34    Ho    Gadella & Kliphuis,
                                                      1966
 58 flavispina Boiss. ex DC.
 59 creticum (Lam.) D'Urv.
    (a) creticum
    (b) triumfetti (Lacaita) Werner
 60 arvense (L.) Scop.                     34    Cz    Májovský et al.,
                                                      1970

119 Picnomon Adanson
  1 acarna (L.) Cass.                      32    Ga    Moore, 1967
                                           34    Lu    Fernandes & Queirós,
                                                      1971

120 Notobasis Cass.
  1 syriaca (L.) Cass.                     34    Bl    Dahlgren et al.,
                                                      1971

121 Ptilostemon Cass.
  1 strictus (Ten.) W. Greuter
  2 niveus (C. Presl) W. Greuter
  3 afer (Jacq.) W. Greuter
  4 chamaepeuce (L.) Less.
  5 gnaphaloides (Cyr.) Sojak
    (a) gnaphaloides
    (b) pseudofruticosus (Pamp.) W. Greuter
  6 echinocephalus (Willd.) W. Greuter
  7 hispanicus (Lam.) W. Greuter
  8 casabonae (L.) W. Greuter               32    Sa    Renzoni-Cela, 1963
  9 stellatus (L.) W. Greuter               24    It    Burdet & Greuter,
                                                      1973

122 Lamyropsis (Charadze) Dittrich
  1 cynaroides (Lam.) Dittrich
  2 microcephala (Moris) Dittrich &
    W. Greuter

123 Galactites Moench
  1 tomentosa Moench                        22    Lu    Fernandes & Queirós, 1971
  2 duriaei Spach ex Durieu                 22    Hs    Gadella & Kliphuis,
                                                      1968

124 Tyrimnus (Cass.) Cass.
  1 leucographus (L.) Cass.

125 Onopordum L.
  1 acaulon L.
    (a) acaulon
    (b) uniflorum (Cav.) Franco
  2 acanthium L.
```

(a) acanthium	34	Po	Skalińska et al., 1959

(b) gautieri (Rouy) Franco
(c) parnassicum (Boiss. & Heldr.) Nyman
3 laconicum Heldr. & Sart. ex Rouy

4 nervosum Boiss.	34	Lu	Fernandes & Queirós, 1971

5 corymbosum Willk.
  (a) corymbosum
  (b) visegradense Franco
6 tauricum Willd.
7 argolicum Boiss.
8 caulescens D'Urv.
  (a) caulescens
  (b) atticum Franco
9 messeniacum Halácsy
10 macracanthum Schousboe
11 bracteatum Boiss. & Heldr.
  (a) bracteatum
  (b) ilex (Janka) Franco
  (c) creticum Franco
  (d) myriacanthum (Boiss.) Franco

12 illyricum L.	34	Hs	Valdés, 1970b

  (a) illyricum
  (b) cardunculus (Boiss.) Franco
  (c) horridum (Viv.) Franco
13 majorii Beauverd

126 Cynara L.			
1 humilis L.	34	Lu	Fernandes & Queirós, 1971

2 alba Boiss. ex DC.
3 algarbiensis Cosson ex Mariz
4 cornigera Lindley

5 cardunculus L.	34	Lu	Fernandes & Queirós, 1971

6 scolymus L.
7 tournefortii Boiss. & Reuter

127 Silybum Adanson			
1 marianum (L.) Gaertner	34	It	Larsen, 1956a

2 eburneum Cosson & Durieu

128 Palaeocyanus Dostál			
1 crassifolius (Bertol.) Dostál	30	Si	Damboldt et al., 1973

129 Cheirolophus Cass.			
1 sempervirens (L.) Pomel	30	Lu	Guinochet & Foissac, 1962
2 uliginosus (Brot.) Dostál	24	Lu	Guinochet & Foissac, 1962
	32	Lu	Fernandes & Queirós, 1971
3 intybaceus (Lam.) Dostál	32	Ga	Gardou, 1972

130 Serratula L.			
1 tinctoria L.	22	Cz	Májovský et al., 1970

2 wolffii Andrae

```
  3 pinnatofida (Cav.) Poiret
  4 abulensis Pau                          90    Lu    Fernandes & Queirós,
                                                       1971
  5 baetica Boiss. ex DC.
  6 alcalae Cosson
  7 pauana Iljin
  8 nudicaulis (L.) DC.                    30    Ga    Guinochet &
                                                       Logeois, 1962
  9 flavescens (L.) Poiret
 10 leucantha (Cav.) DC.
 11 cichoracea (L.) DC.
    (a) cichoracea
    (b) mucronata (Desf.) Jahandiez &
        Maire
    (c) cretica Turrill
 12 lycopifolia (Vill.) A. Kerner         60    Hu    Baksay, 1957
 13 bulgarica Acht. & Stoj.
 14 radiata (Waldst. & Kit.) Bieb.        30    Hu    Baksay, 1956
    (a) radiata
    (b) cetingensis (Rohlena) Hayek
 15 gmelinii Tausch
 16 cardunculus (Pallas) Schischkin
 17 erucifolia (L.) Boriss.


131 Leuzea DC.
  1 centauroides (L.) J. Holub            26
  2 rhapontica (L.) J. Holub              26    Ga    Guinochet, 1957
    (a) rhapontica
    (b) heleniifolia (Gren. & Godron)
        J. Holub
    (c) bicknellii (Briq.) J. Holub
  3 altaica (Fischer ex Sprengel) Link
  4 conifera (L.) DC.                     26    Hs    Lorenzo-Andreu &
                                                      Garcia, 1950
  5 rhaponticoides Graells               26    Hs    Dittrich, 1968
  6 longifolia Hoffmanns. & Link          26    Lu    Fernandes & Queirós,
                                                      1971

132 Amberboa (Pers.) Less.
  1 turanica Iljin

133 Volutaria Cass.
  1 lippii (L.) Maire

134 Cyanopsis Cass.
  1 muricata (L.) Dostál

135 Mantisalca Cass.
  1 salmantica (L.) Briq. & Cavillier    18    It    Chiappini, 1954
                                         20    Ga    Guinochet, 1957
                                         22    Lu    Guinochet & Foissac,
                                                      1962

136 Acroptilon Cass.
  1 repens (L.) DC.

137 Phalacrachena Iljin
  1 inuloides (Fischer ex Janka) Iljin

138 Centaurea L.
  1 centaurium L.
```

```
 2 amplifolia Boiss. & Heldr.
 3 fraylensis Schultz Bip. ex Nyman
 4 africana Lam.                        30      Lu      Fernandes & Queirós,
                                                        1971
 5 ruthenica Lam.
 6 linaresii Laz.-Ibiza
 7 kasakorum Iljin
 8 taliewii Kleopow
 9 alpina L.
10 jankae Brandza
11 collina L.                           20      Lu      Fernandes & Queirós,
                                                        1971
12 salonitana Vis.                      20      Rm      Lungeanu, 1975
                                        40      Gr      Phitos, 1970
   (a) salonitana
   (b) ognianoffii (Urum.) Dostál
13 centauroides L.
14 rumelica Boiss.
15 nicolai Bald.
16 ornata Willd.
   (a) ornata                           20      Lu      Queirós, 1973
                                        40      Lu      Fernandes & Queirós,
                                                        1971
   (b) saxicola (Lag.) Dostál           60      Hs      Gardou, 1972
17 granatensis Boiss. ex DC.
18 tuntasia Heldr. ex Halácsy
19 macedonica Boiss.
   (a) macedonica
   (b) parnonia (Halácsy) Dostál
20 mannagettae Podp.
21 dichroantha A. Kerner                20      Ju      Nilsson & Lassen,
                                                        1971
22 rupestris L.                         20      Ju      Nilsson & Lassen,
                                                        1971
   (a) rupestris
   (b) ceratophylla (Ten.) Gugler
   (c) finazzeri (Adamović) Hayek
   (d) athoa (DC.) Gugler
23 kosaninii Hayek
24 clementei Boiss. ex DC.
25 orientalis L.                        20      Bu      Kuzmanov &
                                                        Kozuharov, 1970
26 neiceffii Degen & H. Wagner
27 tauromenitana Guss.
28 chrysolepis Vis.
29 prolongi Boiss. ex DC.
30 polymorpha Lag.                      40      Hs      Gardou, 1972
31 atropurpurea Waldst. & Kit.          18      Bu      Kozuharov et al.,
                                                        1968
   (a) atropurpurea
   (b) soskae (Stoj. & Acht.) Dostál
32 kotschyana Heuffel ex Koch
33 murbeckii Hayek
34 candelabrum Hayek & Kosanin
35 immanuelis-loewii Degen
36 grbavacensis (Rohlena) Stoj. & Acht.
37 oliverana DC.                        22      Gr      Runemark, 1967c
38 ragusina L.
   (a) ragusina
   (b) lungensis (Ginzberger) Hayek
39 rechingeri Phitos                    22      Gr      Runemark, 1967c
```

```
40 graeca Griseb.                          20      Gr    Phitos, 1970
   (a) graeca
   (b) ceccariniana (Boiss. & Heldr.) Dostál
41 laconica Boiss.
42 achaia Boiss. & Heldr.
43 sibthorpii Halácsy
44 psilacantha Boiss. & Heldr.
45 redempta Heldr.                          20      Cr    Runemark, 1967c
46 cytherea Rech. fil.
47 ebenoides Heldr. ex S. Moore
48 spruneri Boiss. & Heldr.
   (a) spruneri                            110      Gr    Phitos, 1970
   (b) minoa (Heldr. ex Boiss.) Dostál
   (c) guicciardii (Boiss.) Hayek           20      Gr    Runemark, 1967c
                                           100      Gr    Phitos, 1970
                                           110      Gr    Phitos, 1970
   (d) lineariloba (Halácsy & Dörfler)
       Dostál                               20      Gr    Phitos, 1970
49 scabiosa L.                              20      Cz    Májovský et al.,
                                                          1970
                                      20 + 0 - 2B   Ga    Lévèque &
                                                          Gorenflot, 1969
50 cephalariifolia Willk.
51 alpestris Hegetschw.                     20      Po    Skalińska et al.,
                                                          1959
52 sadlerana Janka                          20      Hu    Baksay, 1956
53 badensis Tratt.
54 grinensis Reuter
   (a) grinensis
   (b) fritschii (Hayek) Dostál             20      Ju    Lovka et al., 1971
55 apiculata Ledeb.
   (a) apiculata
   (b) spinulosa (Rochel ex Sprengel) Dostál
   (c) adpressa (Ledeb.) Dostál
56 stereophylla Besser
57 raphanina Sibth. & Sm.
   (a) raphanina                            20      Cr    Runemark, 1967
   (b) mixta (DC.) Runemark                 20      Gr    Runemark, 1967
58 aegialophila Wagenitz                    22      Cr    Runemark, 1967
59 pumilio L.
60 macrorrhiza Willk.                       20      Hs    Gardou, 1972
61 toletana Boiss. & Reuter
62 haenseleri (Boiss.) Boiss.
63 argecillensis Gredilla
64 amblensis Graells
65 lagascana Graells
   (a) lagascana
   (b) podospermifolia (Loscos & Pardo) Dostál
66 acaulis L.
67 loscosii Willk.
68 thracica (Janka) Hayek
69 charrelii Halácsy & Dörfler
70 cineraria L.                             18      It    Gori, 1954
   (a) cineraria                            18      It    Viegi et al., 1972
   (b) busambarensis (Guss.) Dostál         18      Si    Viegi et al., 1972
   (c) cinerea (Lam.) Dostál
   (d) veneris (Sommier) Dostál             18      It    Viegi et al., 1972
71 cuspidata Vis.
72 niederi Heldr.
73 kilaea Boiss.
74 wettsteinii Degen & Dörfler
```

75 argentea L.   18   Cr   Runemark, 1967c
76 pannosa DC.
77 nicopolitana Bornm.
78 cuneifolia Sibth. & Sm.
  (a) cuneifolia
  (b) pallida (Friv.) Hayek
  (c) sublanata (DC.) Hayek
79 ipsaria Stoj. & Kitanov
80 rutifolia Sibth. & Sm.
  (a) rutifolia
  (b) jurineifolia (Boiss.) Nyman
  (c) pseudobovina (Hayek) Dostál
81 varnensis Velen.
82 crithmifolia Vis.
83 friderici Vis.
  (a) friderici
  (b) jabukensis (Ginzberger & Teyber)
      Dostál
84 affinis Friv.
  (a) affinis
  (b) balcanica (Urum. & H. Wagner)
      Dostál
  (c) peloponnesiaca (Halacsy) Dostál
  (d) candida (Velen.) Dostál
  (e) lacerata (Hausskn.) Dostál
85 pallidior Halácsy
  (a) pallidior
  (b) denudata (Halácsy) Dostál
  (c) vatevii (Degen, Urum. & H. Wagner)
      Dostál
86 parlatoris Heldr.
  (a) tenorei (Guss. ex Lacaita) Dostál
  (b) nigra (Fiori) Dostál
  (c) parlatoris
87 horrida Badaro
88 balearica Rodr.
89 arenaria Bieb. ex Willd.
  (a) arenaria
  (b) sophiae (Klokov) Dostál
  (c) majorovii (Dumbadze) Dostál
  (d) odessana (Prodan) Dostál
  (e) borysthenica (Gruner) Dostál
90 ovina Pallas ex Willd.
  (a) besserana (DC.) Dostál
  (b) lavrenkoana (Klokov) Dostál
  (c) steveniana (Klokov) Dostál
  (d) koktebelica (Klokov) Dostál
91 tenuiflora DC.
92 spinosociliata Seenus
  (a) cristata (Bartl.) Dostál
  (b) spinosociliata
  (c) tommasinii (A. Kerner) Dostál
93 incompta Vis.
  (a) incompta
  (b) derventana (Vis. & Pančić) Dostál
94 chalcidicaea Hayek
95 grisebachii (Nyman) Form.
  (a) grisebachii
  (b) confusa (Halácsy) Dostál
  (c) paucijuga (Halácsy) Dostál
96 tauscheri A. Kerner

97 biokovensis Teyber
98 gracilenta Velen.
99 kalambakensis Freyn & Sint.
100 transiens Halácsy
101 subsericans Halácsy
102 attica Nyman
   (a) ossaea (Halácsy) Dostál
   (b) asperula (Halácsy) Dostál
   (c) attica
   (d) megarensis (Halácsy & Hayek) Dostál
   (e) drakiensis (Freyn & Sint.) Dostál
   (f) pentelica (Hausskn.) Dostál
103 soskae Hayek ex Kosanin
104 kartschiana Scop.
105 dalmatica A. Kerner
106 brachtii Reichenb. fil.
107 schousboei Lange
   (a) schousboei
   (b) septentrionalis (J. Arenes) Dostál
108 spinabadia Bubani ex Timb.-Lagr.
   (a) shuttleworthii (Rouy) Dostál
   (b) spinabadia
   (c) isernii (Willk.) Dostál
   (d) hanryi (Jordan) Dostál

109 limbata Hoffmanns. & Link	18	Lu	Queirós, 1973

110 urgellensis Sennen
111 rothmalerana (J. Arenes) Dostál
112 aristata Hoffmanns. & Link
   (a) exilis (J. Arenes) Dostál
   (b) langeana (Willk.) Dostál
   (c) geresensis (J. Arenes) Dostál
   (d) aristata

113 paniculata L.	18	Ga It	Gardou, 1972
	19	Ga	Guinochet, 1957

   (a) paniculata
   (b) rigidula (Jordan) Dostál
   (c) polycephala (Jordan) Nyman
   (d) esterellensis (Burnat) Dostál
   (e) castellana (Boiss. & Reuter) Dostál
   (f) cossoniana (J. Arenes) Dostál
114 micrantha Hoffmanns. & Link
   (a) micrantha
   (b) herminii (Rouy) Dostál
   (c) melanosticta (Lange) Dostál

115 leucophaea Jordan	18	Ga	Delay, 1969

   (a) brunnescens (Briq.) Dostál
   (b) reuteri (Reichenb. fil.) Dostál

(c) leucophaea	18	Ga	Guinochet, 1957
(d) pseudocoerulescens (Briq.) Dostál	18	Ga	Guinochet, 1957

   (e) biformis (Timb.-Lagr.) Dostál
   (f) ochrolopha (Costa) Dostál
116 filiformis Viv.
117 corymbosa Pourret
118 subtilis Bertol.
119 exarata Boiss. ex Cosson

120 maculosa Lam.	18	Ga	Guinochet, 1957

   (a) maculosa
   (b) chaubardii (Reichenb. fil.) Dostál
   (c) albida (Lecoq & Lamotte) Dostál
   (d) subalbida (Jordan) Dostál
121 vallesiaca (DC.) Jordan

122 rhenana Boreau
   (a) rhenana   18   Cz   Májovský et al.,
                                            1970
                      18 + O - 2B Po   Skalińska et al.,
                                            1959
   (b) tartarea (Velen.) Dostál
   (c) pseudomaculosa (Dobrocz.) Dostál
   (d) savranica (Klokov) Dostál
123 glaberrima Tausch
124 triniifolia Heuffel
125 reichenbachii DC.
126 calvescens Pančić
127 peucedanifolia Boiss. & Orph.
128 biebersteinii DC.   36   Hu   Baksay, 1958
   (a) biebersteinii
   (b) australis (Pančić) Dostál
   (c) rhodopaea (Hayek & Wagner) Dostál
   (d) radoslavoffii (Urum.) Dostál
   (e) cylindrocephala (Bornm.) Dostál
129 boissieri DC.
   (a) dufourii Dostál
   (b) resupinata (Cosson) Dostál
   (c) prostrata (Cosson) Dostál
   (d) boissieri
   (e) willkommii (Schultz Bip.) Dostál
   (f) pomeliana (Batt. & Trabut) Dostál
   (g) mariolensis (Rouy) Dostál
   (h) pinae (Pau) Dostál
   (i) paui (Loscos ex Willk.) Dostál
   (j) jaennensis (Degen & Debeaux) Dostál
   (k) spachii (Schultz Bip. ex Willk.) Dostál
130 lagascae Nyman
131 bombycina Boiss. ex DC.
   (a) bombycina
   (b) funkii (Schultz Bip.) Dostál
132 monticola Boiss. ex DC.
133 carratracensis Lange
134 aplolepa Moretti
   (a) aplolepa   18   It   Viegi et al., 1972
   (b) aeolica (Guss. ex Lojac.) Dostál 18   Si   Viegi et al., 1972
   (c) pandataria (Fiori & Béguinot) Dostál
   (d) carueliana (Micheletti) Dostál
   (e) maremmana (Fiori) Dostál
   (f) subciliata (DC.) Arcangeli
   (g) ligustica (Gremli ex Briq.) Dostál
   (h) aetaliae (Sommier) Dostál
   (i) lunensis (Fiori) Dostál
   (j) cosana (Fiori) Dostál
   (k) gallinariae (Briq. & Cavillier) Dostál
135 lactiflora Halácsy
136 laureotica Heldr. ex Halácsy
137 pelia DC.
138 rufidula Bornm.
139 tymphaea Hausskn.
   (a) tymphaea ·
   (b) brevispina (Hausskn.) Dostál
140 orphanidea Heldr. & Sart. ex Boiss.
   (a) orphanidea
   (b) thessala (Hausskn.) Dostál
141 diffusa Lam.
142 bovina Velen.

143 aemulans Klokov
144 zuccariniana DC.
145 spinosa L.   36   Gr   Runemark, 1967 c
  (a) spinosa
  (b) cycladum (Heldr.) Hayek
146 hyalolepis Boiss.
147 iberica Trev. ex Sprengel
  (a) iberica
  (b) holzmanniana (Boiss.) Dostál
148 calcitrapa L.   20   Lu   Fernandes & Queirós,
                                                                   1971
149 pontica Prodan & E.I. Nyárády
150 sonchifolia L.
151 seridis L.
  (a) seridis
  (b) cruenta (Willd.) Dostál
  (c) maritima (Dufour) Dostál
152 sphaerocephala L.
  (a) sphaerocephala
  (b) lusitanica (Boiss. & Reuter) Nyman 20   Lu   Guinochet & Foissac,
                                                                   1962
                                                 22   Lu   Fernandes & Queirós,
                                                                   1971
  (c) malacitana (Boiss.) Dostál
  (d) polyacantha (Willd.) Dostál   22   Lu   Guinochet & Foissac,
                                                                   1962
153 aspera L.   22   Lu   Fernandes & Queirós,
                                                                   1971
  (a) aspera
  (b) stenophylla (Dufour) Nyman
  (c) scorpiurifolia (Dufour) Nyman
  (d) pseudosphaerocephala (R.J. Shuttlew.
       ex Rouy) Kugler
154 napifolia L.
155 micracantha Dufour
156 hermannii F. Hermann
157 solstitialis L.   16   Ju   Gadella & Kliphuis,
                                                                   1972
  (a) solstitialis
  (b) adamii (Willd.) Nyman
  (c) schouwii (DC.) Dostál
  (d) erythracantha (Halácsy) Dostál
158 idaea Boiss. & Heldr.
159 melitensis L.   24   Ga   Guinochet, 1957
160 sulphurea Willd.
161 nicaeensis All.
162 eriophora L.   24   Hs   Björkqvist et al.,
                                                                   1969
163 diluta Aiton
164 margaritacea Ten.
  (a) pseudoleucolepis (Kleopow) Dostál
  (b) breviceps (Iljin) Dostál
  (c) margaritacea
  (d) margaritalba (Klokov) Dostál
  (e) protomargaritacea (Klokov) Dostál
  (f) appendicata (Klokov) Dostál
  (g) konkae (Klokov) Dostál
  (h) protogerberi (Klokov) Dostál
  (i) donetzica (Klokov) Dostál
  (j) pineticola (Iljin) Dostál
  (k) dubjanskyi (Iljin) Dostál

        (l) gerberi (Steven) Dostál
        (m) paczoskii (Kotov ex Klokov) Dostál
165 transcaucasica D. Sossn. ex Grossh.
166 sterilis Steven
        (a) sterilis
        (b) semijusta (Juz.) Dostál
        (c) vankovii (Klokov) Dostál
167 alba L.   18   Hs   Gardou, 1972
        (a) costae (Willk.) Dostál
        (b) latronum (Pau) Dostál
        (c) alba
        (d) splendens (L.) Arcangeli   20   Ju   Nilsson & Lassen,
                                                                     1971
        (e) tenoreana (Willk.) Dostál
        (f) pestalottii (De Not.) Arcangeli
        (g) diomedea (Gasparr.) Dostál
        (h) albanica (Halácsy) Dostál
        (i) ipecensis (Rech. fil.) Dostál
        (j) deusta (Ten.) Nyman   18   Ju   Damboldt et al.,
                                                                     1971
        (k) brunnea (Halácsy) Dostál
        (l) epapposa (Velen.) Dostál
        (m) caliacrae (Prodan) Dostál
        (n) leucomalla (Bornm.) Dostál
        (o) formanekii (Halácsy) Dostál
        (p) vandasii (Velen.) Dostál
        (q) euxina (Velen.) Dostál
        (r) heldreichii (Halácsy) Dostál
        (s) princeps (Boiss. & Heldr.) Gugler
        (t) subciliaris (Boiss. & Heldr.) Dostál
168 deustiformis Adamović
        (a) deustiformis
        (b) ptarmicifolia (Halácsy ex Hayek) Dostál
        (c) pseudocadmea (Wagenitz) Dostál
169 ferulacea U. Martelli   18   Sa   Arrigoni & Mori,
                                                                     1971
170 musarum Boiss. & Orph.
171 haynaldii Borbás ex Vuk
172 bracteata Scop.
173 weldeniana Reichenb.
174 rocheliana (Heuffel) Dostál
175 pannonica (Heuffel) Simonkai   22   Ga   Guinochet, 1957
                                                     44   Rm   Tarnavschi, 1948
        (a) pannonica
        (b) substituta (Czerep.) Dostál
176 vinyalsii Sennen
        (a) vinyalsii
        (b) approximata (Rouy) Dostál
177 dracuncifolia Dufour   22   Hs   Gardou, 1972
178 jacea L.   22   Ga It Ju Gardou, 1972
                                                     44   Au Ga Ju Gardou, 1972
179 decipiens Thuill.
        (a) decipiens
        (b) ruscinonensis (Boiss.) Dostál
180 subjacea (G. Beck) Hayek
181 macroptilon Borbas
        (a) oxylepis (Wimmer & Grab.) Soó
        (b) macroptilon
182 microptilon
        (a) microptilon
        (b) emporitana (Vayr. ex Hayek) Dostál

                                268

```
183 transalpina Schleicher ex DC.            44    Ga        Guinochet, 1957
184 nigrescens Willd.                        44    Ga        Guinochet & Foissac,
                                                               1962
    (a) nigrescens
    (b) ramosa Gugler
    (c) neapolitana (Boiss.) Dostál
    (d) pinnatifida (Fiori) Dostál
    (e) smolinensis (Hayek) Dostál
185 carniolica Host
186 debeauxii Gren. & Godron                 22    Ga It     Gardou, 1972
                                             33    Ga        Gardou, 1972
                                             44    Ga        Gardou, 1972
    (a) endressii (Hochst. & Steudel ex
        Lamotte) Dostál
    (b) nevadensis (Boiss. & Reuter) Dostál
    (c) thuillieri Dostál
    (d) nemoralis (Jordan) Dostál
    (e) debeauxii
187 nigra L.                                 22    Ga Hs Lu  Gardou, 1972
                                             44    Ga        Gardou, 1972
    (a) nigra                                22    Ga        Guinochet, 1957
    (b) carpetana (Boiss. & Reuter) Nyman
    (c) rivularis (Brot.) Coutinho           22    Lu        Fernandes & Queirós,
                                                               1971
188 phrygia L.
    (a) phrygia                              22    Rm        Gardou, 1972
    (b) melanocalathia (Borbás) Dostál
    (c) carpatica (Porc.) Dostál
    (d) moesiaca (Urum. & H. Wagner) Hayek
    (e) nigriceps (Dobrocz.) Dostál
    (f) pseudophrygia (C.A. Meyer) Gugler    22    Au        Favarger & Huynh,
                                                               1964
    (g) abbreviata (C. Koch) Dostál
    (h) ratezatensis (Prodan) Dostál
    (i) rarauensis (Prodan) Dostál
189 stenolepis A. Kerner
    (a) stenolepis                           22    Cz        Májovský et al.,
                                                               1970
    (b) razgradensis (Velen.) Stoj. & Acht.
    (c) bosniaca (Murb.) Dostál
190 indurata Janka
191 uniflora Turra                           22    Ga        Guinochet, 1957
    (a) uniflora
    (b) nervosa (Willd.) Bonnier & Layens    22    Ga It     Gardou, 1972
    (c) ferdinandi (Gren.) Bonnier
    (d) davidovii (Urum.) Dostál
192 kernerana Janka
    (a) kernerana
    (b) gheorghieffii (Halácsy) Dostál
193 pectinata L.                             22    Ga        Guinochet, 1957
    (a) pectinata
    (b) acutifolia (Jordan) Dostál
    (c) supina (Jordan) Dostál
194 antennata Dufour
195 trichocephala Bieb. ex Willd.
    (a) trichocephala
    (b) simonkaiana (Hayek) Dostál
196 janeri Graells
197 linifolia L.                             44    Hs        Gardou, 1972
198 hyssopifolia Vahl
199 parilica Stoj. & Stefanov
```

200 procumbens Balbis
    (a) procumbens
    (b) jordaniana (Gren. & Godron) Rouy
    (c) aemilii (Briq.) Dostál
    (d) verguinii (Briq. & Cavillier) Dostál

201 rhaetica	22	It	Favarger & Huynh, 1964

202 leucophylla Bieb.
203 declinata Bieb.
204 dealbata Willd.
205 sibirica L.
206 carbonata Klokov
207 marschalliana Sprengel
208 sumensis Kalenicz
209 trinervia Stephan ex Willd.

210 montana L.	44	Ga	Guinochet, 1957
211 mollis Waldst. & Kit.	44	Po	Skalińska et al., 1961

212 maramarosiensis (Jav.) Czerep.
213 pinnatifida Schur
214 baldaccii Degen ex Bald.
215 pindicola Griseb.

216 triumfetti All.	22	Ga	Guinochet, 1957
	44	Ga	Guinochet, 1957

    (a) aligera (Gugler) Dostál
    (b) tanaitica (Klokov) Dostál
    (c) pirinensis (Degen, Urum. & H.
        Wagner) Dostál
    (d) semidecurrens (Jordan) Dostál
    (e) dominii Dostál
    (f) novakii (Dostál) Dostál
    (g) stricta (Waldst. & Kit.) Dostál
    (h) angelescui (G. Grint.) Dostál
    (i) adscendens (Bartl.) Dostál
    (j) lugdunensis (Jordan) Dostál

(k) triumfetti	22	Hu	Baksay, 1956
	44	Ga	Guinochet, 1957

    (l) cana (Sibth. & Sm.) Dostál
    (m) lingulata (Lag.) Dostál
217 napulifera Rochel
    (a) pseudaxillaris (Stefanov & Georgiev) Dostál
    (b) tuberosa (Vis.) Dostál
    (c) napulifera
    (d) nyssana (Petrovič) Dostál
    (e) thirkei (Schultz Bip.) Dostál
218 depressa Bieb.
219 pinardii Boiss.

220 cyanus L.	24	Ga	Guinochet, 1957
221 pullata L.	22	Ga	Guinochet, 1957

139 Crupina (Pers.) Cass.			
1 vulgaris Cass.	30	Ju	Larsen, 1956a
2 crupinastrum (Moris) Vis.	28	It	Larsen, 1956a

140 Chartolepis Cass.
  1 glastifolia (L.) Cass.

141 Wagenitzia Dostal
  1 lancifolia (Sieber ex Sprengel) Dostál

142 Cnicus L.

```
  1 benedictus L.                            22      Lu        Fernandes & Queirós,
                                                               1971

143 Carthamus L.
  1 arborescens L.                           24      Hs        Harvey & Knowles,
                                                               1965
  2 tinctorius L.                            24      Europe    Ashri & Knowles,
                                                               1960
  3 dentatus (Forskål) Vahl                  20      Bu        Hanelt, 1963
    (a) dentatus
    (b) ruber (Link) Hanelt
  4 glaucus Bieb.
  5 boissieri Halácsy
  6 leucocaulos Sibth. & Sm.                 20      Gr        Hanelt, 1963
  7 lanatus L.
    (a) lanatus                              44      Lu        Hanelt, 1963
    (b) baeticus (Boiss. & Reuter) Nyman     64      Lu        Khidir & Knowles,
                                                               1970
144 Carduncellus Adanson
  1 monspelliensium All.                     48      Ga        Delay, 1969
  2 mitissimus (L.) DC.                      24      Ga        Delay, 1969
  3 pinnatus (Desf.) DC.
  4 dianius Webb
  5 araneosus Boiss. & Reuter
  6 caeruleus (L.) C. Presl                  24      Lu        Fernandes & Queirós,
                                                               1971


   CICHORIOIDEAE

145 Scolymus L.
  1 maculatus L.                             20      Lu        Fernandes & Queirós,
                                                               1971
  2 hispanicus L.                            20      Lu        Fernandes & Queirós,
                                                               1971
  3 grandiflorus Desf.

146 Cichorium L.
  1 intybus L.                               18      Po        Skalińska et al.,
                                                               1964
  2 endivia L.
    (a) endivia                              18      Lu        Fernandes & Queirós,
                                                               1971
    (b) divaricatum (Schousboe) P.D. Sell
  3 spinosum L.

147 Catananche L.
  1 caerulea L.                              18      Ga        Delay, 1968
  2 lutea L.
    (a) lutea
    (b) carpholepis (Schultz Bip.) Nyman

148 Rothmaleria Font Quer
  1 granatensis (Boiss. ex DC.) Font Quer

149 Hymenonema Cass.
  1 laconicum Boiss. & Heldr.
  2 graecum (L.) DC.

150 Tolpis Adanson
  1 barbata (L.) Gaertner                    18      Lu        Fernandes & Queirós,
                                                               1971
```

```
  2 virgata Bertol.
  3 fruticosa Schrank
  4 azorica (Nutt.) P. Silva
  5 staticifolia (All.) Schultz Bip.        18      He      Favarger, 1949a

151 Arnoseris Gaertner
  1 minima (L.) Schweigger & Koerte         18      Br      Stebbins et al.,
                                                            1953

152 Koelpinia Pallas
  1 linearis Pallas

153 Hyoseris L.
  1 scabra L.                               16      Sa      Martinoli, 1953
  2 radiata L.
    (a) radiata                             16      Sa      Martinoli, 1953
    (b) graeca Halácsy                      16      Sa      Martinoli, 1953
  3 taurina (Pamp.) G. Martinoli            16      Sa      Martinoli, 1953

154 Hedypnois Miller
  1 cretica (L.) Dum.-Courset                8      Lu      Fernandes & Queirós,
                                                            1971
                                           11      Lu      Fernandes & Queirós,
                                                            1971
                                           12      Ga      Fahmy, 1955
                                           13      Lu      Fernandes & Queirós,
                                                            1971
                                           14      Lu      Fernandes & Queirós,
                                                            1971
                                           15      Lu      Fernandes & Queirós,
                                                            1971
                                           16      Ga      Fahmy, 1955
                                           18      Lu      Fernandes & Queirós,
                                                            1971
  2 arenaria (Schousboe) DC.                 6      Lu      Fernandes & Queirós,
                                                            1971

155 Rhagadiolus Scop.
  1 stellatus (L.) Gaertner

156 Aposeris Cass.
  1 foetida (L.) Less.                      16      Cz      Májovský et al.,
                                                            1970

157 Urospermum Scop.
  1 dalechampii (L.) Scop. ex F.W.          14      Bl      Dahlgren et al.,
    Schmidt                                                 1971
  2 picroides (L.) Scop. ex F.W. Schmidt    10      Ga      Delay, 1970

158 Hypochoeris L.
  1 robertia Fiori                           8      Sa      Martinoli, 1953
  2 laevigata (L.) Cesati, Passer. &
    Gibelli
  3 achyrophorus L.                         12      Ga      Kliphuis &
                                                            Wieffering, 1972
  4 cretensis (L.) Bory & Chaub.             6      Gr      Stebbins et al.,
                                                            1953
  5 maculata L.                             10      Ga      Delay, 1968
  6 uniflora Vill.                          10      Po      Skalińska, 1950
  7 tenuiflora (Boiss.) Boiss.
  8 glabra L.                               10      Lu      Fernandes & Queirós,
                                                            1971
```

272

```
  9 radicata L.                            8      Lu      Fernandes & Queirós,
                                                          1971

159 Leontodon L.
    1 pyrenaicus Gouan
      (a) pyrenaicus
      (b) cantabricus (Widder) Finch & P.D. Sell
      (c) helveticus (Merat) Finch & P.D.   12      He      Favarger, 1953
          Sell
    2 croceus Haenke
      (a) croceus                          24      Au      Favarger, 1959a
      (b) rilaensis (Hayek) Finch & P.D.   14      Bu      Kozuharov &
          Sell                                             Kuzmanov, 1964b
    3 montanus Lam.
      (a) montanus                         12      Ga      Delay, 1970
      (b) montaniformis (Widder) Finch &
          P.D. Sell
      (c) pseudotaraxaci (Schur) Finch &   12      Po      Skalińska et al.,
          P.D. Sell                                        1959
    4 autumnalis L.                        12      No      Knaben & Engelskjon,
                                                           1967
                                           24      Scand.  Vaarama, 1948

      (a) autumnalis
      (b) pratensis (Koch) Arcangeli
    5 keretinus F. Nyl.
    6 carpetanus Lange
      (a) carpetanus
      (b) nevadensis (Lange) Finch & P.D.
          Sell
    7 duboisii Sennen ex Widder
    8 microcephalus (Boiss. ex DC.) Boiss.
    9 cichoraceus (Ten.) Sanguinetti
   10 muelleri (Schultz Bip.) Fiori
   11 salzmannii (Schultz Bip.) Ball
   12 hispidulus (Delile) Boiss
   13 repens Schur
   14 schischkinii V. Vassil.
   15 hispidus L.
      (a) hispidus                         14      Cz      Májovský et al.,
                                                           1970
      (b) alpinus (Jacq.) Finch & P.D. Sell
      (c) pseudocrispus (Schultz Bip. ex
          Bischoff) J. Murr
      (d) danubialis (Jacq.) Simonkai      14      Po      Skalińska et al.,
                                                           1964
      (e) opimus (Koch) Finch & P.D. Sell
      (f) hyoseroides (Welw. ex Reichenb.)
          J. Murr.
   16 siculus (Guss.) Finch & P.D. Sell
   17 boryi Boiss. ex DC.                  14      Hs      Küpfer, 1968
   18 hirtus L.                             8      It      Larsen, 1956a
   19 crispus Vill.
      (a) rossianus (Degen & Lengyel) Hayek
      (b) crispus                           8      Al      Strid, 1971
      (c) asperrimus (Willd.) Finch & P.D.
          Sell
      (d) bourgaeanus (Willk.) Finch & P.D.
          Sell
      (e) graecus (Boiss. & Heldr.) Hayek   8      Gr      Phitos & Damboldt,
                                                           1971

   20 hellenicus Phitos
```

21 incanus (L.) Schrank
  (a) incanus   8   Hu   Baksay, 1956
  (b) tenuiflorus (Gaudin) Hegi
22 berinii (Bartl.) Roth
23 tuberosus L.   8   Lu   Fernandes & Queirós,
                                          1971
24 maroccanus (Pers.) Ball   8
25 taraxacoides (Vill.) Mérat
  (a) taraxacoides   8   Ga   Delay, 1969
  (b) longirostris Finch & P.D. Sell   8   Lu   Fernandes & Queirós,
                                          1971
26 filii (Hochst.) Paiva & Ormonde
27 rigens (Aiton) Paiva & Ormonde

160 Picris L.
  1 aculeata Vahl
  2 echioides L.   10   Lu   Fernandes & Queirós,
                                          1971
  3 comosa (Boiss.) B.D. Jackson
  4 algarbiensis Franco
  5 spinifera Franco
  6 hispanica (Willd.) P.D. Sell
  7 hispidissima (Bartl.) Koch
  8 scaberrima Guss.
  9 hieracioides L.   10   Lu   Fernandes & Queirós,
                                          1971
  (a) longifolia (Boiss. & Reuter) P.D.
      Sell
  (b) hieracioides
  (c) spinulosa (Bertol. ex Guss.)
      Arcangeli
  (d) villarsii (Jordan) Nyman
  (e) grandiflora (Ten.) Arcangeli
  10 pauciflora Willd.
  11 sprengerana (L.) Poiret
  12 willkommii (Schultz Bip.) Nyman

161 Scorzonera L.
  1 cana (C.A. Meyer) O. Hoffm.   14   Hu   Pólya, 1948
  2 laciniata L.   14   Lu   Fernandes & Queirós,
                                          1971
  3 mollis Bieb.
  4 idaea (Gand.) Lipsch.
  5 undulata Vahl
  6 purpurea L.   14   Po   Skalińska et al.,
                                          1971
  (a) purpurea
  (b) rosea (Waldst. & Kit.) Nyman   14   It   Favarger, 1965b
  (c) peristerica Form.
  7 graminifolia L.
  8 austriaca Willd.   14   Cz   Májovský et al.,
                                          1970
  (a) austriaca
  (b) crispa (Bieb.) Nyman
  (c) bupleurifolia (Pouzolz) Bonnier
  9 humilis L.   14   Lu   Fernandes & Queirós,
                                          1971
                             15   Lu   Fernandes & Queirós,
                                          1971
  10 parviflora Jacq.   14   Cz   Májovský et al.,
                                          1970

11 baetica (Boiss.) Boiss.
12 transtagana Coutinhò
13 crocifolia Sibth. & Sm.   14   Gr   Damboldt, 1968
14 aristata Ramond ex DC.   14   It   Favarger & Huynh,
                                                           1964
15 hispanica L.   14   Po   Skalińska et al.,
                                                           1964
16 crispatula (Boiss.) Boiss.
17 brevicaulis Vahl
18 scyria M. Gustafsson & Snogerup   14   Gr   Gustaffson &
                                                           Snogerup, 1972
19 fistulosa Brot.
20 pusilla Pallas
21 villosa Scop.
   (a) villosa
   (b) columnae (Guss.) Nyman
   (c) callosa (Moris) Chater
22 ensifolia Bieb.
23 hirsuta L.   12   Ga   Delay, 1968
                                               14   Ga   Delay, 1968
24 doria Degen & Bald.
25 cretica Willd.
26 albicans Cosson
27 lanata (L.) Hoffm.
28 tuberosa Pallas

162 Tragopogon L.
  1 ruber S.G. Gmelin
  2 marginifolius Pawl.
  3 cretaceus S. Nikitin
  4 porrifolius L.
    (a) porrifolius   12   Br   Edmonds et al., 1974
    (b) australis (Jordan) Nyman   12   Gr   Damboldt, 1968
    (c) cupani (Guss. ex DC.) I.B.K.
        Richardson
  5 pterodes Pančić   12   Bu   Kozuharov &
                                                           Kuzmanov, 1968
  6 balcanicus Velen.   12   Bu   Kozuharov &
                                                           Kuzmanov, 1968
  7 longirostris Bischoff ex Schultz Bip.
  8 crocifolius L.
    (a) crocifolius
    (b) samaritani (Heldr. & Sart. ex Boiss.)
        I.B.K. Richardson
  9 kindingeri Adamović
 10 dubius Scop.   12   Hu   Pólya, 1950
 11 pratensis L.
    (a) pratensis   12   Bu   Kozuharov &
                                                           Kuzmanov, 1968
    (b) minor (Miller) Wahlenb.   12   Ga   Delay, 1969
    (c) orientalis (L.) Čelak.   12   Po   Skalińska et al.,
                                                           1961
 12 lassithicus Rech. fil.
 13 hayekii (Soó) I.B.K. Richardson
 14 tommasinii Schultz Bip.
 15 brevirostris DC.
    (a) brevirostris
    (b) volgensis (S. Nikitin) C. Regel
    (c) podolicus (DC.) C. Regel
    (d) bjelorussicus (Artemczuk) C. Regel

(e) longifolius (Heldr. & Sart. ex Boiss.)
    I.B.K. Richardson
16 dasyrhynchus Artemczuk
  (a) dasyrhynchus
  (b) daghestanicus Artemczuk
17 elatior Steven
18 floccosus Waldst. & Kit.

(a) floccosus	12	Cz	Májovský et al., 1970

(b) heterospermus (Schweigger) C. Regel
19 ruthenicus Besser ex Krasch. & S.
   Nikitin
  (a) ruthenicus
  (b) donetzicus (Artemczuk) I.B.K.
     Richardson

20 hybridus L.	14	Lu	Queirós, 1973

163 Reichardia Roth

1 tingitana (L.) Roth	16	Bl	Dahlgren et al., 1971
2 gaditana (Willk.) Coutinho	16	Lu	Fernandes & Queirós, 1971
3 picroides (L.) Roth	14	It	Larsen, 1955b
4 intermedia (Schultz Bip.) Coutinho	14	Lu	Fernandes & Queirós, 1971

164 Launaea Cass.
  1 resedifolia (L.) O. Kuntze
  2 nudicaulis (L.) Hooker fil.

3 pumila (Cav.) O. Kuntze	16	Hs	Lorenzo-Andreu & García, 1950
4 cervicornis (Boiss.) Font Quer & Rothm.	18	Bl	Dahlgren et al., 1971
5 lanifera Pau	16	Hs	Casas, 1973

6 arborescens (Batt.) Murb.

165 Aetheorhiza Cass.

1 bulbosa (L.) Cass.	18	Lu	Fernandes & Queirós, 1971

  (a) bulbosa
  (b) microcephala Rech. fil.
  (c) willkommii (Burnat & W. Barbey)
    Rech. fil.

166 Sonchus L.
  1 asper (L.) Hill

(a) asper	18	Ho	Gadella & Kliphuis, 1966

  (b) glaucescens (Jordan) Ball

2 tenerrimus L.	14	It	Larsen, 1956a
3 oleraceus L.	32	Lu	Rodriguez, 1953
4 maritimus L.	18	Ga	Delay, 1968

  (a) maritimus
  (b) aquatilis (Pourret) Nyman

5 palustris L.	18	Da	Henin, 1960

6 crassifolius Pourret ex Willd.

7 arvensis L.	36	Fe	V. Sorsa, 1962
	54	Ho	Gadella & Kliphuis, 1968

  (a) arvensis
  (b) uliginosus (Bieb.) Nyman
8 pustulatus Willk.

167 Cephalorrhynchus Boiss.
1 tuberosuş (Steven) Schchian

168 Steptorhamphus Bunge
1 tuberosus (Jacq.) Grossh.

169 Lactuca L.
1 viminea (L.) J. & C. Presl
  (a) alpestris (Gand.) Ferakova
  (b) viminea   18   Cz   Feráková, 1968
  (c) chondrilliflora (Boreau) Bonnier
  (d) ramosissima (All.) Bonnier
2 acanthifolia (Willd.) Boiss.   18   Cr   Phitos & Kamari,
                                                              1974
3 longidentata Moris ex DC.
4 tatarica (L.) C.A. Meyer   18   Rm   Tarnavschi, 1935
5 sibirica (L.) Maxim.
6 quercina L.
  (a) quercina   18   Cz   Feráková, 1968b
  (b) wilhelmsiana (Fischer & C.A. Meyer
      ex DC.) Feráková
7 watsoniana Trelease
8 aurea (Schultz Bip. ex Pančič)   16   Bu   Babcock et al.,
  Stebbins   1937
9 serriola L.   18   Hu   Pólya, 1950
10 sativa L.   18   Su   Lindqvist, 1960
11 saligna L.   18   Au   Lindqvist, 1960
12 altaica Fischer & C.A. Meyer
13 virosa L.   18   Lu   Fernandes & Queirós,
                                                              1971
14 livida Boiss. & Reuter
15 perennis L.   18   Ge   Babcock et al.,
                                                              1937
16 tenerrima Pourret   16   Bl   Babcock et al.,
                                                              1937
17 graeca Boiss.

170 Cicerbita Wallr.
1 alpina (L.) Wallr.   18   Bu No   Stebbins et al.,
                                                                1953
2 macrophylla (Willd.) Wallr.
3 plumieri (L.) Kirschleger
4 pancicii (Vis.) Beauverd

171 Prenanthes L.
1 purpurea L.   18   Bu   Kozuharov &
                                                              Kuzmanov, 1967

172 Mycelis Cass.
1 muralis (L.) Dumort   18   Su   Löve & Löve, 1942

173 Taraxacum Weber
1 glaciale Huet ex Hand.-Mazz.
2 bessarabicum (Hornem.) Hand.-Mazz.   16   Hu   Gustaffson, 1932
3 serotinum (Waldst. & Kit.) Poiret   16   Cz   Májovský et al.,
                                                              1970
4 phymatocarpum J. Vahl
5 glabrum DC.
6 pacheri Schultz Bip.   32   Au   Fürnkranz, 1965
7 schroeteranum Hand.-Mazz.   24   He   Hou-Liu, 1963
8 bithynicum DC.

277

minimum (Briganti ex Guss.) N. Terracc.	16	Ju	Richards, 1969
9 obovatum (Willd.) DC.	32	Hs	Bjorkqvist et al., 1969
10 glaucanthum (Ledeb.) DC.			
11 ceratophorum (Ledeb.) DC.			
tornense T.C.E. Fries	32	Fe Su	Gustaffson, 1935
12 spectabile Dahlst.			
eximium Dahlst.	40	No	Fagerlind, 1968
faeroense (Dahlst.) Dahlst.	40	Br	Richards, 1969
spectabile Dahlst.	40	Br	Richards, 1969
13 praestans H. Lindb. fil.			
euryphyllum (Dahlst.) M.P. Christiansen	32	Br	Richards, 1969
lainzii Van Soest	24	Br	Richards, 1969
landmarkii Dahlst.	32	No	Fagerlind, 1968
maculigerum H. Lindb. fil.	32	Br	Richards, 1969
naevosiforme Dahlst.	32	Br	Richards, 1972
naevosum Dahlst.	32	No	Fagerlind, 1968
praestans H. Lindb. fil.	32	Br	Richards, 1969
stoctophyllum Dahlst.	32	No	Fagerlind, 1968
14 unguilobum Dahlst.			
fulvicarpum Dahlst.	32	Br	Richards, 1969
unguilobum Dahlst.	32	Br	Richards, 1969
15 croceum Dahlst.			
craspedotum Dahlst.	32	Br	Richards, 1972
croceum Dahlst.	32	Br	Richards, 1972
cymbifolium H. Lindb. fil. ex Dahlst.	32	Sb	Engelskjon, 1968
pycnostictum M.P. Christiansen	32	Br	Richards, 1969
repletum (Dahlst.) Dahlst.	40	No	Fagerlind, 1968
16 adamii Claire	24	Br	Richards, 1969
litorale Raunk.	24	Da	Richards, 1969
nordstedtii Dahlst.	48	Br	Richards, 1969
17 palustre (Lyons) Symons			
austrinum G. Hagl.	24	Cz	Malecka, 1970
	32	Br	Richards, 1969
balticiforme Dahlst.	24	Po	Malɔcka, 1970
balticum Dahlst.	24		
	31	Su	Richards, 1972
	32	Po	Malecka, 1970
crocodes Dahlst.	40	Su	Richards, 1969
illyricum Dahlst. ex Van Soest	29	Hu	Richards, 1972
palustre (Lyons) Symons	40	Br	Richards, ined.
turfosum (Schultz Bip.) Van Soest	24	Po	Malecka, 1970
18 apenninum (Ten.) Ten.			
helveticum Van Soest	32	Po	Richards, 1972
venustum Dahlst.	32	Au	Fürnkranz, 1965
19 nigricans (Kit.) Reichenb.			
aestivum Van Soest	32	Po	Richards, 1972
nigricans (Kit.) Reichenb.	24	Au Po	Richards, 1972
	32	Po	Malecka, 1962
rhaeticum Van Soest	24	He	Hou-Liu, 1963
20 fontanum Hand.-Mazz.			
fontaneosquameum Van Soest	25	Au	Richards, 1972
pohlii Van Soest	32	Au	Richards, 1972
21 cucullatum Dahlst.			
concullatum A.J. Richards	24	Au	Richards, 1972
22 dissectum (Ledeb.) Ledeb.			
23 obliquum (Fries) Dahlst.	24	Ho	Hou-Liu, 1963
24 erythrospermum Andrz. ex Besser			
austriaeum Van Soest	16	Cz	Richards, 1970
	24	Cz	Richards, 1970

brachyglossum (Dahlst.) Dahlst.	16	Br	Richards, 1969
	24	Br	Richards, 1969
disseminatum G. Hagl.	24	Ge	Richards, 1969
dunense Van Soest	24	Br	Richards, 1969
isophyllum G. Hagl.	24	Su	Richards, 1969
lacistophyllum (Dahlst.) Raunk.	24	Br	Richards, 1969
	25	Su	Richards, 1972
laetum (Dahlst.) Dahlst.	24	Su	Richards, 1972
polyschistum Dahlst.	24	Su	Richards, ined.
proximum (Dahlst.) Dahlst.	24	Br Cz	Richards, 1969
rubicundum (Dahlst.) Dahlst.	24	Br	Richards, 1969
scanicum Dahlst.	25	Po	Malecka, 1967
silesiacum Dahlst. ex G. Hagl.	24	Cz	Richards, 1970
taeniatum G. Hagl. ex Holmgren	24	Ho	Hou-Liu, 1963
tenuilobum (Dahlst.) Dahlst.	24	Rs(B)	Gustaffson, 1935
	25	Po	Malecka, 1969
25 simile Raunk.			
dissimile Dahlst.	24	Su	Gustaffson, 1935
placidum A.J. Richards	24	Ga	Richards, 1969
proximiforme Van Soest	24	Br	Richards, 1969
pseudolacistophyllum Van Soest	24	Br	Richards, 1969
purpureomarginatum Van Soest	24	Cz	Richards, 1969
simile Raunk.	32	Br	Richards, 1969
tortilobum Florström	24	Ho	Hou-Liu, 1963
26 fulvum Raunk.			
fulviforme Dahlst.	32	Br	Richards, 1969
fulvum Raunk.	32	Da	Gustaffson, 1935
oxoniense Dahlst.	32	Br	Richards, 1969
27 gasparrinii Tineo ex Lojac.			
28 hoppeanum Griseb.			
amborum G. Hagl.	24	Gr	Gustaffson, 1935
aquilonare Hand.-Mazz.	24	Au	Furnkranz, 1965
capricum Van Soest	24	Ga	Richards, 1969
pieninicum Pawl.	16	Po	Malecka, 1961
29 crassipes H. Lindb.			
cochleatum Dahlst. & H. Lindb. fil.	24	Fe	Richards, 1969
30 officinale Weber			
aequilobum Dahlst.	24	Su	Richards, 1969
alatum H. Lindb. fil.	24	Br	Richards, 1969
cordatum Palmgren	24	Su	Gustaffson, 1932
cyanolepis Dahlst.	24	Su	Richards, 1969
dahlstedtii H. Lindb. fil.	24	Br	Richards, 1969
duplidens H. Lindb. fil.	24	Su	Gustaffson, 1935
	26	Br	Richards, ined.
ekmanii Dahlst.	24	Su	Gustaffson, 1935
fasciatum Dahlst.	24	Su	Gustaffson, 1935
haematicum G. Hagl.	24	Da	Richards, 1969
haematopus H. Lindb. fil.	24	Da	Gustaffson, 1935
hamatiforme Dahlst.	24	Br	Richards, 1969
hamatum Raunk.	24	Br Ga	Richards, 1969
insigne E.L. Ekman ex Wiinst. & K.	25	Br	Richards, 1972
involucratum Dahlst.	24	Su	Gustaffson, 1932
laciniosifrons Wiinst.	19	Da	Sörensen &
	20	Da	Gudjonsson,
	21	Da	1946
	22	Da	
	23	Da	Sörensen &
	24	Da	Gudjónsson,
	48	Da	1946
laeticolor Dahlst.	24	Su	Gustaffson, 1935
longisquameum H. Lindb. fil.	24	Su	Gustaffson, 1932

279

marklundii Palmgren.	24	Br	Richards, 1969
melanthoides Dahlst.	24	Su	Gustaffson, 1935
mimulum Dahlst.	24	Rs(N)	Gustaffson, 1932
oblongatum Dahlst.	24	Br	Richards, ined.
parvuliceps H. Lindb. fil.	24	Rs(N)	Gustaffson, 1932
polyodon Dahlst.	21	Da	Sørensen &
	22	Da	Gudjónsson,
	23	Da	1946
	24	Br Da	Richards, 1969
	44 45	Da	Sørensen &
	46 47	Da	Gudjónsson,
	48	Da	1946
raunkiaeri Wiinst.	24	Br	Richards, 1969
retroflexum H. Lindb. fil.	24	Su	Gustaffson, 1935
sellandii Dahlst.	24	Br	Richards, ined.
speciosum Raunk.	24	Da	Richards, 1969
stenoschistum Dahlst.	24	Su	Gustaffson, 1932
subcyanolepis M.P. Christiansen	24	Br	Richards, 1969
sublaeticolor Dahlst.	24	Su	Gustaffson, 1935
tenebricans (Dahlst.) Dahlst.	24	No	Fagerlind, 1968

174 Chondrilla L.

1 juncea L.	14 + 1B	Ga	Delay, 1970
	30	Ga	Delay, 1947

2 ramosissima Sibth. & Sm.
3 pauciflora Ledeb.
4 chondrilloides (Ard.) Karsten

175 Calycocorsus F.W. Schmidt

1 stipitatus (Jacq.) Rauschert	10	Ga	Delay, 1970

176 Heteracia Fischer & C.A. Meyer
   1 szovitsii Fischer & C.A. Meyer

177 Lapsana L.
   1 communis L.

(a) communis	14	Br	Edmonds et al., 1974
	16	Fe	V. Sorsa, 1962
(b) adenophora (Boiss.) Rech. fil.			
(c) intermedia (Bieb.) Hayek	14		
(d) alpina (Boiss. & Balansa) P.D. Sell			

178 Crepis L.

1 sibirica L.	10	?Europe	Babcock, 1947
2 geracioides Hausskn.	12	?Europe	Babcock, 1947
3 viscidula Froelich	12	Bu	Kuzmanov et al., 1981
4 paludosa (L.) Moench	12	Cz	Májovský et al., 1970
5 pygmaea L.			
(a) pygmaea	12	Ga	Delay, 1970
(b) anachoretica Babcock			
6 terglouensis (Hacq.) A. Kerner	12	?Europe	Babcock, 1947
7 jacquinii Tausch			
(a) jacquinii			
(b) kerneri (Rech. fil.) Merxm.			
8 rhaetica Hegetschw.			
9 aurea (L.) Cass.			
(a) aurea	10	Ga	Delay, 1970

(b) glabrescens (Caruel) Arcangeli	10	?Europe	Babcock, 1947
10 chrysantha (Ledeb.) Turcz.	8	?Europe	Babcock, 1947
11 lampsanoides (Gouan) Tausch	12	Ga	Delay, 1969
12 smyrnaea DC. ex Froelich			
13 mollis (Jacq.) Ascherson	12	Ga	Delay, 1970
14 fraasii Schultz Bip.			
(a) fraasii	12	?Europe	Babcock, 1947
(b) mungieri (Boiss. & Heldr.) P.D.			
Sell	12	?Europe	Babcock, 1947
15 bocconi P.D. Sell	10	Ga	Delay, 1970
16 conyzifolia (Gouan) A. Kerner	8	Ga	Delay, 1970
17 pyrenaica (L.) W. Greuter	8	Ga	Delay, 1970
18 alpestris (Jacq.) Tausch	8	?Europe	Babcock, 1947
19 albida Vill.			
(a) albida	10	Ga	Delay, 1970
(b) asturica (Lacaita & Pau) Babcock	10	?Europe	Babcock, 1947
(c) grosii (Pau) Babcock			
(d) scorzoneroides (Rouy) Babcock			
(e) macrocephala (Willk.) Babcock	10	?Europe	Babcock, 1947
(f) longicaulis Babcock			
20 tingitana Ball	10	?Europe	Babcock, 1947
21 leontodontoides All.	10	?Europe	Babcock, 1947
22 biennis L.	31 36	Br	Edmonds et al., 1974
	38 39 40	Br	Edmonds et al., 1974
23 pannonica (Jacq.) C. Koch	8	Cz	Májovský et al., 1970
24 lacera Ten.	18	?Europe	Babcock, 1947
25 bertiscea Jav.			
26 chondrilloides Jacq.	8	?Europe	Babcock, 1947
27 auriculifolia Sieber ex Sprengel	10		
28 baldaccii Halácsy	10	?Europe	Babcock, 1947
29 turcica Degen & Bald.			
30 pantocsekii (Vis.) Latzel			
31 triasii (Camb.) Nyman	8	?Europe	Babcock, 1947
32 albanica (Jav.) Babcock			
33 macedonica Kitanov			
34 oporinoides Boiss. ex Froelich	8		
35 sibthorpiana Boiss. & Heldr.			
36 incana Sibth. & Sm.	16	?Europe	Babcock, 1947
37 taygetica Babcock	c.40	?Europe	Babcock, 1947
38 guioliana Babcock			
39 crocifolia Boiss. & Heldr.			
40 athoa Boiss.			
41 schachtii Babcock	10	?Europe	Babcock, 1947
42 bithynica Boiss.	10	Bu	Kuzmanov & Kozuharov, 1970
43 praemorsa (L.) Tausch			
(a) praemorsa	8	Ga	Zickler & Lambert, 1967
(b) corymbosa (Gaudin) P.D. Sell			
(c) dinarica (G. Beck) P.D. Sell			
44 tectorum L.	8	Fe	Sorsa, 1963
(a) tectorum			
(b) nigrescens (Pohle) P.D. Sell			
(c) pumila (Liljeblad) Sterner			
45 purpurea (Willd.) Bieb.			
46 reuterana Boiss.			
47 pulchra L.	8	Cz	Májovský et al., 1970

```
48 stojanovii Georgiev                          8
49 alpina L.                                   10      Rs(K)    Babcock, 1947
50 rubra L.                                    10      ?Europe  Babcock, 1947
51 foetida L.
   (a) foetida                                 10      Lu       Fernandes & Queirós,
                                                                1971
   (b) rhoeadifolia (Bieb.) Čelak.             10      Cz       Májovský et al.,
                                                                1970
   (c) commutata (Sprengel) Babcock            10      Lu       Fernandes & Queirós,
                                                                1971
52 multicaulis Ledeb.                          10      ?Europe  Babcock, 1947
53 sancta (L.) Babcock                         10      Bu       Kuzmanov & Juakova,
                                                                1977
54 dioscoridis L.                               8      ?Europe  Babcock, 1947
55 multiflora Sibth. & Sm.                      8      ?Europe  Babcock, 1947
56 zacintha (L.) Babcock                        6      ?Europe  Babcock, 1947
57 pusilla (Sommier) Merxm.                    10               Merxmüller, 1968
58 nicaeensis Balbis                            8      Ga       Delay, 1970
59 foliosa Babcock
60 capillaris (L.) Wallr.                       6      Ge       Gross, 1966
61 micrantha Czerep.                            8      ?Europe  Babcock, 1947
62 neglecta L.
   (a) neglecta                                 8      ?Europe  Babcock, 1947
   (b) corymbosa (Ten.) Nyman                   8      ?Europe  Babcock, 1947
   (c) fuliginosa (Sibth. & Sm.) Vierh.         6      ?Europe  Babcock, 1947
   (d) cretica (Boiss.) Vierh.                  8      ?Europe  Babcock, 1947
63 suffreniana (DC.) Lloyd
   (a) suffreniana                              8      ?Europe  Babcock, 1947
   (b) apula (Fiori) P.D. Sell                  8
64 spathulata Guss.
65 bourgeaui Babcock ex Maire                   8      ?Europe  Babcock, 1947
66 vesicaria L.
   (a) vesicaria                                8      It       Marchi, 1971
                                               16      Si       Pavone et al.,
   (b) hyemalis (Biv.) Babcock                  8      Si       1981
   (c) haenseleri (Boiss. ex DC.) P.D..Sell     8      Br       Edmonds et al.,
                                               16               1974
   (d) congenita Babcock
67 tybakiensis Vierh.
68 setosa Haller fil.                           6      Br       Edmonds & Walters,
                                                                ined.
                                                8      Al       Strid, 1971
69 bellidifolia Loisel.                         8      ?Europe  Babcock, 1947
70 bursifolia L.                                8      Ga       Van Loon et al.,
                                                                1971

179 Hispidella Barnades ex Lam.
  1 hispanica Barnades ex Lam.                 18      Lu       Fernandes & Queirós,
                                                                1971

180 Andryala L.
  1 integrifolia L.                            18      Ga       Kliphuis &
                                                                Wieffering, 1972
  2 laxiflora DC.                              18      Lu       Fernandes & Queirós,
                                                                1971
  3 ragusina L.
  4 agardhii Haenseler ex DC.
  5 levitomentosa (E.I. Nyárády) P.D. Sell

181 Hieracium L.
  1 castellanum Boiss. & Reuter               18      Hs       Merxmüller, 1975
  2 hoppeanum Schultes                         18      He Ju    Gadella, 1972
                                               45
```

(a) hoppeanum
(b) pilisquamum Naegeli & Peter
(c) testimoniale Naegeli ex Peter
(d) troicum Zahn
(e) cilicicum Naegeli & Peter
3 x viridifolium Peter
4 x hypeuryum Peter
5 x ruprechtii Boiss.
6 x tephrocephalum Vuk.

7 peleteranum Mérat	18	Br	Richards, 1972
	27 36		
	45		
(a) peleteranum	18	Br	Finch, ined.
(b) subpeleteranum Naegeli & Peter	18	Ho	Gadella & Kliphuis, 1968
(c) sabulosorum Dahlst.	18	Su	G. & B. Turesson, 1960
(d) tenuiscapum (Pugsley) P.D. Sell			
(e) ligericum Zahn			
8 x longisquamum Peter	27	Su	G. & B. Turesson, 1960
9 argyrocomum (Fries) Zahn	18	Hs	Merxmüller, 1975
10 pilosella L.	18	It	Gadella & Kliphuis, 1970
	36	Ho	Gadella & Kliphuis, 1966
	45	Ga	Gadella, 1972
	54	It	Scannerini, 1971
	63	Su	G. & B. Turesson,
(a) micradenium Naegeli & Peter	18		1960
	36 45		
(b) euronotum Naegeli & Peter	45	Br	Finch, ined.
(c) pilosella	36	Su	G. & B. Turesson,
(d) trichosoma Peter	36		1960
	45		
	54		
	63		
(e) tricholepium Naegeli & Peter	36	Br	Finch, ined.
(f) melanops Peter	36		
	45		
(g) trichoscapum Naegeli & Peter	45	Su	B. Turesson, 1972
(h) velutinum Fries	45	Ga	Gadella, 1972
	54		

11 x florentoides Arvet-Touvet
12 x brachiatum Bertol. ex Lam.
13 pseudopilosella Ten.

(a) pseudopilosella	18	Hs It	Merxmüller, 1975
(b) banaticola E.I. Nyárády & Zahn			
14 flagellare Willd.			
(a) flagellare	36	Europe	Gustaffson, 1933
	45	Po	Skalińska, 1967
(b) bicapitatum (P.D. Sell & C. West) P.D. Sell	54	Br	Finch, ined.
15 x flagellariforme G. Schneider			
16 lactucella Wallr.	18	Po	Skalińska, 1967
	27	Co	Contandriopoulos, 1962
	36		

(a) nanum (Scheele) P.D. Sell
(b) lactucella
(c) magnauricula (Naegeli & Peter) P.D. Sell

17 x auriculiforme Fries 18
36

18 x schultesii F.W. Schultz
19 x paragogum Naegeli & Peter
20 x sulphureum Doll
21 koernickeanum (Naegeli & Peter) Zahn

22 vahlii Froelich	18	Hs	Merxmüller, 1975
23 glaciale Reyn.	18	Ga	Favarger, 1969

24 x niphostribes Peter
25 sphaerocephalum Froelich
26 breviscapum DC.

27 alpicola Schleicher ex Gaudin	36	He	Favarger, 1959

28 piloselloides Vill.

(a) piloselloides	36	Europe	Finch, ined.

(b) megalomastix (Naegeli & Peter)
P.D. Sell

pavichii	18

29 praealtum Vill. ex Gochnat

(a) praealtum	36
	45

(b) anadenium (Naegeli & Peter) P.D.
Sell

(c) bauhinii (Besser) Petunnikov	45	Cz	Uhríková, 1970

(d) thaumasium (Peter) P.D. Sell

30 cymosum L.	36		
	54 63	Ca	Gadella & Kliphuis, 1970

(a) sabinum (Sebastiani & Mauri)
Naegeli & Peter
(b) cymosum

(c) cymigerum (Reichenb.) Peter	36	Su	G. & B. Turesson, 1963

31 x fallax Willd.
32 x sciadophorum Naegeli & Peter
33 x anchusoides (Arvet-Touvet) Arvet-
Touvet
34 x spurium Chaix ex Froelich
35 x fallacinum F.W. Schultz
36 x zizianum Tausch
37 x densiflorum Tausch

38 caespitosum Dumort.	18	Po	Skalińska & Kubien, 1972
	27	Po	Skalińska, 1967
	36	Po	Skalińska & Kubien, 1972
	45	Po	Skalińska, 1967
(a) caespitosum	45	Su	G. & B. Turesson, 1963

(b) colliniforme (Peter) P.D. Sell

(c) brevipilum (Naegeli & Peter) P.D. Sell	18	Po	Skalińska et al., 1972

39 x ambiguum Ehrh.	36		
	45		
40 x dubium L.	36	Su	G. & B. Turesson,
	45		1963
41 x macranthelum Naegeli & Peter	45	Su	G. & B. Turesson, 1963
42 x floribundum Wimmer & Grab.	27	Po	Skalińska, 1967
islandicum (Lange) Dahlst.	27	Is	Finch, ined.

43 x piloselliflorum Naegeli & Peter
44 x duplex Peter
45 x melinomelas Peter
46 x arvicola Naegeli & Peter
47 x polymastix Peter

48 aurantiacum L.	18	Po	Skalińska, 1970
	27	Po	Skalińska, 1967
	36	Po	Skalińska, 1967
	45	Po	Skalińska, 1967
	54	Po	Skalińska et al., 1968
	63	Po	Skalińska, 1967
	72		
(a) aurantiacum	36	Br	Finch, ined.

(b) carpathicola Naegeli & Peter
49 x fuscatrum Naegeli & Peter
50 x guthnickianum Hegetschw.
51 x plaicense Woloszczak
52 x biflorum Arvet-Touvet

53 x fuscum Vill.	54	Po	Skalińska, 1967
54 x peteranum Kaeser			
55 x stoloniflorum Waldst. & Kit.	45 46	Europe	Finch, ined.

56 x calomastix Peter

57 echioides Lumn.	36	Bot. Gard.	Gentscheff, 1937
(a) echioides	36	Cz	Májovský et al., 1970

(b) procerum (Fries) P.D. Sell
58 x macrotrichum Boiss.
59 x bifurcum Bieb.
60 x heterodoxum (Tausch) Naegeli & Peter
61 x euchaetium Naegeli & Peter
62 x auriculoides A.F. Lang
63 x echiogenes (Naegeli & Peter) Juxip
64 verruculatum Link

65 murorum L.	27 36	Europe	Finch, ined.
grandidens Dahlst.	27	Europe	Finch, ined.
66 glaucinum Jordan	27		
	36		
gougetianum Gren. & Godron	27	Hb	Mills & Stace, 1974
scotostictum Hyl.	27	Europe	Finch, ined.
67 bifidum Kit.	27	Cz	Májovský et al., 1970
	36		

68 subcaesiiforme (Zahn) Zahn
69 fuscocinereum Norrlin
70 incisum Hoppe

71 vulgatum Fries	27	Br	Morton, 1974
72 benzianum J. Murr & Zahn			
73 maculatum Sm.	27	Europe	Finch, ined.
74 caesium (Fries) Fries	27 36	Europe	Finch, ined.
caesiomurorum Lindeb.	36	Europe	Finch, ined.
75 hypastrum Zahn			
76 ramosum Waldst. & Kit. ex Willd.	27	It	Battaglia, 1947
77 argillaceum Jordan	27	Europe	Finch, ined.
acuminatum Jordan	27	Europe	Finch, ined.
78 rotundatum Kit. ex Schultes			
79 diaphanum Fries	27	Br	Morton, 1974

	smolandicum group	27		Gustaffson, 1933
80	schmidtii Tausch	27 36	Europe	Finch, ined.
	schmidtii Tausch	27	Europe	Finch, ined.
81	hypochoeroides Gibson	27 36		
82	sommerfeltii Lindb.			
83	aymericianum Arvet-Touvet			
84	bourgaei Boiss.			
85	stelligerum Froelich			
86	onosmoides Fries	27	Br	Morton, 1974
	subrude (Arvet-Touvet) Arvet-Touvet	27	Europe	Finch, ined.
87	saxifragum Fries			
88	scoticum F.J. Hanb.			
89	caledonicum F.J. Hanb.	27	Br	Mills & Stace, 1974
90	laniferum Cav.	18	Hs	Merxmüller, 1975
91	elisaeanum Arvet-Touvet ex Willk.	18		Merxmüller, ined.
92	candidum Scheele			
93	phlomoides Froelich			
94	rupicaprinum Arvet-Touvet & Gaut.			
95	eriopogon Arvet-Touvet & Gaut.			
96	lawsonii Vill.			
97	briziflorum Arvet-Touvet			
98	subsericeum Arvet-Touvet			
99	cordifolium Lapeyr.			
100	sonchoides Arvet-Touvet			
101	purpurascens Scheele ex Willk.			
102	guadarramense Arvet-Touvet			
103	aragonense Scheele			
104	loscosianum Scheele			
105	cerinthoides L.			
106	alatum Lapeyr.	27 36	Br	Morton, 1974
	anglicum Fries	36	Br	Mills & Stace, 1974
	vogesiacum (Kirschléger) Fries	36	Bot. Gard.	Gentscheff, 1937
107	lamprophyllum Scheele			
108	longifolium Schleicher ex Froelich			
109	villosum Jacq.	27 36	Au He	Polatschek, 1966 Larsen, 1954
110	pilosum Schleicher ex Froelich	27 36		
111	scorzonerifolium Vill.	36		Christoff & Popoff, 1933
112	leucophaeum Gren. & Godron			
113	ctenodon Naegeli & Peter			
114	dentatum Hoppe			
115	chondrillifolium Fries			
116	cryptadenum Arvet-Touvet			
117	valdepilosum Vill.			
118	wilczekianum Arvet-Touvet			
119	chlorifolium Arvet-Touvet			
120	piliferum Hoppe			
121	dasytrichum Arvet-Touvet			
122	aphyllum Naegeli & Peter			
123	cochlearioides Zahn			
124	armerioides Arvet-Touvet			
125	mixtum Froelich			
126	mixtiforme Arvet-Touvet			
127	pictum Schleicher ex Pers.			

128 farinulentiforme Zahn			
129 pulchellum Gren.			
130 caesioides Arvet-Touvet			
131 pseudoprasinops Zahn			
132 cephalotes Arvet-Touvet			
133 leiopogon Gren. ex Verlot			
134 rupestre All.			
135 lanatum Vill.			
136 erioleucum Zahn			
137 jordanii Arvet-Touvet			
138 pellitum Fries			
139 lansicum Arvet-Touvet			
140 verbascifolium Vill.			
141 chaboissaei Arvet-Touvet			
142 pannosum Boiss.	36		Christoff & Popoff, 1933
143 gymnocephalum Griseb. ex Pant.			
144 pichleri A. Kerner			
145 gaudryi Boiss.			
146 waldsteinii Tausch	27 36		
147 dolopicum Freyn & Sint.			
148 guentheri-beckii Zahn			
149 scheppigianum Freyn			
150 mirificissimum Rohlena & Zahn			
151 lazistanum Arvet-Touvet			
152 calophyllum Uechtr.			
153 pilosissimum Friv.			
154 heldreichii Boiss.			
155 sericophyllum Nejc. & Zahn			
156 jankae Uechtr.			
157 sartorianum Boiss. & Heldr.			
marmoreum Pancic & Vis.	27	Bot. Gard.	Gentscheff, 1937
158 alpinum L.	27	Po	Skalińska et al., 1959
159 nigrescens Willd.	36	Br	Mills & Stace, 1974
calenduliflorum Backh.	36	Br	Morton, 1974
hanburyi Pugsley	36	Br	Mills & Stace, 1974
160 pietroszense Degen & Zahn			
161 fritzei F.W. Schultz			
162 arolae J. Murr			
163 senescens Backh.			
164 atratum Fries	27 36		
165 liptoviense Borbas			
166 krasanii Woloszczak			
167 rohacsense Kit. ex Kanitz			
168 bocconei Griseb.			
169 vollmannii Zahn			
lividorubens (Almq.) Elfstr.	27	Da	Jörgensen et al., 1958
paltinae Jav. & Zahn	36		Christoff & Popoff, 1933
170 sudeticum Sternb.			
171 nigritum Uechtr.	27	Da	Rosenberg, 1926
172 chlorocephalum Wimmer			
173 gombense Lagger & Christener			
174 humile Jacq.	27	Da	Rosenberg, 1926
175 cottetii Godet ex Gremli			

176 kerneri Ausserdorfer ex Zahn
177 valoddae (Zahn) Zahn
178 amplexicaule L.   27   It   Gadella & Kliphuis, 1970

176 kerneri Ausserdorfer ex Zahn			
177 valoddae (Zahn) Zahn			
178 amplexicaule L.	27	It	Gadella & Kliphuis, 1970
	36	Ho	Gadella & Kliphuis, 1968
pulmonarioides Vill.	36	Br	Mills & Stace, 1974
speluncarum Arvet-Touvet	36	Lu	Fernandes & Queirós, 1971
179 chamaepicris Arvet-Touvet			
180 pseudocerinthe (Gaudin) Koch			
181 rupicola Jordan			
182 cordatum Scheele ex Costa			
hispanicum Arvet-Touvet	18	Hs	Merxmüller, 1975
183 glaucophyllum Scheele			
184 pardoanum Arvet-Touvet			
185 pedemontanum Burnat & Gremli			
186 scapigerum Boiss., Orph. & Heldr.			
187 urticaceum Arvet-Touvet			
188 viscosum Arvet-Touvet			
189 ramosissimum Schleicher ex Hegetschw.			
190 arpadianum Zahn			
191 picroides Vill.			
192 neopicris Arvet-Touvet			
193 intybaceum All.	27	Da	Rosenberg, 1926
194 pallidiflorum Jordan ex Ascherson			
195 khekianum Zahn			
196 porrifolium L.	18	Ju	Favarger, 1965
197 bupleuroides C.C. Gmelin	27	Cz	Májovský et al., 1970
198 glaucum All.	27	Au	Polatschek, 1966
	36	Au	Tischler, 1950
199 sparsiramum Naegeli & Peter			
200 falcatum Arvet-Touvet			
201 glabratum Hoppe ex Willd.			
202 franconicum Griseb.			
203 oxyodon Fries			
204 fulcratum Arvet-Touvet			
205 neyraeanum Arvet-Touvet			
206 austriacum Brittinger			
207 dollineri Schultz Bip. ex F.W. Schultz			
208 calcareum Bernh. ex Hornem.			
209 saxatile Jacq.			
210 naegelianum Pančić			
211 silesiacum Krause			
212 sparsum Friv.	18		Christoff, 1942
213 heterogynum (Froelich) Guterm.			
214 macrodon Naegeli & Peter			
215 macrodontoides (Zahn) Zahn			
216 tommasinii Reichenb. fil.			
217 olympicum Boiss.			
218 leiocephalum Bartl. ex Griseb.			
219 virgicaule Naegeli & Peter			
220 pseudobupleuroides Naegeli & Peter			
221 umbrosum Jordan			
222 viride Arvet-Touvet			
223 pinicola Arvet-Touvet			
224 rapunculoides Arvet-Touvet			
225 pedatifolium Omang			
226 prenanthoides Vill.	27	Cz	Májovský et al., 1974
	36		
227 juranum Fries			

228 juraniforme Zahn
229 pocuticum Woloszczak
230 cydonifolium Vill.
231 doronicifolium Arvet-Touvet
232 cantalicum Arvet-Touvet
233 turritifolium Arvet-Touvet
234 sergureum Arvet-Touvet

235 epimedium Fries	27	Au	Polatschek, 1966
zetlandicum Beeby	27	Br	Mills & Stace, 1974
236 carpathicum Besser	36	Br	Mills & Stace, 1974
237 dovrense Fries	36	No	Finch, ined.

238 plicatum Lindeb.

239 truncatum Lindeb.	27		
	36		
vinicaule P.D. Sell & C. West	36	Br	Mills & Stace, 1974

240 latifolium Froelich ex Link

241 virosum Pallas	36	Rs( )	Christoff & Popoff, 1933

242 robustum Fries

243 inuloides Tausch	27	Br	Morton, 1974
latobrigorum (Zahn) Roffey	27	Br	Finch, ined.

244 crocatum Fries

245 lucidum Guss.	18	Si	Merxmüller, 1975
246 racemosum Waldst. & Kit. ex Willd.	27	It	Battaglia, 1947
crinitum Sibth. & Sm.	36		Merxmüller, ined.

247 pseuderiopus Zahn
248 compositum Lapeyr.
249 nobile Gren. & Godron
250 rectum Griseb.
251 symphytaceum Arvet-Touvet
252 insuetum Jordan ex Boreau
253 bracteolatum Sibth. & Sm.

254 sabaudum L.	18		
	27	Lu	Queirós, 1973
sabaudum L.	27	Br	Morton, 1974
vagum Jordan	18	Br	Morton, 1974

255 flagelliferum Ravaud
256 lycopsifolium Froelich

257 umbellatum L.	18	Cz	Májovský et al., 1970
	27	Br	Morton, 1974
ssp. bichlorophyllum	18	Br	Morton, 1974
258 laevigatum Willd.	27	He	Gadella & Kliphuis, 1966
tridentatum Fries	27	Rs( )	Christoff & Popoff, 1933
259 eriophorum St.Amans	18	Ga	Merxmüller, 1975

260 prostratum DC.

HELOBIAE

CLXX ALISMATACEAE

1 Sagittaria L.

1 sagittifiolia L.	22	Po	Skalińska et al., 1966

```
2 natans Pallas                         22      Fe      Vaarama, 1941
3 latifolia Willd.
4 rigida Pursh

2 Baldellia Parl.
1 ranunculoides (L.) Parl.              16      Su      Löve & Löve, 1944
                                        30
2 alpestris (Cosson) Vasc.

3 Luronium Rafin.
1 natans (L.) Rafin.                    42      Su      Björkquist, 1961a

4 Alisma L.
1 plantago-aquatica L.                  14      Ho      Gadella & Kliphuis,
                                                        1963
2 lanceolatum With.                     26      Su      Löve & Löve, 1944
                                        27      Su      Pogan, 1968
                                        28      Da      Hagerup, 1944
3 gramineum Lej.                        14      Cz      Májovský et al.,
                                                        1976
4 wahlenbergii (Holmberg) Juz.          14      Fe      Björkquist, 1961b

5 Caldesia Parl.
1 parnassifolia (L.) Parl.

6 Damasonium Miller
1 alisma Miller                         28      Ga      Schotsman, 1970


CLXXI BUTOMACEAE

1 Butomus L.
1 umbellatus L.                         26      Po      Skalińska et al.,
                                                        1961

                                        39      Ho      Gadella & Kliphuis,
                                                        1963


CLXXII HYDROCHARITACEAE

1 Hydrocharis L.
1 morsus-ranae L.                       28      Ho      Gadella & Kliphuis,
                                                        1973

2 Stratiotes L.
1 aloides L.                            24      ?Ge     Schürhoff, 1926

3 Ottelia Pers.
1 alismoides (L.) Pers.                 22   non-Europe Rao, 1951

4 Egeria Planchon
1 densa Planchon

5 Elodea Michx
1 canadensis Michx                      24      ?Br     Heppell, 1945
                                        48      Po      Skalińska &
                                                        Pogan, 1973
2 ernstiae St. John
3 nuttallii (Planchon) St. John
```

6 Hydrilla L.C.M. Richard
  1 verticillata (L. fil.) Royle   16   non-Europe   Larsen, 1963

7 Lagarosiphon Harvey
  1 major (Ridley) Moss

8 Blyxa Noronha ex Thouars
  1 japonica (Miq.) Maxim.

9 Vallisneria L.
  1 spiralis L.   20   ?Br cult.   Jörgensen, 1927

10 Halophila Thouars
   1 stipulacea (Forskål) Ascherson

CLXXIII SCHEUCHZERIACEAE

1 Scheuchzeria L.
  1 palustris L.   22   Br   Manton, 1949

CLXXIV APONOGETONACEAE

1 Aponogeton L. fil.
  1 distachyos L. fil.

CLXXV JUNCAGINACEAE

1 Triglochin L.
  1 maritima L.   24   Ga   Lévèque & Gorenflot,
                                                            1969

                                          30   Po   Skalińska et al.,
                                                            1961
                                          36   Ga   Labadie, 1976
                                          48   Po   Skalińska et al.,
                                                            1961

  2 palustris L.   24   Ga   Lévèque & Gorenflot,
                                                            1969

  3 striata Ruiz & Pavón
  4 bulbosa L.
    (a) barrelieri (Loisel.) Rouy   30   Lu   Castro & Fontes,
                                                            1946
                                          32   Bl   Dahlgren et al.,
                                                            1971
    (b) laxiflora (Guss.) Rouy   18   Lu   Gardé & Malheiros
                                                            Gardé, 1953

CLXXVI LILAEACEAE

1 Lilaea Humb. & Bonpl.
  1 scilloides (Poiret) Hauman

CLXXVII POTAMOGETONACEAE

1 Potamogeton L.

1 natans L.	52	Su	Palmgren, 1939
2 polygonifolius Pourret	26	Su	Palmgren, 1939
3 coloratus Hornem.	26	Da	Palmgren, 1939
4 nodosus Poiret			
5 lucens L.	52	Su	Palmgren, 1939
6 gramineus L.	52	Su	Palmgren, 1939
7 alpinus Balbis	26	Hs	Löve & Kjellqvist, 1973
	52	Is	Löve & Löve, 1956
8 praelongus Wulfen	52	Su	Palmgren, 1939
9 ⁻erfoliatus L.	52	Su	Palmgren, 1939
10 epihydrus Rafin.			
11 friesii Rupr.	26	Su	Palmgren, 1939
12 rutilus Wolfg.			
13 pusillus L.	26	Su	Palmgren, 1939
14 obtusifolius Mert. & Koch	26	Su	Palmgren, 1939
15 berchtoldii Fieber	26	Su	Löve & Löve, 1956
16 trichoides Cham. & Schlecht.	26	Su	Palmgren, 1939
17 compressus L.	26	Su	Palmgren, 1939
18 acutifolius Link	26	Su	Palmgren, 1942
19 crispus L.	52	Su	Palmgren, 1939
20 filiformis Pers.	78	Is	Löve & Löve, 1956
21 vaginatus Turcz.	c.88	Su	Palmgren, 1939
22 pectinatus L.	‾78	Su	Palmgren, 1942
2 Groenlandia Gay			
1 densa (L.) Fourr.	30	?Su	Palmgren, 1939

CLXXVIII RUPPIACEAE

1 Ruppia L.			
1 maritima L.	20	Ge	Löve & Löve, 1961b
2 cirrhosa (Petagna) Grande	40	Ge	Löve & Löve, 1961b

CLXXIX POSIDONIACEAE

1 Posidonia C. Konig
   1 oceanica (L.) Delile

CLXXX ZOSTERACEAE

1 Zostera L.			
1 marina L.	12	Po	Skalińska et al., 1974
2 angustifolia (Hornem.) Reichenb.	12	Ge	Wulff, 1937a
3 noltii Hornem.	12	Ge	Wulff, 1937a

CLXXXI ZANNICHELLIACEAE

1 Althenia Petit
   1 filiformis Petit

2 Zannichellia L.			
1 palustris L.	12	Bu	Kozuharov & Kuzmanov, 1964

| | 28 | Ge | Scheerer, 1940 |
| 24 32 34 36 | Ge | Reese, 1963 |

3 Cymodocea C. Konig
  1 nodosa (Ucria) Ascherson

CLXXXII NAJADACEAE

1 Najas L.
  1 marina L.   12   Scand.   Löve & Löve, 1948
  2 flexilis (Willd.) Rostk. & W.L.E.
    Schmidt   24   Br Su   Löve & Löve, 1958
  3 minor All.   24   Ga   Schotsman, 1970
  4 tenuissima (A. Braun) Magnus
  5 gracillima (A. Braun ex Engelm.)
    Magnus
  6 graminea Delile

LILIIFLORAE

CLXXIII LILIACEAE

1 Tofieldia Hudson
  1 pusilla (Michx) Pers.
    (a) pusilla   30   Br   Elkington, 1974
    (b) austriaca H. Kunz
  2 calyculata (L.) Wahlenb.   30   Ju   Lovka et al.,
                                                            1971

2 Narthecium Hudson
  1 ossifragum (L.) Hudson   26   Lu   Fernandes, 1950
  2 reverchonii Celak.   26   Co   Contandriopoulos,
                                                            1962

  3 scardicum Kosanin

3 Zigadenus L.C.M. Richard
  1 sibiricus (L.) A. Gray

4 Veratrum L.
  1 nigrum L.   16   Ju   Lovka et al.,
                                                            1972

  2 album L.   16   error
    lobelianum   32   Ju   Lovka et al.,
                                                            1971

5 Asphodelus L.
  1 fistulosus L.   28   Hs   Rejón, 1976
                                              56   Hs   Rejón, 1976

  2 albus Miller
    (a) albus   28   Gr   Alden, 1976

                                              56   Hs   Rejón, 1976
    (b) villarsii (Verlot ex Billot)
        I.B.K. Richardson & Smythies   78   Co   Lambert, 1969
  3 ramosus L.   28   Hs   Löve & Kjellqvist,
                                                            1973
                                              56   Lu   Barros Neves, 1973

	84	Lu	Barros Neves, 1973
4 aestivus Brot.	28 56 84	Lu	Barros Neves, 1973
5 bento-rainhae P. Silva			

6 Asphodeline Reichenb.

	28	Bu	Borhidi, 1968
1 lutea (L.) Reichenb.	28	Bu	Borhidi, 1968
2 liburnica (Scop.) Reichenb.	14	Bu	Borhidi, 1968
	28	Bu	Popova, 1972
3 taurica (Pallas ex Bieb.) Kunth	28	Bu	Popova, 1972

7 Anthericum L.

1 liliago L.	32	Lu	Fernandes, 1950
	48		
	64	Ge	Tischler, 1934
2 baeticum (Boiss.) Boiss.	30	Hs	Küpfer, 1968
3 ramosum L.	30	Su	Strandhede, 1963
	32	Ga	Parreaux, 1971

8 Paradisea Mazzuc.

1 liliastrum (L.) Bertol.	16	Ga	Küpfer, 1969
2 lusitanica (Coutinho) Samp.	32	Lu	Fernandes, 1950
	64	Lu	Fernandes, 1950

9 Eremurus Bieb.
  1 caucasicus Steven
  2 tauricus Steven

10 Simethis Kunth

1 planifolia (L.) Gren.	48	Lu	Barros Neves, 1973

11 Aphyllanthes L.

1 monspeliensis L.	32	Hs	Löve & Kjellqvist, 1973

12 Hemerocallis L.
  1 lilioasphodelus L.
  2 fulva (L.) L.

13 Phormium J.R. & G. Forster
  1 tenax J.R. & G. Forster

14 Aloe L.
  1 aristata Haw.
  2 humilis (L.) Miller
  3 brevifolia Miller
  4 vera (L.) Burm. fil.
  5 ciliaris Haw.
  6 perfoliata L.
  7 succotrina All.
  8 arborescens Miller
  9 ferox Miller

15 Androcymbium Willd.

1 europaeum (Lange) K. Richter	18	Hs	Carpio, 1954
2 rechingeri W. Greuter			

16 Colchicum L.

1 triphyllum G. Kunze	20 21	It	D'Amato, 1956
diampolis Delip. & Česchm ex Kuzmanov & Kozuharov	18	Bu	Popova & Česchmedjiev, 1966

2 burttii Meikle			
3 hungaricum Janka	54	Ju	Milan, 1975
4 cupanii Guss.	54	It	D'Amato, 1957a
5 pusillum Sieber			
6 psaridis Heldr. ex Halacsy			
7 boissieri Orph.			
8 parlatoris Orph.			
9 arenarium Waldst. & Kit.	36	Ju	Milan, 1975
	54	Ju	Milan, 1975
10 alpinum DC.	54	It	D'Amato, 1957a
	56	Ga	Perrenoud &
10 x 14	46 47	Ga	Favarger, 1971
11 corsicum Baker			
12 micranthum Boiss.			
macedonicum Kosanin	54	Ju	Milan, 1975
13 umbrosum Steven			
14 autumnale L.	38	Ga He It	D'Amato, 1957b
15 neapolitanum (Ten.) Ten.	140	It	D'Amato, 1957b
16 lusitanum Brot.	106	It	D'Amato, 1957b
17 lingulatum Boiss. & Spruner			
18 parnassicum Sart.			
19 turcicum Janka			
20 callicymbium Stearn & Stefanov			
21 macrophyllum B.L. Burtt			
22 variegatum L.			
23 bivonae Guss.	36 54	Bot.Gard.	Levan, 1940
	90	Ju	Milan, 1975
17 Bulbocodium L.			
1 vernum L.			
2 versicolor (Ker-Gawler) Sprengel			
18 Merendera Ramond			
1 attica (Spruner ex Tommasini) Boiss. & Spruner			
2 androcymbioides Valdes			
3 sobolifera C.A. Meyer	54	Ju	Lovka et al., 1971
4 pyrenaica (Pourret) P. Fourn.	60	Lu	Fernandes, 1950
	54 + 0 - 6B	Hs	Löve & Kjellqvist, 1973
5 filifolia Camb.	54	Bl	Dahlgren et al., 1971
19 Lloydia Salisb. ex Reichenb.			
1 serotina (L.) Reichenb.	24	He	Larsen, 1954
20 Gagea Salisb.			
1 pratensis (Pers.) Dumort.	36	Cz	Májovský et al., 1976
	48 60	Cz	Mešiček & Hrouda, 1974
transversalis	48	Cz	Mešiček & Hrouda, 1974
2 pusilla (F.W. Schmidt) Schultes & Schultes fil.	24	Cz	Mešiček & Hrouda, 1974
3 lutea (L.) Ker-Gawler	72	Cz	Mešiček & Hrouda, 1974
4 minima (L.) Ker-Gawler	24	Cz	Mešiček & Hrouda, 1974
5 reticulata (Pallas) Schultes & Schultes fil.			

```
 6 fibrosa (Desf.) Schultes & Schultes
   fil.
 7 taurica Steven
 8 bulbifera (Pallas) Schultes &
   Schultes fil.
 9 graeca (L.) A. Terracc.
10 trinervia (Viv.) W. Greuter
11 spathacea (Hayne) Salisb.          106      Cz      Mesiček & Hrouda,
                                                       1974
12 fistulosa (Ram. ex DC.) Ker-Gawler
13 arvensis (Pers.) Dumort.            48      Cz      Mesiček & Hrouda,
                                                       1974
14 dubia A. Terracc.
15 granatellii (Parl.) Parl.
16 peduncularis (J. & C. Presl) Pascher
17 nevadensis Boiss.                   36      Sa      Martinoli, 1950
18 amblyopetala Boiss. & Heldr.
19 foliosa (J. & C. Presl) Schultes &
   Schultes fil.
20 polymorpha Boiss.
21 bohemica (Zauschner) Schultes &     60      Cz      Mesiček & Hrouda,
   Schultes fil.                                       1974
   (a) bohemica
   (b) nebrodensis (Tod. ex Guss.) I.B.K.
       Richardson
   (c) gallica (Rouy) I.B.K. Richardson
22 saxatilis (Mert. & Koch) Schultes &
   Schultes fil.
23 szovitzii (A.F. Lang) Besser

21 Erythronium L.
   1 dens-canis L.                     24      Cz      Hruby, 1934

22 Tulipa L.
   1 sylvestris L.
     (a) sylvestris                    36    ?S. Europe  Hall, 1937
     (b) australis (Link) Pamp.        24      Lu      Barros Neves,
                                                       1973
   2 orphanidea Boiss. ex Heldr.   24 36 48   Bu      Popova, 1972
   3 goulimyi Sealy & Turrill
   4 saxatilis Sieber ex Sprengel      24
                                       36      Cr      Hall, 1937
   5 cretica Boiss. & Heldr.           24      Cr      Hall, 1937
   6 biflora Pallas                    24      Ju      Milan, 1975
   7 clusiana DC.                      48
                                       56
   8 gesnerana L.
     hungarica Borbas                  24      Rm      Tarnavschi &
                                                       Lungeanu, 1970
     urumoffii Hayek                   24      Bu      Popova, 1972
   9 boeotica Boiss. & Heldr.          24      Ju      Lovka et al.,
                                                       1971
  10 praecox Ten.                      36      It      Hall, 1937
  11 agenensis DC.                     24      ?Ga     Hall, 1937
                                       36    Bot. Gard.  Woods &
                                                       Bamford, 1937

23 Fritillaria L.
```

1 meleagris L.	24	Ho	Gadella & Kliphuis, 1973

(a) meleagris
(b) burnatii (Planchon) Rix
2 tubiformis Gren. & Godron
  (a) tubiformis
  (b) moggridgei Boiss. & Reuter ex Rix
3 macedonica Bornm.
4 pontica Wahlenb.
5 involucrata All.
6 messanensis Rafin.

(a) messanensis	24	Si	Gori, 1958
(b) gracilis (Ebel) Rix	24	Ju	Milan, 1975

7 epirotica Turrill ex Rix
8 gussichiae (Degen & Dörfler) Rix

9 graeca Boiss. & Spruner	24	Gr	Phitos, ined.

  (a) graeca
  (b) thessala (Boiss.) Rix
10 davisii Turrill

11 rhodocanakis Orph. ex Baker	24	Gr	Phitos, ined.

12 pyrenaica L.

13 lusitanica Wikström	24	Lu	Fernandes, 1950

14 ruthenica Wikström

15 orientalis Adams	18 + 0 - 9B	Ju	Milan, 1975

16 meleagroides Patrin ex Schultes
  fil.

17 obliqua Ker-Gawler	24	Gr	Phitos, ined.

18 tuntasia Heldr. ex Halácsy
19 drenovskii Degen & Stoj.
20 conica Boiss.
21 euboeica Rix

22 stribrnyi Velen.	24	Bu	Popova, 1972

23 ehrhartii Boiss. & Orph.

24 Lilium L.

1 martagon L.	24 + 0 - 2B	Rs(W)	Van Loon & Oudemans, 1976
2 candidum L.	24	Sa	Arrigoni & Mori, 1976
3 bulbiferum L.	24	Po	Skalińska et al., 1964

4 pyrenaicum Gouan
5 pomponium L.

6 carniolicum Bernh. ex Koch	24	It	Zucconi, 1956

7 chalcedonicum L.
8 rhodopaeum Delip.
9 monadelphum Bieb.
10 lancifolium Thunb.

25 Ornithogalum L.

1 pyrenaicum L.	16	It	Chiappini & Scrugli, 1972
	18	Rm	Lungeanu, 1971
2 sphaerocarpum A. Kerner	16	Au	Leute, 1974

3 visianicum Tommasini
4 creticum Zahar.

5 narbonense L.	54	Hs	Rejón, 1976
6 ponticum Zahar.	16	Rs(K)	Agapova, 1976
7 pyramidale L.	24	Ge	Lungeanu, 1971

8 prasinantherum Zahar.

9 reverchonii Lange			
10 fischeranum Krasch.	22	Rs(K)	Agapova, 1976
	24	Rs(W)	Lungeanu, 1971
11 oligophyllum E.D. Clarke	24	Bu	Markova et al., 1972
12 montanum Cyr.	12	Bu	Lungeanu, 1971
	14	Bu	Markova et al., 1972
	18	Gr Si	Ratter, 1967
	20 + 3B	Si	Garbari & Tornadore, 1972
13 atticum Boiss. & Heldr.			
14 oreoides Zahar.	18	Rm	Lungeanu, 1971
15 fimbriatum Willd.	12 + 0 - 3B	Bu	Markova et al., 1972
16 collinum Guss.			
17 costatum Zahar.			
18 woronowii Krasch.	16	Rs(K)	Agapova, 1976
19 amphibolum Zahar.	18	Rm	Lungeanu, 1971
20 wiedemannii Boiss.			
21 sibthorpii W. Greuter	14	Bu	Markova et al., 1972
	18 24	Rm	Lungeanu, 1971
	28	Bu	Markova et al., 1972
22 exscapum Ten.	18	It	Garbari & Tornadore, 1972
23 comosum L.	14	Bu	Markova et al., 1972
	18	It	Garbari & Tornadore, 1972
24 umbellatum L.	18-20	Lu	Mesquita, 1964
	27 28	Br	Czapik, 1968
	35 36	Gr	Ratter, 1967
	42	Cz	Májovský et al., 1976
	44	Bu	Markova et al., 1972
	45	It	Garbari & Tornadore, 1972
	54	Br	Czapik, 1968
	72 90 108	It	Garbari & Tornadore, 1971
25 orthophyllum Ten.	14 + 0 - 3B	It	Garbari & Tornadore, 1972
	18-20 27-29	Po	Czapik, 1965
	36	Bu	Markova et al., 1972
(a) psammophilum (Zahar.) Zahar.			
(b) orbelicum (Velen.) Zahar.			
(c) baeticum (Boiss.) Zahar.	18	Hs	Löve & Kjellqvist,
	27		1971
(d) orthophyllum			
(e) acuminatum (Schur) Zahar.			
(f) kochii (Parl.) Zahar.			
26 armeniacum Baker			
27 refractum Kit. ex Schlecht.	54	Bu	Markova et al., 1972
	56	Rm	Lungeanu, 1971

	72	Bu	Markova et al., 1972
28 divergens Boreau	18	It	Chiappini & Serugli, 1972
	42	Rm	Lungeanu, 1971
	54	Ju	Sušnik & Lovka, 1973
29 exaratum Zahar.			
30 nutans L.	45	It	Garbari & Tornadore, 1972
31 boucheanum Ascherson	28	Rm	Lungeanu, 1971
32 arabicum L.			
33 unifolium (L.) Ker-Gawler	34	Lu	Barros Neves, 1973
34 concinnum (Salisb.) Coutinho	36	Lu	Barros Neves, 1973
26 Urginea Steinh.			
1 undulata (Desf.) Steinh.	20	Hs	Sañudo & Rejón, 1975
	60		
2 fugax (Moris) Steinh.	20 + 1B 4B	Sa	Battaglia, 1957b
3 maritima (L.) Baker	20	Bl It Si	Battaglia, 1957
	30	It	Battaglia, 1957
	40	Bl It Sa	Battaglia, 1957
	60	Hs	Rejón. 1974
	64	Lu	Barros Neves, 1973
27 Scilla L.			
1 bifolia L.	18	Ju	Sušnik & Lovka, 1973
	36	Ju	Sušnik & Lovka, 1973
	54	Gr	Speta, 1976
2 messeniaca Boiss.			
3 siberica Haw.			
4 amoena L.			
5 bithynica Boiss.			
6 verna Hudson	20	Lu	Barros Neves, 1973
7 ramburei Boiss.	20	Lu	Barros Neves, 1973
8 odorata Link			
9 monophyllos Link	20 40	Lu	Barros Neves, 1973
10 litardierei Breistr.	28	Ju	Lovka et al., 1971
11 hyacinthoides L.	20	Lu	Barros Neves, 1973
12 lilio-hyacinthus L.			
13 peruviana L.	16	Lu	Barros Neves, 1973
cupanii Guss.	14	Si	Maugini, 1952
hughii Tineo ex Guss.	16	Si	Maugini, 1952
14 beirana Samp.			
15 autumnalis L.	14	Hs It Si	Battaglia, 1957
	28	Bl Ga It Lu	Battaglia, 1957
	42	Hu	Baksay, 1956
28 Hyacinthoides Medicus			
1 non-scripta (L.) Chouard ex Rothm.	16 24	Br	Wilson, 1956
2 hispanica (Miller) Rothm.	16 24	Br Lu	Wilson, 1959
3 italica (L.) Rothm.	16	Lu	Barros Neves, 1973
reverchonii Degen & Hervier	16	Hs	Löve & Kjellqvist, 1973

29 Chionodoxa Boiss.
1 nana (Schultes & Schultes fil.)   18   Cr   Speta, 1976
  Boiss. & Heldr.
2 cretica Boiss. & Heldr.

30 Hyacinthus L.
1 orientalis L.

31 Brimeura Salisb.
1 amethystina (L.) Chouard
2 fastigiata (Viv.) Chouard   28   Sa   Chiarugi, 1950

32 Strangweia Bertol.
1 spicata (Sibth. & Sm.) Boiss.   8   Gr   Phitos, ined.

33 Bellevalia Lapeyr.
1 trifoliata (Ten.) Kunth
2 dubia (Guss.) Reichenb.   8   It   Chiarugi, 1949
  boissieri   8   Gr   Bothmer & Bentzer, 1973
3 webbiana Parl.   16   It   Chiarugi, 1949
4 hackelii Freyn
5 lipskyi (Miscz.) Wulf
6 ciliata (Cyr.) Nees
7 sarmatica (Pallas ex Georgi) Woronow
8 romana (L.) Reichenb.   8   Ju   Lovka et al., 1971
9 brevipedicellata Turrill

34 Hyacinthella Schur
1 leucophaea (C. Koch) Schur
2 dalmatica (Baker) Chouard
3 pallasiana (Steven) Losinsk

35 Dipcadi Medicus
1 serotinum (L.) Medicus   8 + 0 - 6B   Lu   Fernandes & Queirós, 1971
    16 32   Hs   Gadella et al., 1966

36 Muscari Miller
1 macrocarpum Sweet   18   Gr   Stuart, 1970
2 moschatum Willd.
3 cycladicum P.H. Davis & Stuart   36   Gr   Bentzer, 1973
    54   Gr   Stuart, 1970
4 gussonei (Parl.) Tod.   18   Si   Garbari & Martino, 1972
5 spreitzenhoferi (Heldr.) Vierh.   18   Cr   Stuart, 1970
    36   Cr   Bentzer, 1973
6 weissii Freyn   18 36   Gr   Stuart, 1970
    54   Gr   Bentzer, 1973
  dionysicum Rech. fil.   18 36   Cr   Bentzer, 1973
7 comosum (L.) Miller   18   Gr It   Stuart, 1970

    27   Gr   Bentzer, 1972
8 tenuiflorum Tausch   18   It   Garbari, 1973
9 botryoides (L.) Miller   18   Hu   Pólya, 1950
    36   It   Garbari, 1969
10 armeniacum Leichtlin ex Baker   18   Gr   Stuart, 1970

11 neglectum Guss. ex Ten.	18	Bu	Kozuharov & Kuzmanov, 1962/3
	36	Lu	Barros Neves, 1973
	45	Ju	Sušnik & Lovka, 1973
	54	Ga	Guinochet & Logeois, 1962
	72	Rs(K)	Stuart, 1970
12 commutatum Guss.	18	Gr	Stuart, 1970
13 parviflorum Desf.	18	It	Garbari, 1969

37 Agapanthus L'Hér.
  1 praecox Willd.

38 Allium L.

1 angulosum L.	16	Cz	Májovský et al., 1970
2 rubens Schrader ex Willd.			
3 senescens L.	16	It	Renzoni & Garbari, 1971
	24 32	Hu	Baksay, 1956
	32 + 0 - 4B	Cz	Holub et al., 1970
4 saxatile Bieb.	16	?Bot. Gard.	Levan, 1929
5 horvatii Lovrić	16		
6 albidum Fischer ex Bieb.	16	Ju	Dietrich, 1967b
(a) albidum	16	Bot. Gard.	Hasitschka-Jenscke,
(b) caucasicum (Regel) Stearn			1958
7 stelleranum Willd.			
8 ericetorum Thore	16 + 0 - 1B	Cz	Holub et al., 1970
	32	Ju	Lovka et al., 1972
9 kermesinum Reichenb.	16	Ju	Lovka et al., 1971
10 suaveolens Jacq.	16	Gr	Alden, 1976
11 lineare L.	16	Bot. Gard.	Dietrich, 1970
	48	Cz	Holub et al., 1970
12 palentinum Losa & Montserrat			
13 hymenorhizum Ledeb.			
14 obliquum L.	16	Bot. Gard.	Dietrich, 1967b
15 narcissiflorum Vill.	14	Ga	Delay, 1968
16 insubricum Boiss. & Reuter	14	It	Favarger, 1959
17 inderiense Fischer ex Bunge			
18 schoenoprasum L.	16	Cz	Májovský et al., 1976
19 schmitzii Coutinho	16	Lu	Barros Neves, 1973
20 cepa L.	16	It	Battaglia, 1957
21 fistulosum L.	16	Ga	Battaglia, 1957
22 victorialis L.	16	Po	Mizianty & Frey, 1973
23 roseum L.	16	Ga	Delay, 1970
	32	Ga	Delay, 1970
24 circinnatum Sieber			
25 massaessylum Batt. & Trabut	14	Lu	Barros Neves, 1973
26 phthioticum Boiss. & Heldr. ex Boiss.			
27 breviradium (Halácsy) Stearn			

28 neapolitanum Cyr.	14 21 28	W.Africa	Kollmann, 1973
	31-36	Lu	Barros Neves, 1973
29 longanum Pamp.			
30 subhirsutum L.	14	It Si	Renzoni & Garbari, 1971
	28	Hs	Rejón, 1974
31 trifoliatum Cyr.			
32 subvillosum Salzm. ex Schultes & Schultes fil.	28	Bl	Dahlgren,et al., 1971
33 moly L.	14	Hs	Löve & Kjellqvist, 1973
34 scorzonerifolium Desf. ex DC.	14	Lu	Barros Neves, 1973
35 triquetrum L.	18	Lu	Barros Neves, 1973
36 pendulinum Ten.	14	It	Renzoni & Garbari, 1970
37 paradoxum (Bieb.) G. Don	16	Br	Barling, 1958
38 chamaemoly L.	22	It	Marchi et al., 1974
39 ursinum L.	14	Cz	Májovský et al., 1976
(a) ursinum			
(b) ucrainicum Kleopow & Oxner			
40 moschatum L.	16	It	Billeri, 1954
41 inaequale Janka			
42 bornmuelleri Hayek			
43 rubellum Bieb.			
44 meteoricum Heldr. & Hausskn. ex Halácsy	16	Gr	Stearn, 1978
45 delicatulum Sievers ex Schultes & Schultes fil.			
46 frigidum Boiss. & Heldr.			
47 chrysonemum Stearn	32	Hs	Stearn, 1978
48 grosii Font Quer			
49 cupani Rafin.			
(a) cupani	32	Si	Garbari et al., 1979
(b) hirtovaginatum (Kunth) Stearn	16	Gr	Garbari et al., 1979
50 callimischon Link			
(a) callimischon	16	Gr	Stearn, 1978
(b) haemostictum Stearn	16	Cr	Miceli & Garbari, 1979
51 obtusiflorum DC			
52 parciflorum Viv.	16	Sa	Martinoli, 1955
53 rouyi Gaut.			
54 caeruleum Pallas			
55 sabulosum Steven ex Bunge	16	Asia	Vakhtina, 1964
56 paniculatum L.	16	Hs	Rejón, 1976
	40	It	Marchi et al., 1974
(a) paniculatum	24	Bu	Češchmedjiev, 1970
	32	Co	Contandriopoulos, 1962
(b) fuscum (Waldst. & Kit.) Arcangeli	16	Gr	Diannelidis, 1951
	32	Gr	Johnson, 1982
(c) euboicum (Rech. fil.) Stearn			
(d) villosulum (Halácsy) Stearn	16	Bu	Češchmedjiev, 1970
57 pallens L.	16	Co	Contandriopoulos, 1957
(a) pallens			
(b) tenuiflorum (Ten.) Stearn			
(c) siciliense Stearn			
58 sipyleum Boiss.			

```
59 rupestre Steven
60 favosum Zahar.
61 macedonicum Zahar.
62 podolicum Blocki ex Racib. & Szafer
63 oleraceum L.                              32      Fe      V. Sorsa, 1962
                                            40      It      Gadella & Kliphuis,
                                                            1970
64 parnassicum (Boiss.) Halácsy
65 staticiforme Sibth. & Sm.
66 pilosum Sibth. & Sm.
67 luteolum Halácsy
68 flavum L.
   (a) flavum                            16  32      Gr      Alden, 1976
   (b) tauricum (Besser ex Reichenb.)        32      Bu      Ceschmedjiev, 1970
       Stearn
69 melanantherum Pančić                      16      Bu      Ceschmedjiev, 1976
                                        24 + 1B      Bu      Ceschmedjiev, 1976
                                            32      Bu      Kozuharov &
                                                            Kuzmanov, 1965a
70 tardans W. Greuter & Zahar.
71 carinatum L.
   (a) carinatum                            16      Ga      Parreaux, 1971
                                            24      Ju      Nilsson & Lassen,
                                                            1971
                                            26      Au      Tschermak-Woess, 1947
   (b) pulchellum Bonnier & Layens          16      It      Marchi et al.,
                                                            1974
72 hirtovaginum P. Candargy
73 hymettium Boiss. & Heldr.
74 stamineum Boiss.
75 sativum L.
76 ampeloprasum L.                          16      It      Carpineri, 1971
                                     32 40 48      Gr      Bothmer, 1970
                                            56      Cr      Bothmer, 1975
                                            80      Lu      Barros Neves, 1973
   var. holmense Ascherson & Graebner       48      Br      Stearn, 1978
   porrum L.                                32      Cz      Murin, 1964
77 polyanthum Schultes & Schultes fil.      20      Hs      Gadella et al.,
                                                            1966
                                            32      Hs      Löve & Kjellqvist,
                                                            1973
78 scaberrimum Serres
79 atroviolaceum Boiss.                      16      Tu      Johnson, unpub.
                                            24
                                            32      Bu      Ceschmedjiev, 1970
                                         40 48

80 bourgeaui Rech. fil.
   (a) bourgeaui                            16      Gr      Bothmer, 1970
   (b) cycladicum Bothmer              16 24 32      Gr      Bothmer, 1970
   (c) creticum Bothmer                     32      Cr      Bothmer, 1975
81 commutatum Guss.                    16 24 32      Gr      Bothmer, 1970
82 pardoi Loscos
83 talijevii Klokov
84 baeticum Boiss.                           32      Lu      Barros Neves, 1973
85 acutiflorum Loisel.
86 pyrenaicum Costa & Vayr.
87 scorodoprasum L.
   (a) rotundum (L.) Stearn           16 + 0 - 2B   Bu      Ceschmedjiev, 1976
                                            32      Hs      Rejón, 1976
```

38 + 1B	Bu	Česchmedjiev, 1976
40 + 1 - 4B	Bu	Česchmedjiev, 1970
48 + 1B		

(b) waldsteinii (G. Don) Stearn   16

32

40

48

(c) jajlae (Vved.) Stearn

(d) scorodoprasum   16   Cz   Májovský et al., 1970

    24   Au   Tschermak-Woess, 1947

88 albiflorum Omelczuk

89 pervestitum Klokov

90 sphaerocephalon L.   16   Lu   Barros Neves, 1973

  16 + 2B   Hs   Rejón, 1976

  (a) sphaerocephalon

  (b) arvense (Guss.) Arcangeli

  (c) trachypus (Boiss. & Spruner) Stearn

91 proponticum Stearn & N. Ozhatay   16   Tu   Stearn & Özhatay, 1977

92 melananthum Coincy

93 pruinatum Link ex Sprengel   16   Lu   Barros Neves, 1973

94 regelianum A. Becker

95 vineale L.   16

  32 40   Lu   Barros Neves, 1973

96 amethystinum Tausch   16 24 32   Gr   Bothmer, 1970

97 guttatum Steven

  (a) guttatum   16   Rm   Buttler, 1969

  (b) sardoum (Moris) Stearn   16   Bu   Česchmedjiev, 1970

  32 48   Lu   Barros Neves, 1973

  (c) dalmaticum (A. Kerner ex Janchen   24   Ju   Sušnik, 1967

98 dilatatum Zahar.

99 gomphrenoides Boiss. & Heldr.

100 jubatum Macbride

101 rubrovittatum Boiss. & Heldr.

102 integerrimum Zahar.

103 chamaespathum Boiss.   16

104 heldreichii Boiss.   16   Gr   Alden, 1976

105 atropurpureum Waldst. & Kit.   16   Gr   Dietrich, 1969

106 nigrum L.   16   Bu   Česchmedjiev, 1976

107 cyrilli Ten.

108 decipiens Fischer ex Schultes & Schultes fil.   16   Bu   Česchmedjiev, 1976

109 orientale Boiss.

110 caspium (Pallas) Bieb.

39 Nectaroscordum Lindley

  1 siculum (Ucria) Lindley   18   Si   Garbari & Tornadore, 1970

  (a) siculum

  (b) bulgaricum (Janka) Stearn

40 Nothoscordum Kunth

  1 inodorum (Aiton) Nicholson   18   Ga   Van Loon et al., 1971

41 Ipheion Rafin.

  1 uniflorum (R.C. Graham) Rafin.

42 Convallaria L.

  1 majalis L.   38   Hu   Pólya, 1949

43 Maianthemum Weber
  1 bifolium (L.) F.W. Schmidt   36   Cz   Májovský et al., 1970

44 Smilacina Desf.
  1 stellata (L.) Desf.

45 Streptopus Michx
  1 amplexifolius (L.) DC.

46 Polygonatum Miller
  1 verticillatum (L.) All.   28   Cz   Májovský et al., 1970
  2 latifolium (Jacq.) Desf.   20   Hu   Pólya, 1949
  3 orientale Desf.
  4 multiflorum (L.) All.   18   Cz   Májovský et al., 1970
                                   32   Fe   Sorsa, 1963
  5 odoratum (Miller) Druce   20   Fe Su   Suomalainen, 1947

47 Paris L.
  1 quadrifolia L.   20   Fe   V. Sorsa, 1962

48 Asparagus L.
  1 asparagoides (L.) Druce
  2 albus L.   20   Lu   Barros Neves, 1973
  3 acutifolius L.   40   It   Bozzini, 1959
  4 aphyllus L.   40   It   Bozzini, 1959
  5 stipularis Forskål   20   It   Bozzini, 1959
  6 verticillatus L.
  7 bresleranus Schultes
  8 persicus Baker
  9 brachyphyllus Turcz.   40   ?Rs   Zakharyeva &
 10 kasakstanicus Iljin   Makushenko, 1969
 11 maritimus (L.) Miller   40   It   Bozzini, 1959
 12 litoralis Steven
 13 officinalis L.
   (a) officinalis   20   Ho It Bozzini, 1959
   (b) prostratus (Dumort.) Corb. 40   Br   Davies & Willis, 1959
 14 pseudoscaber Grec.
 15 tenuifolius Lam.   20   It   Bozzini, 1959

49 Ruscus L.
  1 aculeatus L.   40   Lu   Barros Neves, 1973
  2 hypophyllum L.
  3 hypoglossum L.   40   Sa   Martinoli, 1951

50 Smilax L.
  1 aspera L.   32   It   Marchi, 1971
  2 excelsa L.
  3 canariensis Brouss. ex Willd.

CLXXXIV AGAVACEAE

1 Agave L.
  1 americana L.
  2 atrovirens Karwinski ex Salm-Dyck

2 Furcraea Vent.
  1 foetida (L.) Haw.

3 Yucca L.
  1 filamentosa L.
  2 gloriosa L.

CLXXXV AMARYLLIDACEAE

1 Amaryllis L.
  1 bella-donna L.

2 Sternbergia Waldst. & Kit.
  1 colchiciflora Waldst. & Kit.   20   Ju   Bedalov & Sušnik,
                                                                1970
                                                    26   Rm   Tarnavschi &
                                                                Lungeanu, 1970
  2 lutea (L.) Ker-Gawler ex Sprengel
    (a) lutea   22   Ju   Bedalov & Sušnik,
                                                                1970
    (b) sicula (Tineo ex Guss.) D.A. Webb   18   Gr   Kuzmanov et al.,
                                                                1968

3 Leucojum L.
  1 autumnale L.   14   Lu   Barros Neves, 1939
  2 roseum F. Martin   16   Co Sa Stern, 1956
  3 longifolium (Gay ex M.J. Roemer) Gren.   16   Co   Contandriopoulos,
                                                                1962
  4 trichophyllum Schousboe   14   Lu   Barros Neves, 1939
  5 nicaeense Ardoino   18   Ga   Stern, 1956
  6 valentinum Pau   16   Gr   Damboldt & Phitos,
                                                                1975
  7 vernum L.   20   Po   Skalińska et al.,
                                                                1968
                                                    22   Ge   Stern, 1956
  8 aestivum L.
    (a) aestivum   22   Ju   Sušnik & Lovka,
                                                                1973
                                                    24   Hu   Pólya, 1949
    (b) pulchellum (Salisb.) Briq.   22   Sa   Garbari &
                                                                Tornadore, 1970

4 Galanthus L.
  1 nivalis L.
    (a) nivalis   24   Ge   Stern, 1956
    (b) reginae-olgae (Orph.) Gottl.-Tann.
  2 plicatus Bieb.   24   Rs(K) Van Loon &
                                                                Oudemans, 1976
  3 elwesii Hooker fil.   24   Rm   Tarnavschi &
                                                                Lungeanu, 1970
                                                    48   Gr   Stern, 1956
    (a) elwesii
    (b) minor D.A. Webb
  4 ikariae Baker

5 Lapiedra Lag.
  1 martinezii Lag.   22   Hs   Gadella & Kliphuis,
                                                                1966

6 Narcissus L.
  1 humilis (Cav.) Traub   28   Hs   A. & R. Fernandes, 1945
  2 serotinus L.   10   Hs   Fernandes, 1968
                                                    20   It   Garbari et al.,
                                                                1973

			30	It	Scrugli, 1974
3 tazetta L.					
(a) tazetta					
(b) italicus (Ker-Gawler) Baker			22	It	Maugini, 1952b
(c) aureus (Loisel.) Baker			22	It	Maugini, 1953
4 papyraceus Ker-Gawler					
(a) papyraceus					
(b) polyanthus (Loisel.) Ascherson & Graebner					
(c) panizzianus (Parl.) Arcangeli			22	Lu	Fernandes, 1937a
5 dubius Gouan			50	Hs	Fernandes, 1937b
6 elegans (Haw.) Spach					
7 poeticus L.					
(a) poeticus			14	It	Garbari & Tornadore, 1970
(b) radiiflorus (Salisb.) Baker			14	Ju	Sušnik & Lovka, 1973
8 jonquilla L.			14	Lu	Fernandes, 1966
9 requienii M.J. Roemer			14	Hs	Fernandes, 1939
10 willkommii (Samp.) A. Fernandes			14	Lu	Fernandes, 1966
11 gaditanus Boiss. & Reuter			14	Lu	Fernandes, 1939
			21	Lu	Fernandes & Queirós, 1971
12 rupicola Dufour			14	Lu	Fernandes, 1939
13 cuatrecasasii Casas, Laínz & Ruiz Rejón			14	Hs	Casas et al., 1973
14 calcicola Mendonca			14	Lu	Fernandes, 1939
15 scaberulus Henriq.			14	Lu	Fernandes, 1939
16 viridiflorus Schousboe					
17 triandrus L.					
(a) pallidulus (Graells) D.A. Webb			14	Lu Hs	Fernandes, 1950b
(b) triandrus			14	Lu	Fernandes, 1950b
(c) capax (Salisb.) D.A. Webb			14	Ga	Fernandes, 1950b
18 bulbocodium L.					
(a) bulbocodium		14 21 28 35 42 49 56		Lu	Fernandes, 1967
(b) obesus (Salisb.) Maire			26	Lu	Fernandes, 1967
19 hedraeanthus (Webb & Heldr.) Colmeiro			14	Hs	Rejón, 1974
20 cantabricus DC.			14	Hs	Fernandes, 1959
21 longispathus Pugsley			14	Hs	Heywood, 1955
22 pseudonarcissus L.					
(a) pseudonarcissus					
(b) pallidiflorus (Pugsley) A. Fernandes					
(c) moschatus (L.) Baker					
(d) nobilis (Haw.) A. Fernandes			28	Lu	Fernandes, 1950
(e) major (Curtis) Baker					
(f) portensis (Pugsley) A. Fernandes					
(g) nevadensis (Pugsley) A. Fernandes					
23 bicolor L.					
24 minor L.			14	Lu	Fernandes, 1931
25 asturiensis (Jordan) Pugsley			14 15	Lu	Fernandes & Fernandes, 1946
26 cyclamineus DC.			14	Lu	Fernandes & Fernandes, 1946
27 x incomparabilis Miller					
28 x odorus L.					
29 x medioluteus Miller			24	It	Garbari et al., 1973
30 x intermedius Loisel.					

```
7 Pancratium L.
  1 maritimum L.                                    22      Lu      Rodrigues, 1953
  2 illyricum L.                                    22      Sa      Martinoli, 1949


CLXXXVI DIOSCOREACEAE

1 Dioscorea L.
  1 balcanica Košanin

2 Borderea Miégeville
  1 pyrenaica Miégeville                            24      Ga      Heslot, 1953
  2 chouardii (Gaussen) Heslot                      24      Hs      Heslot, 1953

3 Tamus L.
  1 communis L.                                     48      Bl      Dahlgren et al.,
                                                                    1971


CLXXXVII PONTEDERIACEAE

1 Pontederia L.
  1 cordata L.

2 Heteranthera Ruiz & Pavón
  1 reniformis Ruiz & Pavón

3 Eichornia Kunth
  1 crassipes (C.F.P. Mart.) Solms-Laub.

4 Monochoria C. Presl
  1 korsakowii Regel & Maack


CLXXXVIII IRIDACEAE

1 Sisyrinchium L.
  1 bermudiana L.                                   64      Hb      Löve & Löve, 1958b
                                                    88      Hb      Ingram, 1967
  2 montanum E.L. Greene                            96      Cz      Murín & Májovský,
                                                                    1976
  3 californicum (Ker-Gawler) Aiton fil.

2 Hermodactylus Miller
  1 tuberosus (L.) Miller

3 Iris L.
  1 sibirica L.                                     28      Po      Skalińska et al.,
                                                                    1961
  2 tenuifolia Pallas
  3 unguicularis Poiret
  4 ruthenica Ker-Gawler
  5 foetidissima L.                                 40      It      Mori, 1957
  6 pseudacorus L.                                  24      Cz      Májovský et al.,
                                                                    1970
                                                    30      Hu      Pólya, 1949
                                                    32      Ju      Lovka et al.,
                                                                    1971
                                                    34      Hu It   Félfoldy, 1947
                                                    40      It      Mori, 1957
```

308

7 versicolor L.
8 spuria L.
  (a) spuria   22   Da   Westergaard, 1938
  (b) maritima P. Fourn.
  (c) halophila (Pallas) D.A. Webb &
      Chater   20   Rm   Lungeanu, 1973
  (d) ochroleuca (L.) Dykes
9 graminea L.   34   Ju   Sušnik et al.,
                                       1972

10 sintenisii Janka
  (a) sintenisii   16   Bu   Popova &
                    18   Česchmedjiev, 1975
  (b) brandzae (Prodan) D.A. Webb &
      Chater   (not 10) 20 Rm   Tarnavschi, 1938
11 pontica Zapal.
12 humilis Georgi   22   Au   Mitra, 1956
13 pumila L.
  (a) pumila   32   Au   Mitra, 1956
  (b) attica (Boiss. & Heldr.) Hayek 14   Gr   Simonet, 1955
                    16   Ju   Lovka et al.,
                                       1971

14 lutescens Lam.
  (a) lutescens   40   It   Mitra, 1956
  (b) subbiflora (Brot.) D.A. Webb &
      Chater
15 pseudopumila Tineo   16   Si   Simonet, 1955
16 reichenbachiana Heuffel   24   Ju   Lovka et al.,
                                       1971
                    48   Bu   Popova &
                                       Česchmedjiev, 1975

17 suaveolens Boiss. & Reuter 24   Bu   Popova &
                                       Česchmedjiev, 1975

18 aphylla L.   24   Ga   Simonet, 1934
                    40   It   Mitra, 1956
                    48   Hu   Pólya, 1950
19 variegata L.   24   Au   Mitra, 1956
  reginae I. & M. Horvat   24   Ju   Mitra, 1956
20 germanica L.   36   Hs   Löve & Kjellqvist,
                                       1973
                    44   Ju   Lovka et al.,
                                       1971
                    48   Gr   Alden, 1976
21 marsica I. Ricci & Colasante 40   It   Ricci & Colasante,
22 albicans Lange   1975
23 pallida Lam.   24   It   Mitra, 1956
  (a) pallida
  (b) cengialti (Ambrosi) Foster
24 planifolia (Miller) Fiori & Paol. 24   Si   Garbari &
                                       Tornadore, 1972b
25 latifolia (Miller) Voss
26 xiphium L.   34   Lu   Fernandes &
                                       Queirós, 1971
27 filifolia Boiss.
28 juncea Poiret
29 boissieri Henriq.   36   Lu   Fernandes, 1950
30 serotina Willk.

4 Gynandriris Parl.
  1 sisyrinchium (L.) Parl.   24   Lu   Fernandes &
                                       Queirós, 1971

5 Ixia L.
  1 maculata L.
  2 paniculata Delaroche

6 Freesia Ecklon ex Klatt
  1 refracta (Jacq.) Ecklon ex Klatt

7 Crocus L.

1 malyi Vis.	30	Ju	Brighton,et al., 1973
2 imperati Ten.			
(a) imperati	26	It	Brighton et al., 1973
(b) suaveolens (Bertol.) Mathew	26	It	Brighton et al., 1973
3 versicolor Ker-Gawler	26	Ga	Brighton et al., 1973
4 cambessedesii Gay	16	Bl	Brighton et al., 1973
5 minimus DC.	24	Co Sa	Brighton et al., 1973
6 corsicus Vanucci ex Maw	18	Co	Brighton et al., 1973
	22	Co	Contandriopoulos, 1962
7 vernus (L.) Hill			
(a) vernus	8	Au It Ju Si	Brighton, 1976
	10	Rm	Brighton, 1976
	12	Ju	Brighton, 1976
	16	Cz It	Brighton, 1976
	18 19 20 22 23	Ju	Brighton, 1976
(b) albiflorus (Kit.) Ascherson & Graebner	8	Au It Ju	Brighton et al., 1973
x fritschii Derganc	13	Al Ju	Brighton et al., 1973
8 tommasinianus Herbert	16	Ju	Sušnik & Lovka, 1973
9 etruscus Parl.	8	It	Brighton et al., 1973
10 kosaninii Pulević			
11 nevadensis Amo	26	Hs	Rejón, 1974
	28 30	Hs	Brighton et al., 1973
12 carpetanus Boiss. & Reuter	64	Hs Lu	Brighton et al., 1973
13 sieberi Gay	22	Cr Gr	Brighton et al., 1973
14 veluchensis Herbert	26	Bu	Brighton et al., 1973
15 dalmaticus Vis.	24	Ju	Sušnik & Lovka, 1973
16 angustifolius Weston	12	Rs(K)	Brighton et al., 1973
17 reticulatus Steven ex Adams	12	Ju	Brighton et al., 1973
18 cvijicii Košanin	18	Gr	Brighton et al., 1973
19 scardicus Košanin	32	Ju	Sopova, 1972
	34	Ju	Brighton et al., 1973
	35 36	Ju	Sopova, 1972

310

20 pelistericus Pulević
21 flavus Weston   8 Bu Gr Ju Rm Brighton et al.,
                                   1973

No.	Taxon	2n	Distribution	Reference
20	pelistericus Pulević			
21	flavus Weston	8	Bu Gr Ju Rm	Brighton et al., 1973
22	olivieri Gay	6	Bu Gr	Brighton et al., 1973
23	chrysanthus (Herbert) Herbert	8	Bu Gr Ju	Brighton, ined.
		12	Bu Gr	Brighton, ined.
		20	Gr Ju	Brighton, ined.
24	biflorus Miller	8	It Ju	Brighton et al., 1973
		12	Rs(K)	Brighton et al., 1973
		16	Bu	Brighton, ined.
25	crewei Hooker fil.	12	Gr	Brighton et al., 1973
26	cartwrightianus Herbert	16	Cr Gr	Brighton et al., 1973
	sativus L.	24	Bl	Brighton, ined.
27	pallasii Goldb.	14	Bu Rs(K)	Brighton et al., 1973
28	thomasii Ten.	16	It Ju	Brighton et al., 1973
29	hadriaticus Herbert	16	Gr	Brighton et al., 1973
30	niveus Bowles	28	Gr	Brighton et al., 1973
31	longiflorus Rafin.	28	Si	Brighton et al., 1973
32	goulimyi Turrill	12	Gr	Brighton et al., 1973
33	serotinus Salisb.			
	(a) serotinus	22 23	Lu	Brighton et al., 1973
	(b) clusii (Gay) Mathew	22	Hs Lu	Brighton et al., 1973
		24	Hs Lu	Brighton et al., 1973
	(c) salzmannii (Gay) Mathew	22 24	Hs	Brighton et al., 1973
34	nudiflorus Sm.	48	Ga	Brighton et al., 1973
35	medius Balbis	24 + 0 - 5B	It	Brighton, 1976
36	laevigatus Bory & Chaub.	26	Cr Gr	Brighton et al., 1973
37	boryi Gay	30	Gr	Brighton et al., 1973
38	tournefortii Gay	30	Gr	Brighton et al., 1973
39	cancellatus Herbert	16	Ju	Lovka et al., 1971
40	robertianus C.D. Brickell	20	Gr	Brighton et al., 1973
41	speciosus Bieb.	18	Tu	Brighton et al., 1973
42	pulchellus Herbert	12	Gr Ju Tu	Brighton et al., 1973
43	banaticus Gay	26	Rm	Brighton et al., 1973

8 Romulea Maratti
  1 bulbocodium (L.) Sebastiani & Mauri   34   Lu   Fernandes et al., 1948

	36	Hs	Löve & Kjellqvist, 1973
	42	Ju	Sušnik & Lovka, 1973
2 ligustica Parl.			
3 requienii Parl.	34	Co	Contandriopoulos, 1962
4 revelierei Jordan & Fourr.			
5 linaresii Parl.			
6(a) linaresii			
(b) graeca Béguinot			
6 melitensis Béguinot			
7 ramiflora Ten.	36	It	Ricci, 1961
(a) ramiflora			
(b) gaditana (G. Kunze) Marais			
8 columnae Sebastiani & Mauri			
(a) columnae			
(b) rollii (Parl.) Marais			
9 rosea (L.) Ecklon			

9 Tritonia Ker-Gawler
  1 crocata (L.) Ker-Gawler
  2 x crocosmiflora (Lemoine) Nicholson

10 Gladiolus L.			
1 illyricus Koch	60	Br	Hamilton, ined.
	90	Lu	Fernandes, 1950
2 communis L.			
(a) communis	120	Hs	Hamilton, ined.
(b) byzantinus (Miller) A.P.	90 120	Hs	Hamilton, ined.
Hamilton			
3 palustris Gaudin	60	Po	Skalińska et al., 1974
4 imbricatus L.	60	Po	Skalińska et al., 1964
5 italicus Miller	120	Ju	Sušnik & Lovka, 1973
	171 ± 2	Hs	Hamilton, 1968

JUNCALES

CLXXXIX JUNCACEAE

1 Juncus L.			
1 maritimus Lam.	48	Gr Su	Snogerup, 1963
2 rigidus Desf.			
3 acutus L.	48	Ga Gr It	Snogerup, 1963
(a) acutus			
(b) leopoldii (Parl.) Snogerup			
4 littoralis C.A. Meyer	48	Europe	Snogerup, ined.
5 heldreichianus Marsson ex Parl.	48	Europe	Snogerup, ined.
6 jacquinii L.	c.170	Europe	Snogerup, ined.
7 filiformis L.	84	Scand.	Snogerup, 1971
8 arcticus Willd.	84	Scand.	Snogerup, 1971
9 balticus Willd.	?80	Da	Jørgensen et al., 1958
	84		
10 pyrenaeus Timb.-Lagr. & Jeanb.			
11 inflexus L.	40	Cz	Májovský et al., 1974
	42		

12 effusus L.	40	Cz	Májovský et al., 1976
	42	Su	Snogerup, 1963
13 conglomeratus L.	42	Su	Snogerup, 1963
14 subulatus Forskål	42	Ga	Labadie, 1976
15 trifidus L.	30	Is	Löve & Löve, 1956
(a) trifidus			
(b) monanthos (Jacq.) Ascherson & Graebner			
16 squarrosus L.	42	Da	Snogerup, 1963
17 compressus Jacq.	44	Da Su	Snogerup, 1963
18 gerardi Loisel.	84	Da	Snogerup, 1963
(a) gerardi			
(b) atrofuscus (Rupr.) Printz			
(c) montanus Snogerup			
(d) soranthus (Schrenk) Snogerup			
(e) libanoticus (J. Thieb.) Snogerup			
19 tenuis Willd.	84	Scand.	Snogerup, 1971
20 dudleyi Wieg.	84	non-Europe	
21 imbricatus Laharpe			
22 tenageia L. fil.			
23 sphaerocarpus Nees			
24 foliosus Desf.	26	Br Hb	Cope, 1976
25 sorrentini Parl.	28	Madeira	Cope, 1976
26 bufonius L.	108(not 100-110)	Be Ga Hb	Cope, 1976
27 minutulus Albert & Jahandiez	c.70	Europe	Snogerup, ined.
28 ranarius Song. & Perr.	$\overline{34}$	Hb	Cope, 1976
29 hybridus Brot.	34	Az Bl Hs Si	Cope, 1976
30 planifolius R. Br.			
31 capitatus Weigel	18	Ga	Snogerup, 1963
32 ensifolius Wikström	40	N. America	Snogerup, 1963
33 canadensis Gay	80	Canada	Snogerup, 1963
34 subnodulosus Schrank	40	Su	Snogerup, 1963
35 pygmaeus L.C.M. Richard	40	Scand.	Snogerup, 1972
36 tingitanus Maire & Weiller			
37 heterophyllus Dufour			
38 emmanuelis A. Fernandes & Garcia			
39 bulbosus L.	40	Po	Skalińska et al., 1971
40 atratus Krocker	40	Cz	Snogerup, 1963
41 thomasii Ten.			
42 acutiflorus Ehrh. ex Hoffm.	40	Scand.	Snogerup, 1972
43 valvatus Link			
44 striatus Schousboe ex E.H.F. Meyer			
45 fontanesii Gay			
(a) fontanesii			
(b) pyramidatus (Laharpe) Snogerup	40	Anatolia	Snogerup, ined.
46 alpinus Vill.			
(a) alpinus	40	Br	Elkington, 1974
(b) nodulosus (Wahlenb.) Lindman	40	Fe	Hamet-Ahti & Virrankoski, 1970
(c) alpestris (Hartman) A. & D. Löve	40	Scand.	Snogerup, 1972
47 anceps Laharpe	40	Ge	Wulff, 1937
48 requienii Parl.			
49 articulatus L.	80	Br	Fearn, 1977
50 biglumis L.	60	No	Knaben & Engelskjon, 1967
51 triglumis L.	50 c.134	Is	Löve & Löve, 1956
52 stygius L.			
53 castaneus Sm.	40	Is	Löve & Löve, 1956

2 Luzula DC.

1 campestris (L.) DC.	12	Su	Nordenskjold, 1951
2 multiflora (Retz.) Lej.			
(a) multiflora	24	Ga	Kliphuis & Wieffering, 1972
	36	Ga	Lambert & Giesi, 1967
(b) frigida (Buchenau) V. Krecz.	36	No	Knaben & Engelskjon, 1967
(c) congesta (Thuill.) Hyl.	48	Is	Löve & Löve, 1956
3 pallescens Swartz	12	Fe	Hamet-Ahti & Virrankoski, 1970
4 sudetica (Willd.) DC.	36	Ga	Lambert & Giesi, 1967
	48	Fe	Halkka, 1964
5 arcuata Swartz	36	No	Knaben, 1950
	42		
6 confusa Lindeb.	48	Is	Löve & Löve, 1956
7 arctica Blytt	24	No	Knaben, 1950
8 spicata (L.) DC.			
(a) spicata	24	Su	Nordenskjold, 1951
(b) mutabilis Chrtek & Křísa	12 14	Au	Nordenskjold, 1951
9 italica Parl.			
10 pindica (Hausskn.) Chrtek & Křísa			
11 hispanica Chrtek & Křísa			
12 nutans (Vill.) Duval-Jouve	12	Ga	Lambert & Giesi, 1967
13 caespitosa Gay			
14 nodulosa (Bory & Chaub.) E.H.F. Meyer			
15 sylvatica (Hudson) Gaudin			
(a) sylvatica	12	Ga	Lambert & Giesi, 1967
(b) henriquesii (Degen) P. Silva	c.84	Lu	Fernandes, 1950
16 sieberi Tausch			
17 lutea (All.) DC.	12	Hs	Noronha-Wagner, 1949
18 purpureospoendens Seub.			
19 nivea (L.) DC.	12	Ga	Lambert & Giesi, 1967
20 lactea (Link) E.H.F. Meyer	12	Hs	Noronha-Wagner, 1949
21 pedemontana Boiss. & Reuter			
22 luzuloides (Lam.) Dandy & Wilmott	12	Ga	Lambert & Giesi, 1967
(a) luzuloides			
(b) cuprina (Rochel ex Ascherson & Graebner) Chrtek & Křísa			
23 alpinopilosa (Chaix) Breistr.	12	Au	Nordenskjold, 1951
(a) alpinopilosa			
(b) candollei (E.H.F. Meyer) Rothm.			
(c) obscura Fröhner	12	Cz	Májovský et al., 1970
(d) velenovskyi (Kozuharov) Chrtek & Křísa			
24 glabrata (Hoppe) Desv.			
25 desvauxii Kunth	12	Ga	Lambert & Giesi, 1967
26 wahlenbergii Rupr.	24	No	Knaben & Engelskjon, 1967
	36	Su	Löve & Löve, 1944
27 parviflora (Ehrh.) Desv.	24	No	Knaben & Engelskjon, 1967

```
28 elegans Lowe                          6     Lu    Malheiros et al.,
                                                        1947
29 pilosa (L.) Willd.                   66     Fe    Halkka, 1964
                                        72     Fe    V. Sorsa, 1962
30 forsteri (Sm.) DC.                   24     Ga    Lambert & Giesi, 1967
31 luzulina (Vill.) Dalla Torre & Sarnth.  24  Au    Nordenskjold, 1951
```

BROMELIALES

CXC BROMELIACEAE

1 Fascicularia Mez
  1 pitcairniifolia (Verlot) Mez

COMMELINALES

CXCI COMMELINACEAE

1 Tradescantia L.
  1 virginiana L.
  2 fluminensis Velloso

2 Tripogandra Rafin.
  1 multiflora (Swartz) Rafin.

3 Commelina L.
  1 communis L.

CXCII ERIOCAULACEAE

1 Eriocaulon L.
  1 aquaticum (Hill) Druce   64   Hb   Löve & Löve, 1958
  2 cinereum R. Br.

GRAMINALES

CXCIII GRAMINEAE

Tribe Arundinarieae Steudel

1 Sasa Makino & Shibata
  1 palmata (Burbidge) E.G. Camus

2 Arundinaria Michx
  1 anceps Mitf.
  2 fastuosa (Latour-Marliac ex Mitf.)
      Lehaie
  3 japonica Siebold & Zucc. ex Steudel
  4 simonii (Carriere) A. & C. Riviere
  5 vagans Gamble

3 Phyllostachys Siebold & Zucc.

Tribe Poeae

4 Festuca L.
  1 triflora Desf.   14
  2 caerulescens Desf.   14
```
```

3 paniculata (L.) Schinz & Thell.
(a) paniculata	14	Ga	Küpfer, 1971
(b) spadicea (L.) Litard.	42	Ga	Van Loon et al.,
(c) baetica (Hackel) Markgr.-Dannenb.			1971

4 durandii Clauson	28	Lu	Fernandes & Queirós,
5 spectabilis Jan	42		1969

  (a) spectabilis
  (b) affinis (Boiss. & Heldr. ex Hackel)
    Hackel

6 altissima All.	14	Cz	Váchová & Feráková, 1977
	42	Br	Hubbard, 1954
7 drymeja Mert. & Koch	14	Po	Mizianty & Frey, 1973
8 gigantea (L.) Vill.	42	Cz	Májovský et al., 1970
9 pratensis Hudson	14	Cz	Májovský et al., 1974
	42		
(a) pratensis	14		

  (b) apennina (De Not.) Hegi

10 arundinacea Schreber	28	Ge	Tischler, 1934
	42	Cz	Uhríkova & Májovský, 1977
	70	Ge	Tischler, 1934
(a) arundinacea	42	Hs	Talavera, 1978

  (b) uechtritziana (Wiesb.) Hegi

(c) atlantigena (St-Yves) Auquier	56		
(d) fenas (Lag.) Arcangeli	28		

  (e) orientalis (Hackel) Tzvelev

11 laxa Host	14		
	28	Au	Gervais, 1965
	42	Ju	Lovka et al., 1972
12 dimorpha Guss.	28	Ga	Tombal, 1968

13 calabrica Huter
14 pseudeskia Boiss.

15 scariosa (Lag.) Ascherson & Graebner	14		
16 carpatica F.G. Dietr.	28	Cz	Májovský et al., 1974
17 pulchella Schrader	14	Au	Gervais, 1965
18 eskia Ramond ex DC.	14	Hs	Küpfer, 1969

19 alpestris Roemer & Schultes

20 elegans Boiss.	28	Lu	Fernandes & Queirós, 1969
21 burnatii St-Yves	14	Hs	Küpfer, 1971

22 valida (Uechtr.) Penzes
  (a) valida
  (b) leilaensis Markgr.-Dannenb.

23 gautieri (Hackel) K. Richter	14		
	28		
24 flavescens Bellardi	14	It	Gervais, 1965

25 scabriculmis (Hackel) K. Richter
  (a) scabriculmis
  (b) luedii Markgr.-Dannenb.

26 sardoa (Hackel ex Barbey) K. Richter	14	Co	Litardière, 1950
27 quadriflora Honckeny	14		

28 versicolor Tausch
  (a) versicolor

(b) brachystachys (Hackel) Markgr.-Dannenb.	14	Au	Litardière, 1950b

  (c) pallidula (Hackel) Markgr.-Dannenb.
  (d) dominii Krajina
29 acuminata Gaudin
30 varia Haenke

31 calva (Hackel) K. Richter
32 xanthina Roemer & Schultes
33 balcanica (Acht.) Markgr.-Dannenb.
  (a) balcanica
  (b) neicevii (Acht.) Markgr.-Dannenb.
  (c) rhodopensis Markgr.-Dannenb.
34 bosniaca Kummer & Sendtner
  (a) bosnaica
  (b) chlorantha (G. Beck) Markgr.-
     Dannenb.
  (c) pirinensis (Acht.) Markgr.-
     Dannenb.
35 penzesii (Acht.) Markgr.-Dannenb.
36 pindica (Markgr.-Dannenb.) Markgr.-
   Dannenb.
37 graeca (Hackel) Markgr.-Dannenb.
  (a) graeca
  (b) pawlowskiana Markgr.-Dannenb.
38 rechingeri Markgr.-Dannenb.
39 cyllenica Boiss. & Heldr.
  (a) cyllenica
  (b) pangaei Markgr.-Dannenb.
  (c) thasia Markgr.-Dannenb.
40 adamovicii (St-Yves) Markgr.-
   Dannenb.
41 galicicae I. Horvat ex Markgr.-
   Dannenb.

42 ampla Hackel	28	Lu	Fernandes & Queirós,
	42		1969
	56		

43 petraea Guthnick ex Seub.
44 jubata Lowe
45 morisiana Parl.
46 henriquesii Hackel
47 querana Litard.
48 capillifolia Dufour

49 porcii Hackel	14		
50 tatrae (Csakő) Degen	14	Cz	Baksay, 1957b

51 peristerea (Vetter) Markgr.-Dannenb.
52 korabensis (Jav. ex Markgr.-Dannenb.)

53 borderi (Hackel) K. Richter	14	Ga	Litardière, 1950b

54 norica (Hackel) K. Richter
55 amethystina L.

(a) amethystina	14	?He	Litardière, 1950b

  (b) kummeri (G. Beck) Markgr.-Dannenb.
  (c) ritschlii (Hackel) Lemke ex Markgr.-
     Dannenb.
  (d) orientalis Krajina

56 heterophylla Lam.	28	Cz	Májovský et al.,
			1974

57 plicata Hackel
58 violacea Schleicher ex Gaudin

(a) violacea	14	Ga	Zickler, 1968

  (b) macrathera (Hackel ex G. Beck)
     Markgr.-Dannenb.
  (c) handelii Markgr.-Dannenb.
59 nitida Kit.
  (a) nitida
  (b) flaccida (Schur) Markgr.-Dannenb.
60 iberica (Hackel) K. Richter
61 puccinellii Parl.

```
62 clementei Boiss.                          14      Hs     Küpfer, 1968
63 picta Kit.                                14      Cz     Májovský et al.,
                                                            1974
64 baffinensis Polunin                       28      As     Zhukova, 1965
65 nigrescens Lam.
   (a) nigrescens                            28
                                             42      Br     Maude, 1940
   (b) microphylla (St-Yves) Markgr.-
       Dannenb.
66 rubra L.                                  14      Lu     Rodrigues, 1953
                                          28 70      Is     Löve & Löve, 1942
                                             42      Su     Löve & Löve, 1944
                                             56      Hu     Pólya, 1949
   (a) rubra
   (b) litoralis (G.F.W. Meyer) Auquier      42      Be     Auquier &
                                                            Rammeloo, 1973
   (c) arenaria (Osbeck) Syme                56      Da     Kjellqvist, 1964
   (d) pruinosa (Hackel) Piper               42
   (e) juncea (Hackel) Soó                   42      Be     Auquier &
                                                            Rammeloo, 1973
   (f) thessalica Markgr.-Dannenb.
   (g) asperifolia (St-Yves) Markgr.-Dannenb.
67 richardsonii Hooker                       42      Is     Löve & Löve, 1956
68 pyrenaica Reuter                          28
69 trichophylla (Ducros ex Gaudin) K.        42
   Richter
70 pseudotrichophylla Patzke                 42
71 cyrnea (Litard. & St-Yves) Markgr.-Dannenb.
72 oelandica (Hackel) K. Richter             42
73 juncifolia St-Amans                       56      Be     Auquier &
                                                            Rammeloo, 1973
74 diffusa Dumort.                           42
                                             56      Be     Auquier &
                                                            Rammeloo, 1973
75 rivularis Boiss.                       14 42
76 rothmaleri (Litard.) Markgr.-Dannenb.
77 nevadensis (Hackel) Markgr.-Dannenb.      70
78 cretacea T. Popov & Proskorj.
79 brachyphylla Schultes                     28
                                             42      Asia   Zhukova, 1965
80 hyperborea Holmen                         28      Is     Löve & Löve, 1956
81 pirinica I. Horvat ex Markgr.-Dannenb.
82 intercedens (Hackel) Ludi ex Becherer
83 riloensis (Hackel ex Hayek) Markgr.-
   Dannenb.
84 halleri All.
   (a) halleri                               14      He     Scholte, 1977
   (b) scardica (Griseb.) Markgr.-Dannenb.
85 glacialis (Miégeville ex Hackel) K.       14
   Richter
86 rupicaprina (Hackel) A. Kerner            14  Bot. Gard. Brandberg, 1948
87 bucegiensis Markgr.-Dannenb.
88 frigida (Hackel) K. Richter
89 alpina Suter                              14      Ga     Parreaux, 1972
90 olympica Vetter
91 vizzavonae Ronniger
92 horvatiana Markgr.-Dannenb.
93 jeanpertii (St-Yves) Markgr.
   (a) jeanpertii
```

(b) achaica (Markgr.-Dannenb.) Markgr.-
    Dannenb.
(c) campana (N. Terracc.) Markgr.-
    Dannenb.

94 stenantha (Hackel) K. Richter	14		
95 pseudodura Steudel	28		
96 circummediterranea Patzke	14	Ga	Parreaux, 1972
97 costei (St-Yves) Markgr.-Dannenb.	28	Ga	Parreaux, 1972
98 hervieri Patzke	14	Ga	Litardière, 1950
99 patzkei Markgr.-Dannenb.			
100 hystrix Boiss.			
101 reverchonii Hackel			
102 tenuifolia Sibth.	14 28	Ga	Bidault, 1972
103 ovina L.	14	Is	Löve & Löve, 1956
104 ophioliticola Kerguélen	28	Ga	Kerguélen, 1975
105 niphobia (St-Yves) Kerguélen	28	Ga	Kerguélen, 1976
106 airoides Lam.	14	Cz	Hadde & Haskova, 1956
	28		
	35		
107 armoricana Kerguélen	28	Ga	Kerguélen, 1975
108 eggleri Tracey	28	Au	Tracey, 1977
109 sipylea (Hackel) Markgr.-Dannenb.			
110 lapidosa (Degen) Markgr.-Dannenb.			
111 occitanica (Litard.) Auquier & Kerguélen	14	Ga	Kerguélen, 1975
112 keteropachys (St-Yves) Patzke ex Auquier	28	Be	Auquier & Rammeloo, 1973
113 guestfalica Boenn. ex Reichenb.	28		
114 lemanii Bast.	42	Be	Auquier & Rammeloo, 1973
115 arvernensis Auquier, Kerguélen & Markgr.-Dannenb.	28	Ga	Auquier & Kerguélen, 1977
116 longifolia Thuill.	14	Br	Hubbard, 1954
117 inops De Not.			
118 vaginata Waldst. & Kit. ex Willd.	14	Hu	Baksay, 1956
(a) vaginata			
(b) dominii (Krajina) Soó			
119 psammophila (Hackel ex Čelak.) Fritsch			
120 pannonica Wulfen ex Host			
121 beckeri (Hackel) Trautv.			
122 polesica Zapal.	14	Da	Bücher, 1947
123 javorkae Májovský			
124 arenicola (Prodan) Soó			
125 cinerea Vill.	28	Ga	Bidault, 1968
126 pallens Host			
(a) pallens	14 28	Cz	Májovský et al., 1974
(b) scabrifolia (Hackel ex Rohlena) Zielonk.			
(c) treskana Markgr.-Dannenb.			
127 huonii Auquier	42	Ga	Auquier, 1973
128 gracilior (Hackel) Markgr.-Dannenb.	14	Ga	Parreaux, 1972
129 liviensis (Verguin) Markgr.-Dannenb.	14	Ga	Auquier & Kerguélen, 1977
130 ochroleuca Timb.-Lagr.	28		

(a) ochroleuca
(b) gracilior (Verguin) Kerguélen
(c) bigorrensis (St-Yves) Kerguélen
(d) heteroidea (Verguin) Markgr.-Dannenb.
131 brigantina (Markgr.-Dannenb.) Markgr.-
    Dannenb.

132 valentina (St-Yves) Markgr.-Dannenb.
133 degenii (St-Yves) Markgr.-Dannenb.   28   Ga   Bidault, 1968
134 oviniformis Vetter
135 hirtovaginata (Acht.) Markgr.-Dannenb.
136 ticinensis (Markgr.-Dannenb.
137 grandiaristata Markgr.-Dannenb.
138 thracica (Acht.) Markgr.-Dannenb.
139 hercegovinica Markgr.-Dannenb.
140 robustifolia Markgr.-Dannenb.
141 panciciana (Hackel) K. Richter
142 apuanica Markgr.-Dannenb.
143 koritnicensis Hayek & Vetter
144 pohleana Alexeev
145 macedonica Vetter
146 polita (Halácsy) Tzvelev
147 curvula Gaudin
   (a) curvula   56 + 0 - 1B Ga   Bidault, 1968
   (b) crassifolia (Gaudin) Markgr.-   56 + 1B
       Dannenb.
   (c) cagiriensis (Timb.-Lagr.) Markgr.
       Dannenb.
148 pachyphylla Degen ex E.I. Nyárády
149 igoschiniae Tzvelev
150 vivipara (L.) Sm.   28   Is   Löve & Löve, 1956
                      21 35 42 49
                      56
151 glauca Vill.   42
152 vasconcensis (Markgr.-Dannenb.)   42   Ga Hs Kerguélen, 1976
   Auquier & Kerguélen
153 durissima (Hackel) Kerguélen   42 56   Ga   Kerguélen, 1975

154 indigesta Boiss.
   (a) indigesta   42   Hs   Küpfer, 1968
   (b) hackeliana (St-Yves) Markgr.-   42
       Dannenb.
   (c) aragonensis (Willk.) Kerguélen
   (d) alleizettei (Litard.) Kerguélen
   (e) molinieri (Litard.) Kerguélen
   (f) litardierei (St-Yves) Kerguélen
155 pseudovina Hackel ex Wiesb.   14   Hu   Pólya, 1948
                      28   Hu   Felföldy, 1947
156 illyrica Markgr.-Dannenb.
157 wagneri (Degen, Thaisz & Flatt)
   Krajina   28   Hu   Horánszky et al.,
158 makutrensis Zapal   1972
159 valesiaca Schleicher ex Gaudin   14   Cz   Baksay, 1958
                      28   Hu   Felföldy, 1947b
160 carnuntina Tracey   42   Au   Tracey, 1977
161 rupicola Heuffel   42   Hu   Polya, 1949
   (a) rupicola
   (b) saxatilis (Schur) Rauschert
162 taurica (Hackel) A. Kerner ex Trautv.
163 dalmatica (Hackel) K. Richter   28
164 pseudodalmatica Krajina ex Domin   28   Cz   Baksay, 1957
165 callieri (Hackel ex St-Yves) Markgraf
166 wolgensis P. Smirnov
167 stricta Host   42   Hu   Felföldy, 1947
168 trachyphylla (Hackel) Krajina   42   Ga   Bidault, 1972
169 brevipila Tracy   42   Au   Tracey, 1977
170 duvalii (St-Yves) Stohr   28

x Festulolium Ascherson & Graebner
  1 loliaceum (Hudson) P. Fourn.   14   Ho   Gadella & Kliphuis,
                                                                1966
  2 braunii (K. Richter) A. Camus   14   Br   Peto, 1933

5 Lolium L.
  1 perenne L.   14 + 1 - 3B
  2 multiflorum Lam.   14   Lu   Fernandes & Queirós,
                                                                1969

  3 rigidum Gaudin
    (a) rigidum   14   Lu   Fernandes & Queirós,
                                                                1969
    (b) lepturoides (Boiss.) Sennen &
        Mauricio
  4 temulentum L.   14   Bl   Dahlgren et al.,
                                                                1971
  5 remotum Schrank   14 + 1B   Br   Hovin & Hill, 1966

6 Vulpia C.C. Gmelin
  1 sicula (C. Presl) Link   14   It Si   Cotton & Stace,
                                                                1976
  2 geniculata (L.) Link   14   Lu   Cotton & Stace,
                                                                1976
  3 ligustica (All.) Link   14   Si   Cotton & Stace,
                                                                1976
  4 alopecuros (Schousboe) Dumort.   14   Lu   Fernandes & Queirós,
                                                                1969
  5 fasciculata (Forskål) Samp.   28   Bl   Dahlgren et al.,
                                                                1971
  6 membranacea (L.) Dumort.   14   Ga Hs   Cotton & Stace,
                                                                1976
  7 fontquerana Melderis & Stace   14   Hs   Cotton & Stace,
                                                                1976
  8 bromoides (L.) S.F. Gray   14   Br   I. & O. Hedberg, 1961
  9 muralis (Kunth) Nees   14   Lu   Fernandes & Queirós,
                                                                1969
 10 myuros (L.) C.C. Gmelin   42 Be Br Co Ga Cotton & Stace,
                                           Gr It Ju Si   1976
 11 ciliata Dumort.
    (a) ciliata   28 Br Co Cr Ga Cotton & Stace,
                                           Gr Hs Ju Lu   1976
                                           Rm Si
    (b) ambigua (Le Gall) Stace & Auquier 28   Be Br Ga Cotton & Stace,
                                                            1976
 12 unilateralis (L.) Stace   14   Bu   Kozuharov &
                                                              Petrova, 1973

7 Ctenopsis De Not.
  1 delicatula (Lag.) Paunero   14   Lu   Fernandes & Queirós,
                                                                1969
  2 gypsophila (Hackel) Paunero

8 Wangenheimia Moench
  1 lima (L.) Trin.

9 Micropyrum Link
  1 tenellum (L.) Link   14   Lu   Fernandes & Queirós,
                                                                1969
  2 patens (Brot.) Rothm. ex Pilger   14   Lu   Fernandes & Queirós,
                                                                1969

10 Castellia Tineo
  1 tuberculosa (Moris) Bor

11 Narduroides Rouy
  1 salzmannii (Boiss.) Rouy   14

12 Desmazeria Dumort.
  1 sicula (Jacq.) Dumort.
  2 marina (L.) Druce   14 28   Lu   Fernandes &
                                                       Queirós, 1969
  3 rigida (L.) Tutin   14   Bu   Kozuharov &
                                                       Petrova, 1974
    (a) rigida
    (b) hemipoa (Delile ex Sprengel) Stace

13 Cutandia Willk.
  1 maritima (L.) W. Barbey   14   Lu   Fernandes &
                                                       Queirós, 1969
  2 stenostachya (Boiss.) Stace
  3 divaricata (Desf.) Ascherson ex W.
    Barbey
  4 memphitica (Sprengel) K. Richter

14 Sphenopus Trin.
  1 divaricatus (Gouan) Reichenb.   12 N. Africa Litardière, 1950b

15 Vulpiella (Trabut) Burollet
  1 tenuis (Tineo) Kerguelen

16 Poa L.
  1 annua L.   28   Br   Tutin, 1957
  2 informa Kunth   14   Ga   Tutin, 1957
  3 supina Schrader   14   He   Tutin, 1957
  4 trivialis L.   14   Hs   Løve & Kjellqvist,
                                                       1973
                                  14 + 0 - 5B   Su   Bosemark, 1957
                                        28   Bu   Kozuharov &
                                                       Kuzmanov, 1970
    (a) trivialis
    (b) sylvicola (Guss.) H. Linb. fil.
  5 feratiana Boiss. & Reuter
  6 alpigena (Fries) Lindman   35   Sb   Flovik, 1940
                                        43   Sb   Flovik, 1940
                                        56   Rs(N) Sokolovskaya &
                                        70-72   Rs(N)   Strelkova, 1960
                                        84   Rs(N) Sokolovskaya &
                                                       Strelkova, 1941
  7 subcaerulea Sm.   38-117   Is   Løve & Løve, 1956
  8 pratensis L.   42   Su   Almgaard, 1960
                                        50-78   Is   Løve & Løve, 1956
                                        91   Is   Løve & Løve, 1942
                                        98   Bu   Christov & Nikolov,
                                                       1962
  9 angustifolia L.   46-63   Is   Løve & Løve, 1956
 10 arctica R. Br.   38   No   Nannfeldt, 1940
                                        56   Sb   Flovik, 1940
                                        72   Rs(N) Sokolovskaya &
                                                       Strelkova, 1941
                                        76   Is   Løve & Løve, 1956
 11 granitica Br.-Bl.

(a) granitica	64 67 71 72 94	Po	Skalińska et al., 1957
(b) disparilis (E.I. Nyárády) E.I. Nyárády)			
12 cenisia All.			
(a) sardoa E. Schmid	56	Co	Contandriopoulos, 1962
(b) cenisia	28	He	Favarger, 1959
	49	Ga	Bajon & Greber, 1971
	50-56	Ga Ge	Favarger, 1959
13 longifolia Trin.			
14 sibirica Roshev.			
15 chaixii Vill.	14	Cz	Májovský et al., 1974
16 hybrida Gaudin	14	He	Favarger, 1959
17 remota Forselles	14	Su	Nannfeldt, 1942
18 stiriaca Fritsch & Hayek	14	Cz	Edmondson, ined.
19 flexuosa Sm.	42	Is	Löve & Löve, 1956
20 laxa Haenke	21 28	He	Nygren, 1956
21 pirinica Stoj. & Acht.	14	Bu	Stoeva, 1977
22 minor Gaudin	28	He	Nygren, 1956
23 compressa L.	42	Cz	Májovský et al., 1974
	45 49 56	Is	Löve & Löve, 1948
24 flaccidula Boiss. & Reuter			
25 palustris L.			
26 glauca Vahl	42	Is	Löve & Löve, 1956
	44	No	Knaben & Engelskjon, 1967
	49 56 63	Is	Löve & Löve, 1956
	65	Su	Aakerberg, 1942
	c.70	Sb	Flovik, 1940
27 nemoralis L.	28	Po	Skalińska et al., 1957
	33	Ga	Guinochet, 1943
	42	Is	Löve & Löve, 1956
	56	Br	Hooper, 1961
28 rehmannii (Ascherson & Graebner) Woloszczak	14	Rm	Lungeanu, 1972
29 pannonica A. Kerner			
(a) pannonica			
(b) scabra (Ascherson & Graebner) Soó			
30 versicolor Besser			
31 sterilis Bieb.			
32 stepposa (Krylov) Roshev.			
33 abbreviata R. Br.			
34 trichophylla Heldr. & Sart. ex Boiss.	14	Gr	Edmondson, ined.
35 bulbosa L.	21	Lu	Fernandes & Queirós, 1969
	28	Al Co Ga	Strid, 1971; Litardière, 1949
	39 42	Po	Skalińska et al., 1957
	42	It	Tutin, 1961
36 perconcinna J.R. Edmondson	14	Bu Ga He	Nygren, 1962
37 ligulata Boiss.	14	Hs	Löve & Kjellqvist, 1973
38 timoleontis Heldr. ex Boiss.	14	Ju	Nygren, 1957
39 media Schur	14 + 0 - 2B	Ju	Edmondson, ined.
40 pumila Host	14	Au	Nygren, 1962

41 molinerii Balbis	14	He	Nygren, 1955
42 badensis Haenke ex Willd.	14 + 2 - 8B	Au Ge He	Muntzing & Nygren, 1955
	18	Cz	Holub et al., 1971
	21 28 Ga	Ge Hu	Nygren, 1962
43 alpina L.	22 26 28 33-35	Po	Skalińska, 1950
	37-40	Su	Muntzing, 1966
	42	Rs(N)	Sokolovskaya & Strelkova, 1960
	43-45	Su	Muntzing, 1966
	33-46	No	Knaben & Engelskjon, 1967
	56	Bu	Stoeva, 1977
44 jubata A. Kerner			
45 flabellata (Lam.) Hooker fil.			
46 x hartzii Gand.			
47 x jemtlandica (Almq.) K. Richter	c.36	Br	Hedberg, 1958
48 x nobilis Skalinska	c.80	Po	Skalińska, 1950
49 x taurica H. Pojark.			

17 Bellardiochloa Chiov.

1 violacea (Bellardi) Chiov.	14	He	Gervais, 1965

18 Eremopoa Roshev.
  1 altaica (Trin.) Roshev.

19 Catabrosella (Tzvelev) Tzvelev
  1 humilis (Bieb.) Tzvelev

20 Arctophila Rupr. ex N.J. Andersson

1 fulva (Trin.) N.J. Andersson	42	Rs(N)	Sokolovskaya & Strelkova, 1960

21 Puccinellia Parl.
  1 distans (L.) Parl.

(a) distans	42	Cz	Váchová & Feráková, 1978
(b) limosa (Schur) Jáv.	28	Hu	Pólya, 1948
(c) hauptiana (Trin. ex V. Krecz.) W.E. Hughes	28	Asia	
	42	Asia	
(d) borealis (Holmberg) W.E. Hughes	42	Is	Löve & Löve, 1956
2 tenella (Lange) Holmberg svarlbadensis Rønning	42	Sb	Rønning, 1962
3 fasciculata (Torrey) E.P. Bicknell			
(a) fasciculata	28	Br	Rutland, 1941
(b) pungens (Pau) W.E. Hughes			
4 maritima (Hudson) Parl.	56	Ga	Delay, 1969
5 phryganodes (Trin.) Scribner & Merr.	28	Rs(N)	Sokolovskaya & Strelkova, 1960
6 festuciformis (Host) Parl.	35	Ga	Labadie, 1976

  (a) festuciformis
  (b) convoluta (Hornem.) W.E. Hughes
  (c) intermedia (Schur) W.E. Hughes
  (d) tenuifolia (Boiss. & Reuter) W.E. Hughes
  7 dolicholepis (V. Krecz.) V. Krecz. ex Pavlov
  8 gigantea (Grossh.) Grossh.

```
 9 poecilantha (C. Koch) Grossh.
10 tenuissima (V. Krecz.) V. Krecz. ex Pavlov
11 angustata (R. Br.) Rand & Redf.        42
12 rupestris (With.) Fernald & Weatherby  42      Br      Rutland, 1941
13 vahliana (Liebm.) Scribner & Merr.
```

x Pucciphippsia Tzvelev
  1 vacillans (Th. Fries) Tzvelev

22 Phippsia R. Br.
   1 algida (Solander) R. Br.   28   No   Knaben, 1950
   2 concinna (Th. Fries) Lindeb.   28   No   Knaben, 1950

23 Schlerochloa Beauv.
   1 dura (L.) Beauv.   14   Cz   Vachova &
                                                     Ferakova, 1977

24 Dupontia R. Br.
   1 fisheri R. Br.   88   Rs(N)   Sokolovskaya &
                                                     Strelkova, 1960

   2 psilosantha (Rupr.) Griseb.   44   Rs(N)   Sokolovskaya &
                                                     Strelkova, 1960

25 Arctagrostis Griseb.
   1 latifolia (R. Br.) Griseb.   56 62   Rs(N)   Sokolovskaya &
                                                     Strelkova, 1960

26 Dactylis L.
   1 glomerata L.   14   Su   Lundqvist, 1965
                                     28   Is   Løve & Løve, 1956
                             28 + 0 - 4B   Lu   Fernandes & Queirós,
                                                     1969

27 Beckmannia Host
   1 eruciformis (L.) Host   14   Cz   Májovský et al.,
                                                     1970
   (a) eruciformis
   (b) borealis Tzvelev
   2 syzigachne (Steudel) Fernald

28 Cynosurus L.
   1 cristatus L.   14   Lu   Fernandes & Queirós,
                                                     1969
   2 echinatus L.   14   Bu   Kozuharov &
                                                     Kuzmanov, 1970
   3 elegans Desf.   14   Lu   Fernandes & Queirós,
                                                     1969
   4 callitrichus W. Barbey

29 Lamarckia Moench
   1 aurea (L.) Moench   14   Lu   Fernandes & Queirós,
                                                     1969

30 Catabrosa Beauv.
   1 aquatica (L.) Beauv.   20   Is   Løve & Løve, 1956

31 Cinna L.
   1 latifolia (Trev.) Griseb.

32 Apera Adanson
   1 spica-venti (L.) Beauv.   14   Ho   Gadella & Kliphuis,
                                                     1968
```
```

(a) spica-venti
(b) maritima (Klokov) Tzvelev
2 interrupta (L.) Beauv.   14   Br   Maude, 1940
3 intermedia Hackel

33 Psilurus Trin.
1 incurvus (Gouan) Schinz & Thell.   28   Lu   Fernandes & Queirós,
                                                          1969

34 Mibora Adanson
1 minima (L.) Desv.

35 Briza L.
1 media L.   14   Lu   Fernandes & Queirós,
                                                          1969
                                            28   Bu   Kozuharov et al.,
(a) media   1974
(b) elatior (Sibth. & Sm.) Rohlena
2 maxima L.   14   Lu   Fernandes & Queirós,
                                                          1969
3 minor L.   10   Lu   Fernandes & Queirós,
                                                          1969
4 humilis Bieb.   14   Bu   Kozuharov et al.,
                                                          1974

Tribe Seslerieae Koch

36 Sesleria Scop.
1 ovata (Hoppe) A. Kerner
2 leucocephala DC.
3 sphaerocephala Ard.   14   It   Favarger & Huynh,
                                                          1964
4 tenuifolia Schrader
(a) tenuifolia   28   Al   Ujhelyi, 1959
                                            42
(b) kalnikensis (Jav.) Deyl   56   Ju   Ujhelyi, 1960b
5 comosa Velen.   28   Bu   Ujhelyi, 1959
6 tenerrima (Fritsch) Hayek   28   Gr   Ujhelyi, 1960b
7 sadlerana Janka
(a) sadlerana   56   Hu   Ujhelyi, 1959
(b) tatrae (Degen) Deyl   56   Po   Rychlewski, 1955
8 heuflerana Schur
(a) heuflerana   28   Cz   Murín &
                                                          Májovský, 1976
(b) hungarica (Ujhelyi) Deyl   56   Hu   Ujhelyi, 1959
9 bielzii Schur   56   Rm   Ujhelyi, 1959
10 coerulans Friv.   56   Bu   Ujhelyi, 1959
11 korabensis (Kummerle & Jav.) Deyl
12 vaginalis Boiss. & Orph.
13 doerfleri Hayek
14 wettsteinii Dörfler & Hayek   56   Al   Ujhelyi, 1959
15 robusta Schott, Nyman & Kotschy
(a) robusta   56   Al   Ujhelyi, 1959
(b) skanderbeggii (Ujhelyi) Dehl
16 nitida Ten.   28   Bot. Gard.   Ujhelyi, 1960b
17 alba Sibth. & Sm.
18 latifolia (Adamovic) Degen   28   Ju   Ujhelyi, 1959
19 autumnalis (Scop.) F.W. Schultz   28   Ju   Ujhelyi, 1959
20 anatolica Deyl
21 argentea (Savi) Savi   28   Hs   Löve & Kjellqvist,
                                                          1973
22 albicans Kit. ex Schultes   28   Cz   Murín & Májovský,
                                                          1976

```
    (a) albicans
    (b) angustifolia (Hackel & G. Beck) Deyl
 23 caerulea (L.) Ard.                    28    Ga    Parreaux, 1971
 24 taygetea Hayek                        28    Gr    Ujhelyi, 1960
    serbica                               28    Ju    Ujhelyi, 1959
 25 rigida Heuffel ex Reichenb.
    (a) rigida
    (b) achtarovii (Deyl) Deyl
 26 insularis Sommier                     28    Co    Litardière, 1949
    (a) insularis                         28    Co    Litardière, 1949
    (b) italica (Pamp.) Deyl
    (c) sillingeri (Deyl) Deyl


 37 Oreochloa Link
    1 disticha (Wulfen) Link              14    Au    Reese, 1953
    2 blanka Deyl                         14
    3 seslerioides (All.) K. Richter      14    Ga    Gervais, 1965
    4 confusa (Coincy) Rouy


 38 Ammochloa Boiss.
    1 palaestina Boiss.


 39 Echinaria Desf.
    1 capitata (L.) Desf.                 18    Bu    Kozuharov &
                                                      Petrova, 1974


    Tribe Meliceae Reichenb.


 40 Melica L.
    1 nutans L.                           18    Fe    V. Sorsa, 1962
    2 picta C. Koch                       18          Doulat, 1943
    3 uniflora Retz.                      18    Ho    Gadella & Kliphuis,
                                                      1963
    4 rectiflora Boiss. & Heldr.
    5 minuta L.                           18    Ga    Delay, 1969
                                          36    Lu    Fernandes & Queirós,
                                                      1969
    6 altissima L.                        18          Doulat, 1943
    7 ciliata L.
    (a) ciliata                           18    Ga    Delay, 1969
    (b) taurica (C. Koch) Tzvelev         18
    (c) monticola (Prokudin) Tzvelev
    (d) magnolii (Gren. & Godron)
    8 transsilvanica Schur                18    Po    Frey, 1971
    (a) transsilvanica
    (b) klokovii Tzvelev
    9 cupanii Guss.
   10 bauhinii All.


 41 Schizachne Hackel
    1 callosa (Turcz. ex Griseb.) Ohwi


    Tribe Glycerieae Endl.


 42 Glyceria R. Br.
    1 triflora (Korsh.) Komarov
    2 arundinacea Kunth
    3 maxima (Hartman) Holmberg           60    Ho    Gadella & Kliphuis,
                                                      1963
    4 lithuanica (Gorski) Gorski
    5 striata (Lam.) A.S. Hitchc.
```

```
 6 nemoralis (Uechtr.) Uechtr. &          20     Cz     Holub et al.,
     Koernicke                                            1970
 7 declinata Breb.                        20     Lu     Fernandes & Queirós,
                                                         1969
 8 fluitans (L.) R. Br.                   40     Hu     Pólya, 1950
 9 spicata (Biv.) Guss.
10 plicata (Fries) Fries                  40     Hs     Löve & Kjellqvist,
                                                         1973
11 x pedicellata Townsend

43 Pleuropogon R. Br.
 1 sabinei R. Br.


Tribe Bromeae Dumort.

44 Bromus L.
 1 diandrus Roth                          56     Hs     Löve & Kjellqvist,
                                                         1973
 2 rigidus Roth                           42     Ga     Cugnac & Simonet, 1941b
 3 sterilis L.                            14     Cz     Májovský et al.,
                                                         1974
 4 tectorum L.                            14     Hs     Löve & Kjellqvist,
                                                         1973
 5 madritensis L.                         28     Bu     Kozuharov et al.,
                                                         1974
                                          42
 6 fasciculatus C. Presl                  14
 7 rubens L.                              28     Hs     Björkqvist et al.,
                                                         1969
 8 inermis Leysser                        56     Su     Löve & Löve, 1944
 9 sibiricus Drobov
10 ramosus Hudson                         14     Br     Hubbard, 1954
                                          28     Hu     Pólya, 1950
                                          42     Cz     Májovský et al.,
                                                         1976
11 benekenii (Lange) Trimen               28     Cz     Májovský et al.,
                                                         1976
12 erectus Hudson                         42     Hu     Baksay, 1961
                                          56     Hu     Pólya, 1949
    (a) erectus
    (b) condensatus (Hackel) Ascherson &
        Graebner
    (c) stenophyllus (Link) Ascherson &
        Graebner
    (d) transsilvanicus (Steudel) Ascherson
        & Graebner
13 pannonicus Kummer & Sendtner
    (a) pannonicus
    (b) monocladus (Domin) P.M. Sm.       28     Cz     Májovský et al.,
                                                         1976
14 moellendorffianus (Ascherson & Graebner)
    Hayek
15 pindicus Hausskn.
16 tomentellus Boiss.
17 moesiacus Velen.
18 riparius Rehmann                       70
19 cappadocicus Boiss. & Balansa          42
    (a) cappadocicus
    (b) lacmonicus (Hausskn.) P.M. Sm.
20 arvensis L.                            14     Bu     Kozuharov et al.,
                                                         1974
```

21 pseudosecalinus P.M. Sm.	14	Br	Smith, 1967
22 secalinus L.	28	Su	Löve & Löve, 1944
23 bromoideus (Lej.) Crépin	28	Ga	Cugnac & Simonet, 1941a
24 grossus Desf. ex DC.	28	Be	Tournay, 1968
25 commutatus Schrader	28	Bu	Kozuharov et al., 1974
(a) commutatus			
(b) neglectus (Parl.) P.M. Sm.			
26 racemosus L.	28	Is	Löve & Löve, 1956
27 hordeaceus L.	28	Hs	Löve & Kjellqvist, 1973
(a) hordeaceus			
(b) molliformis (Lloyd) Maire & Weiller			
(c) ferronii (Mabille) P.M. Sm.			
(d) thominii (Hard.) Maire & Weiller			
28 interruptus (Hackel) Druce	28	Br	Hubbard, 1954
29 lepidus Holmberg	28	Br	Hubbard, 1954
30 scoparius L.	14	Bu	Kozuharov et al., 1974
31 alopecuros Poiret	14		
(a) alopecuros			
(b) caroli-henrici (W. Greuter) P.M. Sm.			
32 lanceolatus Roth	28	Lu	Gardé & Malheiros-Gardé, 1954
33 intermedius Guss.	14		
34 japonicus Thunb.	14	Bu	Kozuharov et al., 1974
(a) japonicus			
(b) anatolicus (Boiss. & Heldr.) Penzes			
35 squarrosus L.	14	Bu	Jasiewicz & Mizianty, 1975
36 willdenowii Kunth	42	Lu	Fernandes & Queirós, 1969
37 carinatus Hooker & Arnott	56	Br	Hubbard, 1954

Tribe Brachypodieae C.O. Harz

45 Brachypodium Beauv.			
1 sylvaticum (Hudson) Beauv.	18	Br	Hedberg & Hedberg, 1961
	28 42 + 2B 56	Bu	Kozuharov et al., 1974
(a) sylvaticum			
(b) glaucovirens Murb.	16	Bu	Kozuharov et al., 1974
2 pinnatum (L.) Beauv.	14 16 28	Bu	Kozuharov et al., 1974
(a) pinnatum	28	Bu	Kozuharov et al., 1974
(b) rupestre (Host) Schübler & Martens	14	Bu	Kozuharov & Nicolova, 1975
3 retusum (Pers.) Beauv.	18		
4 phoenicoides (L.) Roemer & Schultes	28	Lu	Fernandes & Queirós, 1969
5 distachyon (L.) Beauv.	10	Bu	Kozuharov et al., 1974
	28	Ga	Kliphuis & Wieffering, 1972
	30	Lu	Fernandes & Queiros, 1969

Tribe Triticeae Dumort.

46 Festucopsis (C.E. Hubbard) Melderis
  1 serpentini (C.E. Hubbard) Melderis   14   Al   Jones, 1955
  2 sancta (Janka) Melderis

47 Leymus Hochst.
  1 arenarius (L.) Hochst.   56   Is   Löve & Löve, 1956
    mollis   28   Is   Löve & Löve, 1956
  2 racemosus (Lam.) Tzvelev
    (a) racemosus   28   ?Europe Bowden, 1957
    (b) sabulosus (Bieb.) Tzvelev 28   Asia   Bowden, 1957
    (c) klokovii Tzvelev
  3 karelinii (Turcz.) Tzvelev   56
  4 akmolinensis (Drobov) Tzvelev 28   Rs(E)
  5 paboanus (Claus) Pilger
  6 multicaulis (Kar. & Kir.) Tzvelev
  7 ramosus (Trin.) Tzvelev

48 Elymus L.
  1 sibiricus L.   28
  2 panormitanus (Parl.) Tzvelev   28   It   Schulz-Schaeffer
                                         & Jurasits, 1962
  3 trachycaulus (Link) Gould ex Shinners 28
  4 mutabilis (Drobov) Tzvelev   28
  5 caninus (L.) L.   28   Lu   Fernandes & Queirós,
                                         1969
  6 uralensis (Nevski) Tzvelev
    (a) uralensis
    (b) viridiglumis (Nevski) Tzvelev
  7 fibrosus (Schrenk) Tzvelev   28
  8 macrourus (Turcz.) Tzvelev
    (a) macrourus   28
    (b) turuchanensis (Reverdatto) Tzvelev 28   Rs(N)   Sokolovskaya, 1970
  9 alaskanus (Scribner & Merrill) A. Löve
    (a) borealis (Turcz.) Melderis 28
    (b) scandicus (Nevski) Melderis 28   Su   Runemark & Heneen,
                                         1968
    (c) subalpinus (Neuman) Melderis
 10 reflexiaristatus (Nevski) Melderis
    (a) reflexiaristatus
    (b) strigosus (Bieb.) Melderis 14 27 Rs(K)   Petrova, 1968
 11 stipifolius (Czern. ex Nevski)
    Melderis   28   Rs(K)   Petrova, 1968
 12 bungeanus (Trin.) Melderis
    (a) scythicus (Nevski) Melderis
    (b) pruiniferus (Nevski) Melderis
 13 nodosus (Nevski) Melderis
    (a) nodosus   28
    (b) corsicus (Hackel) Melderis 28   Co   Contandriopoulos,
                                         1962
 14 flaccidifolius (Boiss. & Heldr.) Melderis
 15 curvifolius (Lange) Melderis
 16 elongatus (Host) Runemark
    (a) elongatus   14   Bu   Kozuharov &
                                         Kuzmanov, 1970
    (b) ponticus (Podp.) Melderis   70
                             56   Rm   Tarnavschi, 1948
 17 pycnanthus (Godron) Melderis 42   Br   Löve & Löve, 1961b
 18 repens (L.) Gould
    (a) repens   42   Hs   Löve & Kjellqvist,
                                          1973

	28		
	56		
(b) pseudocaesius (Pacz.) Melderis	42		
(c) arenosus (Petif) Melderis			
(d) elongatiformis (Drobov) Melderis	42		
(e) calcareus (Černjavski) Melderis			
repens x truncatus	42	Rs(K)	Petrova, 1968
repens x Hordeum secalinum	49		
19 pungens (Pers.) Melderis			
(a) pungens			
(b) campestris (Godron & Gren.) Melderis	56		
(c) fontqueri Melderis			
20 hispidus (Opiz) Melderis			
(a) hispidus	42	Cz	Májovský et al., 1974
	28	Co	Litardière, 1947
	43		
(b) barbulatus (Schur) Melderis	42		
	28		
(c) graecus Melderis			
(d) varnensis (Velen.) Melderis			
(e) pouzolzii (Godron) Melderis			
21 lolioides (Kar. & Kir.) Melderis	58	Iran	
22 farctus (Viv.) Runemark ex Melderis			
(a) farctus	42	Ga	Labadie, 1976
	56		
(b) boreali-atlanticus (Simonet & Guinochet) Melderis	28	Ho Su	Heneen, 1962
(c) bessarabicus (Savul. & Rayss) Melderis	14		
(d) rechingeri (Runemark) Melderis	28	Gr	Heneen & Runemark, 1962
22b x 17	35		
22b x 18	35		
	49		
49 Agropyron Gaertner			
1 cimmericum Nevski			
2 dasyanthum Ledeb.	28	Rs(W)	Petrova, 1968
3 tanaiticum Nevski			
4 fragile (Roth) P. Candargy	28	Rs(E)	Knowles, 1955
	14		
5 desertorum (Fischer ex Link) Schultes	28		Knowles, 1955
	29		Knowles, 1955
	32		Knowles, 1955
6 cristatum (L.) Gaertner			
(a) pectinatum (Bieb.) Tzvelev	14	Au	Titz, 1966
	28	Rm	Lungeanu, 1972
	42		
(b) sabulosum Lavrenko			
(c) brandzae (Panţu & Solac.) Melderis			
(d) ponticum (Nevski) Tzvelev			
(e) sclerophyllum Novopokr. ex Tzvelev			
50 Eremopyrum (Ledeb.) Jaub. & Spach			
1 triticum (Gaertner) Nevski	14		
2 orientale (L.) Jaub. & Spach	28		
3 distans (C. Koch) Nevski	14		
	28		

51 Crithopsis Jaub. & Spach
   1 delileana (Schultes) Roshev.

52 Aegilops L.

1 speltoides Tausch	14	cult. Br	Percival, 1926
2 ventricosa Tausch	28	Bl	Dahlgren et al., 1971
3 cylindrica Host	28	Rm	Priadcencu et al., 1967
4 dichasians (Zhuk.) Humphries	14		
5 comosa Sibth. & Sm.	14		
(a) comosa			
(b) heldreichii (Boiss.) Eig			
6 uniaristata Vis.	28	Asia	
7 triuncialis L.	28	Lu	Fernandes & Queirós, 1969
8 lorentii Hochst.	28	Asia	
9 geniculata Roth	28	Bl	Dahlgren et al., 1971
10 neglecta Req. ex Bertol.	28 42	Al	Strid, 1971

53 Triticum L.

1 baeoticum Boiss.			
2 monococcum L.	14	Lu	Câmara &
3 dicoccon Schrank	28	It Lu	Coutinho, 1939
4 durum Desf.	28	It	Giorgi & Bozzini, 1969
5 turgidum L.	28	It	Giorgi & Bozzini, 1969
6 polonicum L.	28	Lu	Câmara & Coutinho, 1939
7 spelta L.	42		Upadhya & Swaminathan, 1963
8 aestivum L.	42	Lu	Câmara & Coutinho, 1939
9 compactum Host			

54 Dasypyrum (Cosson & Durieu) T. Durand

1 villosum (L.) P. Candargy	14	Al	Strid, 1971
2 hordeaceum (Cosson & Durieu) P. Candargy			

55 Secale L.
   1 sylvestre Host
   2 montanum Guss.
   3 cereale L.

56 Psathyrostachys Nevski

1 juncea (Fischer) Nevski	14	Rs(N)	Bowden, 1964

57 Hordeum L.

1 spontaneum C. Koch			
2 distichon L.	14	Br	Gale & Rees, 1970
3 vulgare L.	14	Br	Gale & Rees, 1970
4 murinum L.			
(a) murinum L.	28	Al	Strid, 1971
(b) glaucum (Steudel) Tzvelev	14		
(c) leporinum (Link) Arcangeli	28	Bl	Dahlgren et al., 1971
5 marinum Hudson	14	Hu	Kliphuis, 1977
	28	Lu	Castro & Fontes, 1946

6 hystrix Roth	14	Hu	Pólya, 1948
	28		
7 bulbosum L.	14	It	Morrison, 1959
	28	Ju	Katznelson &
8 bogdanii Wilensky			Zohary, 1967
9 brevisubulatum (Trin.) Link			
10 secalinum Schreber	14	Ga	Labadie, 1976
	28	Lu	Fernandes & Queiros, 1969
11 jubatum L.	28 Br Ge Lu Po		Morrison, 1959

58 Hordelymus (Jessen) C.O. Harz

1 europaeus (L.) C.O. Harz	28	Cz	Májovský et al., 1970

59 Taeniatherum Nevski

1 caput-medusae (L.) Nevski	14	Lu	Fernandes & Queirós, 1969

Tribe Aveneae

60 Avena L.
  1 clauda Durieu
  2 eriantha Durieu

3 longiglumis Durieu	14	Hs It	Ladizinsky &
4 barbata Pott ex Link			Zohary, 1971
(a) barbata	14 28	Lu	Fernandes & Queiros, 1969
(b) atherantha (C. Presl) Rocha Afonso	14		
5 saxatilis (Lojac.) Rocha Afonso			
6 prostrata Ladizinsky	14		
7 strigosa Schreber			
(a) strigosa	14	Hs It	Ladizinsky & Zohary,
(b) agraria (Brot.) Tab. Mor.			1971
8 brevis Roth	14		
9 nuda L.	14		

  10 fatua L.
  11 sativa L.
    (a) sativa
    (b) praegravis (Krause) Tab. Mor.
    (c) macrantha (Hackel) Rocha Afonso
  12 byzantina C. Koch
  13 sterilis L.

(a) sterilis	42	Bl	Dahlgren et al., 1971
(b) ludoviciana (Durieu) Nyman	42	Lu	Fernandes & Queirós, 1969
14 murphyi Ladizinsky	28	Hs	Ladizinsky, 1971

61 Helictotrichon Besser

1 decorum (Janka) Henrard	14	Rm	Litardière, 1950
2 sedenense (DC.) J. Holub	14 + 0 - 2B	Ga	Gervais, 1966
	28 + 0 - 2B	Ga	Gervais, 1966
3 planifolium (Willk.) J. Holub	28 + 0 - 2B		
4 parlatorei (J. Woods) Pilger	14	Ga It	Gervais, 1966
5 setaceum (Vill.) Hendrard	14	Ga	Gervais, 1965
6 petzense Melzer	14	Alps	Sauer, 1971
7 sempervirens (Vill.) Pilger	42 + 0 - 5B	Ga Gr It	Gervais, 1966
8 convolutum (C. Presl) Henrard			
(a) convolutum	14	Gr	Gervais, 1968
(b) heldreichii (Parl.) Gervais			

9 sarracenorum (Gand.) J. Holub   14 28 70   Hs   Gervais, 1972

10 filifolium (Lag.) Henrard   70   Hs   Löve & Kjellqvist,
                                                 1973
                              84 + 0 - 1B   Hs   Gervais, 1968, 1972
11 murcicum J. Holub
12 cantabricum (Lag.) Gervais   84 98   Ga   Gervais, 1973
                                    98
13 desertorum (Less.) Nevski
   (a) desertorum
   (b) basalticum (Podp.) J. Holub   14   Cz   Holub et al.,
                                                1971

62 Avenula (Dumort.) Dumort.
  1 pubescens (Hudson) Dumort.   14   Br No Su   Hedberg & Hedberg,
                                                 1961
                              14 + 2B   Cz   Gervais, 1966
   (a) pubescens
   (b) laevigata (Schur) J. Holub
  2 versicolor (Vill.) Lainz
   (a) versicolor   14 + 0 - 3B Au Ga   Gervais, 1966
   (b) pretutiana (Parl. ex   14
      Arcangeli) J. Holub
  3 schelliana (Hackel) W. Sauer &   14   ?Rs   Sauer & Chmelitschek,
    Chmelitschek   1976
  4 marginata (Lowe) J. Holub   14   Ga   Gervais, 1966
   (a) marginata
   (b) sulcata (Gay ex Delastre) Franco
   (c) myrenaica J. Holub
  5 albinervis (Boiss.) Laínz   28   Lu   Fernandes & Queirós,
                                                 1969
  6 occidentalis (Gervais) J. Holub
   (a) occidentalis   42
   (b) stenophylla Franco
  7 delicatula Franco
  8 levis (Hackel) J. Holub   14   Hs
  9 compressa (Heuffel) W. Sauer &   14   Hu   Gervais, 1966
    Chmelitschek
10 aetolica (Rech. fil.) J. Holub
11 blavii (Ascherson & Janka) W. Sauer &
    Chmelitschek
12 mirandana (Sennen) J. Holub   98
13 gonzaloi (Sennen) J. Holub
14 requienii (Mutel) J. Holub
15 pungens (Sennen ex St-Yves) J. Holub
16 pratensis (L.) Dumort.   126   Br Su   Hedberg & Hedberg,
                                                 1961
   (a) pratensis
   (b) hirtifolia (Podp.) J. Holub
17 praeusta (Reichenb.) J. Holub   126   It   Sauer & Chmelitschek,
18 planiculmis (Schrader) W. Sauer &   1976
    Chmelitschek   126   Po   Gervais, 1966
   (a) planiculmis
   (b) angustior J. Holub
19 bromoides (Gouan) H. Scholz   14 + 0 - 2B Ga Hs   Gervais, 1968
20 murcica J. Holub
21 cincinnata (Ten.) J. Holub   14   Bu   Kozuharov &
22 peloponnesiaca J. Holub   Petrova, 1975
23 cycladum (Rech. fil. & Scheffer) W.
    Greuter
24 gervaisii J. Holub

25 pruinosa (Batt. & Trabut) J. Holub
26 crassifolia (Font Quer) J. Holub
27 hackelii (Henriq.) J. Holub

63 Danthoniastrum (J. Holub) J. Holub
 1 compactum (Boiss. & Heldr.) J.
   Holub

64 Arrhenatherum Beauv.
 1 elatius (L.) Beauv. ex J. & C.
   Presl
   (a) elatius   28   Is   Löve & Löve, 1956
   (b) bulbosum (Willd.) Schübler &   28   Lu   Fernandes & Queirós,
       Martens   1969
       sardoum (Schmid) Gamisans   14   Co   Gamisans, 1974
 2 palaestinum Boiss.
 3 album (Vahl) W.D. Clayton   28   Lu   Fernandes & Queirós,
                                                          1971

65 Pseudarrhenatherum Rouy
 1 longifolium (Thore) Rouy   14
 2 pallens (Link) J. Holub

66 Gaudinia Beauv.
 1 fragilis (L.) Beauv.   14 + 0 - 2B Lu   Fernandes & Queirós,
                                                          1969
 2 hispanica Stace & Tutin   14
 3 coarctata (Link) T. Durand &
   Schinz

67 Ventenata Koeler
 1 dubia (Leers) Cosson   14   Ju   Larsen, 1960
 2 macra (Bieb.) Boiss.

68 Koeleria Pers.
 1 vallesiana (Honckeny) Gaudin   14 + 0 - 2B Ga   Küpfer &
                                                          Favarger, 1967
                                       28 42 Ga Hs   Küpfer, 1971
                                          43
                                      112 126
 2 cenisia Reuter ex Reverchon   14 + 0 - 2B Ga   Favarger, 1969
 3 hirsuta Gaudin   14 + 0 - 5B He   Favarger, 1969
 4 asiatica Domin   28   Rs(C)   Zhukova & Tikhonova,
                                                          1971
 5 macrantha (Ledeb.) Schultes   14   Hu   Ujhelyi, 1961
                                 28 + 0 - 6B Au   Larsen, ined.
                                       42 70 Hu   Ujhelyi, 1962

 6 glauca (Schrader) DC.   14   Ga   Delay, 1969
                                         28   Da   Böcher, 1943
                                       42 70 Ge Hu   Ujhelyi, 1960b
 7 lobata (Bieb.) Roemer & Schultes   28   Da   Larsen, ined.
 8 splendens C. Presl   28   Al Hu   Ujhelyi, 1962
 9 caudata (Link) Steudel
10 crassipes Lange   14   Lu   Fernandes & Queirós,
                                                          1969
11 nitidula Velen.
12 eriostachya Pančić
13 pyramidata (Lam.) Beauv.   42   Europe   Ujhelyi, 1970
                                         56   Au Ju   Larsen, ined.
                                         70   Europe   Ujhelyi, 1972
                                         84   Da   Böcher, 1943

335

69 Lophochloa Reichenb.
1 cristata (L.) Hyl.   26   Lu   Fernandes & Queirós,
2 hispida (Savi) Jonsell   1969
3 pumila (Desf.) Bor
4 pubescens (Lam.) H. Scholz
5 salzmannii (Boiss. & Reuter) H. Scholz

70 Trisetum Pers.
1 velutinum Boiss.
2 macrotrichum Hackel
3 laconicum Boiss. & Orph.
4 distichophyllum (Vill.) Beauv.   28   Ga   Favarger, 1959
                                            56   He   Favarger, 1959
5 argenteum (Willd.) Roemer & Schultes   28   It   Favarger, 1959
6 glaciale (Bory) Boiss.   14   Hs   Küpfer, 1968
7 gracile (Moris) Boiss.   14   Co   Gamisans, 1973
8 bertolonii Jonsell
9 hispidum Lange
10 spicatum (L.) K. Richter
   (a) spicatum   28   Is No Su Jonsell et al.,
                                                                 1975
   (b) ovatipaniculatum Hultén ex Jonsell   28   Au   Jonsell et al.,
                                                                 1975
   (c) pilosiglume (Fernald) Hultén   42   Is   Jonsell et al.,
                                                                 1975
11 subalpestre (Hartman) Neuman   28   No   Knaben & Engelskjon,
                                                                 1967
12 baregense Laffitte & Miégeville   14
13 flavescens (L.) Beauv.
   (a) flavescens   28   Cz   Májovský et al.,
   (b) splendens (C. Presl) Arcangeli   1974
   (c) purpurascens (DC.) Arcangeli   12
14 alpestre (Host) Beauv.   14   Po   Skalińska et al.,
                                                                 1971
15 tenuiforme Jonsell
16 fuscum (Kit. ex Schultes) Schultes
17 sibiricum Rupr.   14   Rs(N)   Sokolovskaya &
                                                                 Strelkova, 1960
   (a) sibiricum
   (b) litorale (Rupr.) Roshev.
18 paniceum (Lam.) Pers.   14   Lu   Fernandes & Queirós,
                                                                 1969
19 aureum Ten.
20 parviflorum (Desf.) Pers.
21 scabriusculum (Lag.) Cosson ex Willk.
22 ovatum (Cav.) Pers.   14   Lu   Fernandes & Queirós,
                                                                 1969
23 loeflingianum (L.) C. Presl
24 macrochaetum Boiss.

71 Trisetaria Forskål
1 dufourei (Boiss.) Paunero

72 Parvotrisetum Chrtek
1 myrianthum (Bertol.) Chrtek

73 Lagurus L.
1 ovatus L.   14   Co Gr   Gadella & Kliphuis,
                                                                 1968

74 Deschampsia Beauv.

1 cespitosa (L.) Beauv.
  (a) cespitosa

	24 26	Bu	Kovuharov & Kuzmanov, 1962
	26 + 0 - 7B	Ge	Albers, 1972

  (b) alpina (L.) Tzvelev

52	Is	Löve & Löve, 1956
56	Br	Tutin, 1961

  (c) obensis
  (d) bottnica (Wahlenb.) G.C.S.   26 26 + 2B   Ge   Albers, 1972
      Clarke
  (e) borealis (Trautv.) Tzvelev   28   Rs(N)   Sokolovskaya & Strelkova, 1960
  (f) orientalis Hultén
  (g) paludosa (Schübler & Martens) 26 + 0 - 2B Ge   Albers, 1972
      G.C.S. Clarke
      littoralis (Gaudin) Reuter   56
2 media (Gouan) Roemer & Schultes   26
3 setacea (Hudson) Hackel   14   Da   Hagerup, 1939
4 foliosa Hackel
5 flexuosa (L.) Trin.

26		Ga	Delay, 1970
28	Fe Ge No		Albers, 1972
56		Lu	Albers, 1972

75 Vahlodea Fries
  1 atropurpurea (Wahlenb.) Fries ex   14   No   Knaben & Engelskjon, 1967
    Hartman

76 Aira L.
  1 praecox L.   14   Br   Hedberg & Hedberg, 1961

  2 caryophyllea L.
    (a) caryophyllea   28   Ga   Albers, 1973
    (b) multiculmis (Dumort.) Bonnier & 28 Da Ga Lu Böcher & Larsen, 1958
      Layens
  3 cupaniana Guss.   14   Co   Albers, 1973
  4 uniaristata Lag. & Rodr.   14
  5 scoparia Adamović
  6 elegantissima Schur   14   Bl   Dahlgren et al., 1971

  7 tenorii Guss.   14   Ga   Albers & Albers, 1973

  8 provincialis Jordan   28   Ga   Albers, 1973

77 Molineriella Rouy
  1 minuta (L.) Rouy
  2 laevis (Brot.) Rouy   8   Lu   Fernandes & Queirós, 1969

78 Airopsis Desv.
  1 tenella (Cav.) Ascherson & Graebner 8   Lu   Fernandes & Queirós, 1969

79 Periballia Trin.
  1 involucrata (Cav.) Janka   14   Lu   Fernandes & Queirós, 1969

80 Antinoria Parl.
  1 agrostidea (DC.) Parl.
  2 insularis Parl.

81 Hierochloë R. Br.

337

1 australis (Schrader) Roemer & Schultes 14   Au Fe Ge   Weimarck, 1971
                                                It Rm
2 odorata (L.) Beauv.
  (a) odorata   28   Po   Skalińska et al.,
                                                              1974
                                         42   Fe   Sorsa, 1963
  (b) baltica G. Weim.   42   Fe Su   Weimarck, 1971
3 hirta (Schrank) Borbás   56   Fe No Su   Weimarck, 1971
  (a) hirta
  (b) arctica (C. Presl) G. Weim.
4 repens (Host) Simonkai   28   Cz Rm   Weimarck, 1971
                                   28 + 2B   Cz   Weimarck, 1971
5 stepporum P. Smirnov
6 pauciflora R.Br.
7 alpina (Willd.) Roemer & Schultes   56   Rs(N)   Sokolovskaya &
                                                              Strelkova, 1960

82 Anthoxanthum L.
   1 odoratum L.   10   Au He No Su   I. Hedberg, 1970
                                         20   Cz   Májovský et al.,
                                                              1970
     alpinum   10   Is   Löve & Löve, 1956
   2 pauciflorum Adamović
   3 amarum Brot.   90   Lu   Fernandes & Queirós,
                                                              1969
   4 aristatum Boiss.
     (a) aristatum   10 + 0 - 2B   Hs   Valdés, 1973
     (b) macranthum Valdés   10 + 0 - 1B   Hs   Valdés, 1973
   5 ovatum Lag.   10   Hs   Valdés, 1973
   6 gracile Biv.

83 Holcus L.
   1 lanatus L.   14 + 0 - 2B   Po   Skalińska et al.,
                                                              1968
   2 setiglumis Boiss. & Reuter   14   Lu   Queirós, 1973
     (a) setiglumis
     (b) duriensis P. Silva
   3 notarisii Nyman
   4 mollis L.   14 28   Lu   Fernandes & Queirós,
                                                              1969
                                         35   Po   Walter, 1973
                                      42 49   Br   Jones, 1958
     1 x 4   21   Br   Jones, 1958
   5 rigidus Hochst.
   6 gayanus Boiss.   8   Lu   Fernandes, 1950
   7 grandiflorus Boiss. & Reuter
   8 caespitosus Boiss.   14   Hs   Küpfer, 1969

84 Corynephorus Beauv.
   1 canescens (L.) Beauv.   14   Lu   Fernandes & Queirós,
                                                              1969
   2 divaricatus (Pourret) Breistr.   14   Co   Litardière, 1949
   3 fasciculatus Boiss. & Reuter   14   Lu   Fernandes & Queirós,
                                                              1969
   4 macrantherus Boiss. & Reuter

85 Avellinia Parl.
   1 michelii (Savi) Parl.

86 Triplachne Link
   1 nitens (Guss.) Link

87 Agrostis L.

1 canina L.	14	Br	Hubbard, 1954
	16	Ge	Heitz, 1967
	28	Bu	Kozuharov & Nicolova, 1975
2 vinealis Schreber	28	No	Knaben & Engelskjon, 1967
3 curtisii Kerguelen	14	Br Ga Lu	Björkman, 1960
4 mertensii Trin.	56	Fe No Su	Björkman, 1960
5 alpina Scop.	14	Ga He	Björkman, 1960
6 schleicheri Jordan & Verlot	42	Ga He	Björkman, 1960
7 rupestris All.	14 + 0 - 3B	Ga Po	Björkman, 1960
	21	Ga	Björkman, 1960
	28	Au Ga He	Björkman, 1960
8 nevadensis Boiss.	42 + 0 - 10B	Hs	Björkman, 1960
9 agrostiflora (G. Beck) Rauschert	28	He	Scholte, 1977
10 clavata Trin.	42	Su	Björkman, 1960
11 scabra Willd.			
12 pourretii Willd.	14 + 0 - 2B	Lu	Björkman, 1960
	16 18 19 21	Lu	Fernandes & Queirós, 1969
13 tenerrima Trin.	14	Lu	Björkman, 1960
14 juressi Link	14	Lu	Björkman, 1960
15 nebulosa Boiss. & Reuter			
16 delicatula Pourret ex Lapeyr.	14	Lu	Björkman, 1960
17 durieui Boiss. & Reuter ex Willk.			
18 reuteri Boiss.	14	Lu	Björkman, 1960
19 capillaris L.	28	Su	Hedberg, 1964
20 gigantea Roth	42 + 0 - 2B	Su	Björkman, 1954
(a) gigantea			
(b) maeotica (Klokov) Tzvelev			
21 salsa Korsh.			
22 stolonifera L.	28	Hu	Pólya, 1948
	30 32 35		
	42	Lu	Fernandes & Queirós, 1969
	44 46	Ge	Juhl, 1952
23 congestiflora Tutin & E.F. Warburg			
(a) congestiflora			
(b) oreophila Franco			
24 gracililaxa Franco			
25 castellana Boiss. & Reuter	28 + 0 - 4B	Lu	Björkman, 1954
	42 + 0 - 3B	Lu	Fernandes & Queirós, 1969

88 Gastridium Beauv.

1 ventricosum (Gouan) Schinz & Thell.	14	Lu	Fernandes & Queirós, 1969

89 Polypogon Desf.

1 monspeliensis (L.) Desf.	28	Lu	Fernandes & Queirós, 1969
2 maritimus Willd.	14	Lu	Fernandes & Queirós, 1969
	28	Lu	Rodriguez, 1953
(a) maritimus			
(b) subspathaceus (Req.) Bonnier & Layens	14	Bl	Dahlgren et al., 1969
3 viridis (Gouan) Breistr.	28	Lu	Fernandes & Queirós, 1969

```
x Agropogon P. Fourn.
  1 littoralis (Sm.) C.E. Hubbard          28      Br      Rutland, 1941

90 Ammophila Host
  1 arenaria (L.) Link                     14      Da      Westergaard, 1941
                                           28      Ga      Wulff, 1937a
                                           56      Po      Skalińska et al.,
     (a) arenaria                                          1957
     (b) arundinacea H. Lindb. fil.

x Ammocalamagrostis P. Fourn.
  1 baltica (Flugge ex Schrader) P. Fourn. 28 42   Po      Skalińska et al.,
                                                           1964

91 Calamagrostis Adanson
  1 epigejos (L.) Roth                      28      Ho      Gadella & Kliphuis,
                                                           1967

                                           42      Su      Nygren, 1946
                                           56      Hu      Polya, 1949
  2 pseudophragmites (Haller fil.) Koeler  28      Po      Kubién, 1968
  3 villosa (Chaix) J.F. Gmelin            56      Cz      Murín & Májovský,
                                                           1976
  4 canescens (Weber) Roth                 28      Su      Nygren, 1946
     (a) canescens
     (b) vilnensis (Besser) H. Scholz
  5 purpurea (Trin.) Trin.                  28      Fe      V. Sorsa, 1962
                                           56-91   Su      Nygren, 1946
     (a) purpurea
     (b) phragmitoides (Hartman) Tzvelev
     (c) langsdorfii (Link) Tzvelev
     (d) pseudopurpurea (Gerstlauer ex R.
         Heine) G.C.S. Clarke
  6 stricta (Timm) Koeler                   28      Su      Nygren, 1946
  7 scotica (Druce) Druce
  8 holmii Lange
  9 deschampsioides Trin.
 10 lapponica (Wahlenb.) Hartman            28      Rs(N)   Sokolovskaya &
                                                           Strelkova, 1960
                                        42-112             Nygren, 1946
 11 arundinacea (L.) Roth          28 + 0 - 2B  Cz         Holub et al.,
                                                           1972
 12 varia (Schrader) Host
     (a) varia                              28      Hu      Baksay, 1956
     (b) corsica (Hackel) Rouy              28      Co      Litardière, 1949
 13 chalybaea (Laest.) Fries                42      Su      Nygren, 1946
 14 obtusata Trin.

92 Chaetopogon Janchen
  1 fasciculatus (Link) Hayek              14      Lu      Fernandes & Queirós,
                                                           1969

93 Phleum L.
  1 pratense L.
     (a) pratense                          42      Hu      Polya, 1950
                               21 35 36 49 63 Su          Muntzing, 1935, 1937
                                     56 70 84  Su          Muntzing & Prakken,
                                                           1940
     (b) bertolonii (DC.) Bornm.           14      Su      Muntzing, 1935
  2 alpinum L.
     (a) alpinum                           28      Su      Muntzing, 1935
     (b) rhaeticum Humphries               14      He      Muntzing, 1935

                                   340
```

```
 3 echinatum Host                      10       It       Ellerstrom & Tjio,
                                                          1950
 4 phleoides (L.) Karsten              14       Su       Muntzing, 1935
                                28 + 1 - 5B     Su       Bosemark, 1956
 5 montanum C. Koch
 6 hirsutum Honckeny                   14       It       Nordenskjold, 1937
 7 paniculatum Hudson                  28       Rs(K)    Avdulov, 1928
 8 arenarium L.                        14       Br Da Hb Bücher & Larsen,
                                                Ho Ga    1958
 9 graecum Boiss. & Heldr.
   (a) graecum
   (b) aegaeum (Vierh.) W. Greuter
10 crypsoides (D'Urv.) Hackel
   (a) crypsoides
   (b) sardoum (Hackel) Horn af Rantzien
11 subulatum (Savi) Ascherson & Graebner
   (a) subulatum
   (b) ciliatum (Boiss.) Humphries

94 Alopecurus L.
   1 pratensis L.
     (a) pratensis                     28       Fe       V. Sorsa, 1962
     (b) laguriformis (Schur) Tzvelev
   2 arundinaceus Poiret               28       Su       Johnsson, 1941
   3 geniculatus L.                    28       Su       Johnsson, 1941
   4 aequalis Sobol.                   14       Cz       Májovský et al.,
                                                          1970
   5 bulbosus Gouan                    14       Br       Maude, 1940
   6 himalaicus Hooker fil.
   7 alpinus Sm.
     (a) alpinus                    100 117     Br       Fearn, 1977
                                    c.120       Sb       Flovik, 1938
     (b) glaucus (Less.) Hultén
   8 utriculatus Solander
   9 setarioides Gren.
  10 myosuroides Hudson               14        Su       Johnsson, 1941
  11 creticus Trin.                   14
  12 rendlei Eig                      14        Ju       Gadella & Kliphuis,
                                                          1972
  13 gerardii Vill.                   14        Ga       Favarger & Huynh,
                                                          1964
  14 vaginatus (Willd.) Boiss.

95 Cornucopiae L.
   1 cucullatum L.

Tribe Hainardieae W. Greuter

96 Pholiurus Trin.
   1 pannonicus (Host) Trin.          14        Hu       Polya, 1948

97 Parapholis C.E. Hubbard
   1 incurva (L.) C.E. Hubbard        38        Lu       Fernandes & Queirós,
                                                          1969
   2 marginata Runemark               14        Gr       Runemark, 1962
   3 filiformis (Roth) C.E. Hubbard   14        Lu       Rodriguez, 1953
   4 strigosa (Dumort.) C.E. Hubbard  28        Lu       Fernandes & Queirós,
                                                          1969
   5 pycnantha (Hackel) C.E. Hubbard

98 Hainardia W. Greuter
```

1 cylindrica (Willd.) W. Greuter	26	Lu	Fernandes & Queirós, 1969

Tribe Phalarideae Kunth

99 Phalaris L.

1 arundinacea L.			
(a) arundinacea	28	Br	Hedberg & Hedberg, 1961
	42 + 6B	Lu	Fernandes & Queirós, 1969
(b) rotgesii (Fouc. & Mandon ex Husnot) Kerguelen	14	Co	Litardière, 1949
2 truncata Guss. ex Bertol.			
3 aquatica L.	28	Lu	Fernandes & Queirós, 1969
4 canariensis L.	12	Lu	Fernandes & Queirós, 1969
5 minor Retz.	28	Lu	Fernandes & Queirós, 1969
6 brachystachys Link	12	Lu	Fernandes & Queirós, 1969
7 paradoxa L.	14	Lu	Fernandes & Queirós, 1969
8 coerulescens Desf.	14	Lu	Fernandes & Queirós, 1969
	42	Hs	Löve & Kjellqvist, 1973

Tribe Coleantheae Ascherson & Graebner

100 Coleanthus Seidl
    1 subtilis (Tratt.) Seidl

Tribe Scolochloeae Tzvelev

101 Scolochloa Link			
1 festucacea (Willd.) Link	28	Su	Löve & Löve, 1944

Tribe Milieae Endl.

102 Milium L.			
1 effusum L.	14	Rs(N)	Sokolovskaya & Strelkova, 1960
	28	Fe	V. Sorsa, 1962
2 vernale Bieb.	8	Ga	Tutin, ined.

103 Zingeria P. Smirnov			
1 biebersteiniana (Claus) P. Smirnov	4	Rs(E)	Tzvelov & Zhukova, 1974
2 pisidica (Boiss.) Tutin			

Tribe Stipeae Dumort.

104 Piptatherum Beauv.			
1 miliaceum (L.) Cosson			
2 virescens (Trin.) Boiss.			
3 paradoxum (L.) Beauv.	24	Hs	Löve & Kjellqvist, 1973
4 coerulescens (Desf.) Beauv.			
5 holciforme (Bieb.) Roemer & Schultes			

105 Stipa L.
 1 pennata L.
   (a) austriaca (G. Beck) Martinovsky &
       Skalicky
   (b) lithophila (P. Smirnov) Martinovsky
   (c) eriocaulis (Borbás) Martinovsky   44   Rm   Tarnavschi &
                                                      Lungeanu, 1970
   (d) pennata
   (e) kiemii Martinovsky
 2 pulcherrima C. Koch   44
 3 mayeri Martinovsky
 4 dasyvaginata Martinovsky
 5 crassiculmis P. Smirnov
 6 bavarica Martinovsky & H. Scholz
 7 epilosa Martinovsky
 8 endotricha Martinovsky
 9 cretacea P. Smirnov
10 syreistschikovii P. Smirnov
11 iberica Martinovsky
   (a) iberica
   (b) pauneroana Martinovsky
12 apertifolia Martinovsky
13 rechingeri Martinovsky
14 austroitalica Martinovsky
15 novakii Martinovsky
16 dasyphylla (Lindem.) Trautv.
17 zalesskii Wilensky
18 ucrainica P. Smirnov   44
19 pontica P. Smirnov
20 joannis Čelak.
   (a) puberula (Podp. & Suza) Martinovsky
   (b) joannis   44   Rm   Tarnavschi, 1948
   (c) balcanica Martinovsky
21 borysthenica Klokov ex Prokudin
   (a) borysthenica
   (b) germanica (Endtmann) Martinovsky &
       Rauschert
22 styriaca Martinovsky
23 anomala P. Smirnov
24 danubialis Dihoru & Roman   44   Rm   Tarnavschi &
                                                      Lungeanu, 1970
25 tirsa Steven   44
   (a) tirsa
   (b) albanica Martinovsky
26 lessingiana Trin. & Rupr.   44   Rm   Lungeanu, 1975
   (a) lessingiana
   (b) brauneri Pacz.
27 capensis Thunb.   36   Lu   Fernandes & Queirós,
                                                      1969
28 barbata Desf.
29 arabica Trin. & Rupr.
30 orientalis Trin.
31 capillata L.   44   Po   Rychlewski, 1968
32 sareptana A. Becker   44   Rs(E) Titova, 1935
   (a) sareptana
   (b) praecapillata (Alechin) Tzvelev
33 fontanesii Parl.
34 lagascae Roemer & Schultes
35 celakovskyi Martinovsky
36 offneri Breistr.   44   Bl   Nilsson & Lassen,
                                                      1971

343

```
37 korshinskyi Roshev
38 parviflora Desf.                      28      Hs      Lorenzo-Andreu,
                                                          1950
39 tenacissima L.                        40      Hs      Fernandes & Queirós,
                                                          1969
40 gigantea Link                         96      Lu      Fernandes & Queirós,
                                                          1969
41 bromoides (L.) Dörfler                28      Hu      Ujhelyi, 1961
42 trichotoma Nees

106 Achnatherum Beauv.
    1 calamagrostis (L.) Beauv.          24      Ga He   Gervais, 1966
    2 splendens (Trin.) Nevski
```

Tribe Ampelodesmeae (Concert) Tutin

```
107 Ampelodesmos Link
    1 mauritanica (Poiret) T. Durand &   48      Bl      Nilsson & Lassen,
      Schinz                                            1971
```

Tribe Arundineae Dumort.

```
108 Arundo L.
    1 donax L.                           110 112  Ga     Gorenflot et al.,
                                                          1972a
    2 plinii Turra                       72       Lu     Fernandes & Queirós,
                                                          1969
109 Phragmites Adanson
    1 australis (Cav.) Trin. ex Steudel 36 44 46 48 Ge He Gorenflot et al.,
                                        49 50 51 52  Rm    1972b
                                        96
```

Tribe Cortaderieae Zotov

```
110 Cortaderia Stapf
    1 selloana (Schultes & Schultes fil.)
      Ascherson & Graebner
```

Tribe Danthonieae (G. Beck) C.E. Hubbard

```
111 Danthonia DC.
    1 decumbens (L.) DC.                 24      Ge      Schwarz & Bässler,
                                                          1964
                                         36      Cz      Májovský et al.,
                                                          1974
                                         124     Br      Maude, 1940
    2 alpina Vest                        36      Europe  Packer, 1961

112 Schismus Beauv.
    1 barbatus (L.) Thell.
    2 arabicus Nees
```

Tribe Molinieae Jirasek

```
113 Molinia Schrank
    1 caerulea (L.) Moench
      (a) caerulea                       36      Hu      Pólya, 1950
      (b) arundinacea (Schrank) H. Paul  ?36     Ga      Guinochet & Lemée,
                                                          1950
                                         90      Hu      Ujhelyi, 1961
```

Tribe Aristideae C.E. Hubbard ex Bor

114 Aristida L.
   1 adscensionis L.

115 Stipagrostis Nees
   1 pennata (Trin.) De Winter
   2 karelinii (Trin. & Rupr.) Tzvelev

Tribe Nardeae Reichenb.

116 Nardus L.			
1 stricta L.	26	Br	Hedberg & Hedberg, 1961

Tribe Lygeeae Lange

117 Lygeum Loefl. ex L.			
1 spartum L.	40	Hs	Lorenzo-Andreu & Garcia, 1951

Tribe Pappophoreae Kunth

118 Enneapogon Desv. ex Beauv.
   1 persicus Boiss.

Tribe Aeluropodeae Nevski ex Bor

119 Aeluropus Trin.			
1 littoralis (Gouan) Parl.	20	Bot. Gard.	Litardière, 1950
2 lagopoides (L.) Trin. ex Thwaites			

Tribe Eragrostideae Stapf

120 Cleistogenes Keng			
1 serotina (L.) Keng			
(a) serotina	40	Cz	Májovský et al., 1970
(b) bulgarica (Bornm.) Tutin	40	Bu	Kozuharov & Petrova, 1974
2 squarrosa (Trin.) Keng			

121 Eragrostis N.M. Wolf			
1 pilosa (L.) Beauv.	40	Lu	Fernandes & Queirós, 1969
2 pectinacea (Michx) Nees			
3 papposa (Dufour) Steudel			
4 cilianensis (All.) F.T. Hubbard	20	Lu	Fernandes & Queirós, 1969
5 minor Host	40	Cz	Májovský et al., 1974
	80	Lu	Fernandes & Queirós, 1969
6 barrelieri Daveau			
7 aegyptiaca (Willd.) Link			
8 diarrhena (Schultes & Schultes fil.) Steudel			
9 collina Trin.			

122 Sporobolus R. Br.			
1 pungens (Schreber) Kunth			
2 indicus (L.) R. Br.	36	Lu	Fernandes & Queirós, 1969

123 Crypsis Aiton
    1 alopecuroides (Piller & Mitterp.)
        Schrader
    2 acuminata Trin.
    3 aculeata (L.) Aiton   16   Hu   Pólya, 1948
    4 schoenoides (L.) Lam.   36   Lu   Fernandes & Queirós,
                                                        1969

    5 turkestanica Eig.

124 Eleusine Gaertner
    1 indica (L.) Gaertner
    2 tristachya (Lam.) Lam.

125 Dactyloctenium Willd.
    1 aegyptium (L.) Beauv.

    Tribe Chlorideae J.G. Agardh

126 Cynodon L.C.M. Richard
    1 dactylon (L.) Pers.   36   Hu   Pólya, 1948
                                        40   Lu   Fernandes & Queirós,
                                                        1969

    Tribe Spartineae (Prat) C.E. Hubbard

127 Spartina Schreber
    1 maritima (Curtis) Fernald   60   Lu   Fernandes & Queirós,
                                                        1969
    2 x townsendii H. & J. Groves   62   Br   Marchant, 1968
    3 anglica C.E. Hubbard   120 122 124   Br   Marchant, 1968
    4 alterniflora Loisel.   62   Br   Marchant, 1968
    5 versicolor Fabre
    6 densiflora Brongn.

    Tribe Zoysieae Miq.

128 Tragus Haller
    1 racemosus (L.) All.

    Tribe Oryzeae Dumort.

129 Oryza L.
    1 sativa L.

130 Leersia Swartz
    1 oryzoides (L.) Swartz   48   Lu   Fernandes & Queirós,
                                                        1969

131 Zizania L.
    1 aquatica L.
    2 latifolia (Griseb.) Stapf.

    Tribe Paniceae R. Br.

132 Panicum L.
    1 miliaceum L.
    2 capillare L.
    3 dichotomiflorum Michx
    4 implicatum Scribner ex Britton & A. Br.
    5 repens L.   54   Lu   Fernandes & Queirós,
                                                        1969

6 maximum Jacq.

133 Oplismenus Beauv.
1 undulatifolius (Ard.) Roemer & Schultes

134 Echinochloa Beauv.
1 colonum (L.) Link
2 crus-galli (L.) Beauv.   54   Lu   Fernandes & Queirós,
1969
3 oryzoides (Ard.) Fritsch

135 Brachiaria (Trin.) Griseb.
1 eruciformis (Sibth. & Sm.) Griseb.

136 Eriochloa Kunth
1 succincta (Trin.) Kunth

137 Digitaria Haller
1 debilis (Desf.) Willd.   36   Lu   Fernandes & Queirós,
1969
2 sanguinalis (L.) Scop.   36   Lu   Fernandes & Queirós,
1969
3 ciliaris (Retz.) Koeler
4 ischaemum (Schreber) Muhl.

138 Paspalum L.
1 dilatatum Poiret   50   Lu   Fernandes & Queirós,
1969
2 urvillei Steudel   20   Lu   Fernandes & Queirós,
1969
3 paspalodes (Michx) Scribner   60   Lu   Fernandes & Queirós,
1969
4 vaginatum Swartz   20   Lu   Fernandes & Queirós,
1969

139 Stenotaphrum Trin.
1 secundatum (Walter) O. Kuntze

140 Setaria Beauv.
1 pumila (Poiret) Schultes   36   Cz   Májovský et al.,
1976
     72   Hs   Björkqvist et al.,
1969
geniculata (Lam.) Beauv.   72   Lu   Fernandes & Queirós,
1969
2 verticillata (L.) Beauv.   18   Lu   Fernandes & Queirós,
1969
     36   Cz   Májovský et al.,
1974
     $\underline{c}$.54   Al   Strid, 1971
3 viridis (L.) Beauv.   18
2 x 3   36   Lu   Fernandes & Queirós,
1969
4 italica (L.) Beauv.

141 Pennisetum L.C.M. Richard
1 setaceum (Forskål) Chiov.
2 villosum R. Br. ex Fresen.   45   Lu   Fernandes & Queirós,
1969

142 Cenchrus L.

1 ciliaris L.
2 incertus M.A. Curtis

143 Tricholaena Schrader
   1 teneriffae (L. fil.) Link

Tribe Andropogoneae Dumort.

144 Imperata Cyr.
   1 cylindrica (L.) Raeuschel   50 60   Lu   Fernandes & Queirós,
                                                          1969

145 Saccharum L.
   1 spontaneum L.
   2 officinarum L.
   3 ravennae (L.) Murray
   4 strictum (Host) Sprengel

146 Sorghum Moench
   1 halepense (L.) Pers.   40 41 42 43 Lu   Fernandes & Queirós,
                                                          1969
   2 sudanense (Piper) Stapf
   3 bicolor (L.) Moench

147 Chrysopogon Trin.
   1 gryllus (L.) Trin.

148 Dichanthium Willemet
   1 insculptum (A. Richard) W.D. Clayton
   2 ischaemum (L.) Roberty

149 Andropogon L.
   1 distachyos L.   40   Ga   Saura, 1943

150 Hyparrhenia N.J. Andersson ex E. Fourn.
   1 hirta (L.) Stapf   30   Ga   Larsen, 1954
                                         40   Lu   Fernandes & Queirós,
                                                           1969

151 Heteropogon Pers.
   1 contortus (L.) Beauv. ex Roemer &
       Schultes

152 Phacelurus Griseb.
   1 digitatus (Sibth. & Sm.) Griseb.   20   Bu   Kozuharov &
                                                           Nicolova, 1975

153 Hemarthria R. Br.
   1 altissima (Poiret) Stapf & C.E. Hubbard

154 Coix L.
   1 lacryma-jobi L.

155 Zea L.
   1 mays L.

PRINCIPES

CXCIV PALMAE

1 Chamaerops L.

1 humilis L.

2 Phoenix L.
  1 dactylifera L.
  2 canariensis hort. ex Chabaud
  3 theophrasti W. Greuter

SPATHIFLORAE

CXCV ARACEAE

1 Acorus L.

1 calamus L.	24	?Fe	Vaarama, 1948
	36	Ho	Gadella & Kliphuis, 1973

2 Calla L.

1 palustris L.	36	Ge	Wulff, 1939
	72	Su	Löve & Löve, 1944

3 Lysichiton Schott
  1 americanus Hulten & St. John

4 Zantedeschia Sprengel
  1 aethiopica (L.) Sprengel

5 Colocasia Schott
  1 esculenta (L.) Schott

6 Arum L.

1 italicum Miller			
(a) italicum	84	It	Marchi, 1971
(b) neglectum (Townsend) Prime	84	Br	Prime, 1954
(c) albispathum (Steven) Prime			
(d) byzantinum (Blume) Nyman			
2 maculatum L.	56	Br	Prime, 1954
3 orientale Bieb.			
(a) orientale	28	Au Cz Hu Po Rm	Beuret, 1972
(b) lucanum (Cavara & Grande) Prime	28	It	Beuret, 1972
(c) danicum (Prime) Prime	28	Da	Hagerup, 1944
4 elongatum Steven			
5 petteri Schott			
nigrum var. apulum Carano	56	It	Gori, 1957
6 creticum Boiss. & Heldr.	28	Cr	Marchant, 1972
7 pictum L. fil.	28	Co	Contandriopoulos, 1962

7 Biarum Schott

1 davisii Turrill			
2 tenuifolium (L.) Schott	16	It	Caldo, 1971
	20	Si	Monti & Garbari, 1974
arondandum Boiss. & Reut.	22	Hs	Talavera, 1976
galiani Talavera	26	Hs	Talavera, 1976
3 carratracense (Haenseler ex Willk.) Font Quer	98	Hs	Messia & Rejon, 1976
	c.96	Hs	Talavera, 1976
4 dispar (Schott) Talavera	74	Sa	Chiappini & Scrugli, 1972

5 spruneri Boiss.

8 Arisarum Miller
1 vulgare Targ.-Tozz.
  (a) vulgare   56   Cr   Marchant, 1972
  (b) simorrhinum (Durieu) Maire &
    Weiller
2 proboscideum (L.) Savi   28   Hs   Marchant, 1972

9 Ambrosina Bassi
1 bassii L.

10 Dracunculus Miller
1 vulgaris Schott   32   Bu   Popova &
                                         Ceschmedjiev, 1978
2 muscivorus (L. fil.) Parl.   56   Sa   Scrugli &
                                         Bocchieri, 1977

CXCVI LEMNACEAE

1 Wolffia Horkel ex Schleiden
1 arrhiza (L.) Horkel ex Wimmer   50   Po   Wcislo, 1970

2 Lemna L.
1 trisulca L.   40 Br Ga Ge Hb   Urbanska-
                                       Worytkiewicz, 1975
                 44   Br   Blackburn, 1933
2 gibba L.   40 50 Br Ga Ge   Urbanska-
                        Hs It Lu   Worytkiewicz, 1975
                 64   Br   Blackburn, 1933
3 minor L.   40 42 Br Gr Hb   Urbanska-
                      Hs It Lu   Worytkiewicz, 1975
                 50   Cz   Murín & Májovský,
                                         1978
4 perpusilla Torrey   40   Ga   Urbanska-
                                        Worytkiewicz, 1975
5 valdiviana Philippi   40   Ga   Urbanska-
                                        Worytkiewicz, 1975

3 Spirodela Schleiden
1 polyrhiza (L.) Schleiden   40   Po   Wcislo, 1970

PANDANALES

CXCVII SPARGANIACEAE

1 Sparganium L.
1 erectum L.
  (a) erectum
  (b) microcarpum (Neuman) Domin
  (c) neglectum (Beeby) Schinz & Thell. 30   Da   Larsen, 1965
2 gramineum Georgi   30   Scand.   Lohammon, 1942
3 emersum Rehmann   30   Cz   Májovský et al.,
                                       1976
4 angustifolium Michx   30   Is   Löve & Löve, 1956
5 glomeratum Laest. ex Beurl.   30   Scand.   Lohammon, 1942
6 minimum Wallr.   30   Is   Löve & Löve, 1956
7 hyperboreum Laest. ex Beurl.   30   Is   Löve & Löve, 1956

CXCVIII TYPHACEAE

1 Typha L.
  1 angustifolia L.   30   Ho   Gadella & Kliphuis,
                                                                1963
  2 domingensis (Pers.) Steudel   30   Hs   Löve & Kjellqvist,
                                                                1973
  3 minima Funck
  4 laxmannii Lepechin
  5 latifolia L.   30   Da   Larsen, 1954
  6 shuttleworthii Koch & Sonder

CYPERALES

CXCIX CYPERACEAE

1 Scirpus L.
  1 sylvaticus L.   62   Cz   Májovský et al.,
                                                                1976
                                                    64   Su   Ehrenberg, 1945
  2 radicans Schkuhr
  3 maritimus L.   80   Lu   Rodriguez, 1953
                                                 c.104   Cz   Májovský et al.,
                                                                1976
    (a) maritimus
    (b) affinis (Roth) T. Norlindh
  4 lacustris L.
    (a) lacustris   42   Fe   Sorsa, 1963
    (b) tabernaemontani (C.C. Gmelin) Syme   38   Hu   Pólya, 1949
                                                    42   Cz   Májovský et al.,
                                                                1976
                                                    44   Ge   Wulff, 1938
  5 mucronatus L.
  6 pungens Vahl
  7 litoralis Schrader
  8 triqueter L.
  9 supinus L.
 10 juncoides Roxb.
 11 holoschoenus L.
 12 setaceus L.   28   Ge   Scheerer, 1940
 13 pseudosetaceus Daveau
 14 cernuus Vahl
 15 prolifer Rottb.
 16 fluitans L.   60   Ge   Scheerer, 1940
 17 hudsonianus (Michx) Fernald
 18 cespitosus L.   104   Ge   Scheerer, 1940
    (a) cespitosus
    (b) germanicus (Palla) Broddeson
 19 pumilus Vahl

2 Blysmus Panzer
  1 compressus (L.) Panzer ex Link
  2 rufus (Hudson) Link

3 Eriophorum L.
  1 angustifolium Honckeny   58   Ho   Gadella & Kliphuis,
                                                                1966
    triste (Th. Fries) A. Löve & Hadac   60   Is   Löve & Löve, 1950
  2 latifolium Hoppe   54   Ge   Scheerer, 1940
                                                    72   non-Europe Löve & Löve, 1948

3 gracile Koch ex Roth
4 vaginatum L.   58   Fe   Sorsa, 1963
5 brachyantherum Trautv. & C.A. Meyer
6 russeolum Fries ex Hartman
7 scheuchzeri Hoppe

4 Eleocharis R. Br.
  1 quinqueflora (F.X. Hartmann) O.
    Schwarz   132   Su   Strandhede &
                                              134   Su   Dahlgren, 1968
  2 parvula (Roemer & Schultes) Link ex
    Bluff, Nees & Schauer   10   Ge   Wulff, 1937a
  3 acicularis (L.) Roemer & Schultes   20   Is   Löve & Löve, 1956
  4 bonariensis Nees
  5 ovata (Roth) Roemer & Schultes
  6 obtusa (Willd.) Schultes
  7 atropurpurea (Retz.) C. Presl
  8 palustris (L.) Roemer & Schultes
    (a) palustris   16   Cz   Májovský et al.,
                                                                          1974
    (b) vulgaris Walters   38   Fe No   Strandhede, 1965
                                              37 39   Fe Su   Strandhede, 1965
                                               40   Po   Pogan, 1972
  9 mitracarpa Steudel
 10 austriaca Hayek   16   Au Ge No   Strandhede, 1961
 11 mamillata H. Lindb. fil.   16   Fe No Su   Strandhede, 1965
 12 uniglumis (Link) Schultes   46   Fe No Su   Strandhede, 1965
                                               40   Br   Davies, 1956a
                                               44   Br   Walters, 1949
                                             54 92   Su   Strandhede, 1965
 13 oxylepis (Meinsh.) B. Fedtsch.
 14 multicaulis (Sm.) Desv.   20   Br   Davies, 1956a
 15 carniolica Koch   20   Hu   Baksay, 1958

5 Fuirena Rottb.
  1 pubescens (Poiret) Kunth

6 Fimbristylis Vahl
  1 bisumbellata (Forskål) Bubani
  2 annua (All.) Roemer & Schultes
  3 squarrosa Vahl
  4 ferruginea (L.) Vahl

7 Cyperus L.
  1 papyrus L.
  2 auricomus Sieber ex Sprengel
  3 longus L.
  4 rotundus L.
  5 esculentus L.
  6 glomeratus L.
  7 glaber L.
  8 eragrostis Lam.   42   Ga   Kliphuis &
                                                                         Wieffering, 1972
  9 textilis Thunb.
 10 alternifolius L.
 11 fuscus L.   72   Su   Löve & Löve, 1944
 12 difformis L.
 13 capitatus Vandelli   72   Lu   Rodriguez, 1953
 14 michelianus (L.) Link
    (a) michelianus
    (b) pygmaeus (Rottb.) Ascherson & Graebner

352

15 cerotinus Rottb.
16 pannonicus Jacq.
17 laevigatus L.
  (a) laevigatus
  (b) distachyos (All.) Maire & Weiller
18 flavidus Retz.
19 polystachyos Rottb.
20 mundtii (Nees) Kunth
21 flavescens L.
22 strigosus L.
23 congestus Vahl
24 hamulosus Bieb.
25 squarrosus L.
26 ovularis (Michx) Torrey
27 brevifolius (Rottb.) Hassk.

8 Cladium Browne
  1 mariscus (L.) Pohl   36   Europe   Löve, 1954a
                                               c.60   Da   Larsen, 1965

9 Rhynchospora Vahl
  1 alba (L.) Vahl   26   Ge   Scheerer, 1940
                                                 42   Su   Löve & Löve, 1942
  2 fusca (L.) Aiton fil.   32   Ge   Scheerer, 1940
  3 rugosa (Vahl) S. Gale

10 Schoenus L.
  1 nigricans L.   44   Ga   Labadie, 1976
                                                 54   Lu   Rodriguez, 1953
  2 ferrugineus L.   76   Scand.   Löve & Löve, 1944

11 Kobresia Willd.
  1 myosuroides (Vill.) Fiori   60   Is   Löve & Löve, 1956
  2 simpliciuscula (Wahlenb.) Mackenzie   76   No   Knaben & Engelskjon,
                                                                      1967

12 Carex L.
  1 phyllostachys C.A. Meyer
  2 distachya Desf.   74   Hs   Löve & Kjellqvist,
                                                                      1973
  3 paniculata L.   60 62 64   Su   Lövqvist, 1963
  (a) paniculata
  (b) lusitanica (Schkuhr) Maire
  4 appropinquata Schumacher   64   Su   Lövqvist, 1963
  5 diandra Schrank   60   Is   Löve & Löve, 1956
  6 vulpina L.   68   Su   Lövqvist, 1963
  7 otrubae Podp.   58   It   Dietrich, 1972
                                                 60   Su   Lövqvist, 1963
  8 vulpinoidea Michx
  9 spicata Hudson   58   Ge   Dietrich, 1972
                                                 60   Ju   Druskevič, ined.
 10 muricata L.   c.56   Su   Lövqvist, 1963
                                                 58   Ge   Dietrich, 1972

  (a) muricata
  (b) lamprocarpa Čelak.
 11 divulsa Stokes   56   It   Dietrich, 1972
                                                 58   Br   Davies, 1956b
  (a) divulsa
  (b) leersii (Kneucker) Walo Koch
 12 arenaria L.   58   Lu   Rodriguez, 1953

		64	Ge	Wulff, 1937b
13	ligerica Gay	58	Ge	Dietrich, 1972
14	reichenbachii Bonnet			
15	disticha Hudson	62	Br	Davies, 1956
16	repens Bellardi	70	Au	Dietrich, 1972
17	praecox Schreber	58	Ge	Dietrich, 1972
18	brizoides L.	58	Ge	Dietrich, 1964
19	chordorrhiza L. fil.	60	Su	Lövqvist, 1963
		c.66	No	Knaben & Engelskjon, 1967
20	divisa Hudson	60 62	Hs	Löve & Kjellqvist, 1973
21	stenophylla Wahlenb.	60	It	Dietrich, 1972
22	physodes Bieb.			
23	foetida All.	58	He	Favarger, 1959
24	maritima Gunnerus	60	Is	Löve & Löve, 1956
25	remota L.	62	It	Dietrich, 1972
26	bohemica Schreber	80	Ge	Dietrich, 1972
27	ovalis Good.	64	Hs	Kjellqvist & Löve, 1963
		66	Ge	Dietrich, 1972
		68	Ge	Wulff, 1939
28	macloviana D'Urv.	86	Is	Löve & Löve, 1956
29	echinata Murray	56 58	Su	Lövqvist, 1963
30	dioica L.	52	Su	Lövqvist, 1963
		60		
31	parallela (Laest.) Sommerf.			
	(a) parallela	44	Sb	Flovik, 1943
	(b) redowskiana (C.A. Meyer) Egorova			
32	davalliana Sm.	46	Ge	Dietrich, 1972
33	elongata L.	56	Ju	Druskevic, ined.
34	lachenalii Schkuhr	58	Su	Davies, 1956
		62	Is	Löve & Löve, 1956
		74		
35	glareosa Wahlenb.	66	Sb	Flovik, 1943
36	ursina Dewey	64	Sb	Flovik, 1943
37	mackenziei V. Krecz.	64	Is	Löve & Löve, 1956
38	heleonastes L. fil.	c.64	Is	Löve & Löve, 1956
39	marina Dewey			
40	curta Good.	54	Su	Lövqvist, 1963
		56	No	Knaben & Engelskjon, 1967
41	lapponica O.F. Lang			
42	brunnescens (Pers.) Poiret	56	Is	Löve & Löve, 1956
43	loliacea L.	54	Su	Lövqvist, 1963
44	tenuiflora Wahlenb.			
45	disperma Dewey			
46	baldensis L.	88	It	Favarger, 1959
		90	It	Dietrich, 1972
47	curvula All.	c.86	Au	Reese, 1953
	(a) curvula			
	(b) rosae Gilomen			
48	hirta L.	112	Br	Davies, 1956
49	atherodes Sprengel			
50	lasiocarpa Ehrh.	c.56	Ge	Reese, 1952
51	acutiformis Ehrh.	c.38	Ge	Reese, 1952
		78	Ge	Dietrich, 1972
52	melanostachya Bieb. ex Willd.			
53	riparia Curtis	72	Cz	Májovský et al., 1976

54 pseudocyperus L.	66	Su	Heilborn, 1924
55 rostrata Stokes	60	Ju	Druškevič, ined.
	72-74	Su	Lövqvist, 1963
	76	Cz	Murín & Májovský, 1976
	82	Su	Lövqvist, 1963
56 rhynchophysa Fischer, C.A. Meyer & Ave-Lall.			
57 vesicaria L.	70	Ju	Druskevic, ined.
	74	Su	Lövqvist, 1963
	82	Su	Heilborn, 1924
	86	Ge	Dietrich, 1972
	88	Ju	Druškevič, ined.
58 saxatilis L.	80	No	Knaben & Engelskjon, 1967
59 rotundata Wahlenb.			
60 stenolepis Less.			
61 mollissima Christ			
62 pendula Hudson	58	Hs	Kjellqvist & Löve, 1963
	60	Su	Lövqvist, 1963
	62	Ju	Druškevič, ined.
63 microcarpa Bertol. ex Moris	60	Co	Contandriopoulos, 1962
64 sylvatica Hudson			
(a) sylvatica	58	Cz	Májovský et al., 1976
(b) paui (Sennen) A. & O. Bolos			
65 arnellii Christ			
66 debilis Michx			
67 capillaris L.			
(a) capillaris	54	Su	Löve et al., 1957
(b) fuscidula (V. Krecz. ex Egorova) A. & D. Löve			
68 krausei Boeckeler	36	Is	Löve & Löve, 1956
69 ledebourana C.A. Meyer ex Trev.			
70 strigosa Hudson	66	Su	Lövqvist, 1963
71 vulcani Hochst.			
72 flacca Schreber			
(a) flacca	76	Su	Lövqvist, 1963
	90	Hs	Kjellqvist & Löve, 1963
(b) serratula (Biv.) W. Greuter	76	Ju	Druškevič, ined.
	90	Hs	Löve & Kjellqvist, 1973
73 hispida Willd.	38	Hs	Love & Kjellqvist, 1973
	42	Hs	Kjellqvist & Löve, 1963
74 panicea L.	32	Is	Löve & Löve, 1956
75 vaginata Tausch	32	No	Knaben & Engelskjon, 1967
76 livida (Wahlenb.) Willd.	32	Su	Lövqvist, 1963
77 asturica Boiss.	46	Hs	Löve & Kjellqvist, 1973
78 depauperata Curtis ex With.	44	Ge	Dietrich, 1972
79 brevicollis DC.			
80 michelii Host	62	Ju	Druškevič, ined.
	c.70	Cz	Murín & Májovský, 1976

81 pilosa Scop.	44	Ge	Dietrich, 1972
82 hordeistichos Vill.			
83 secalina Willd. ex Wahlenb.			
84 laevigata Sm.	72	Ge	Dietrich, 1972
85 camposii Boiss. & Reuter	68	Lu	Dietrich, 1964
86 hochstetterana Gay ex Seub.			
87 binervis Sm.	74	Be	Dietrich, 1972
88 distans L.	72	It	Dietrich, 1972
	74	Hs	Löve & Kjellqvist, 1973
89 cretica Gradstein & Kern			
90 punctata Gaudin	68	Ju	Druskević, ined.
91 diluta Bieb.			
92 extensa Good.	60	Ga	Labadie, 1976
93 mairii Cosson & Germ.	68	Hs	Kjellqvist & Löve, 1963
	70	It	Dietrich, 1972
94 hostiana DC.	56	Ju	Druskević, ined.
95 durieui Steudel	52	Hs	Dietrich, 1972
96 flava L.	30 33	Ga	Bidault, 1974
	60	Br	Davies, 1956b
	64	?Su	Heilborn, 1918
97 jemtlandica (Palmgren) Palmgren			
98 lepidocarpa Tausch	58	Ju	Druskević, ined.
	68	Br	Davies, 1956b
99 nevadensis Boiss. & Reuter	68	Hs	Löve & Kjellqvist, 1963
100 bergrothii Palmgren			
101 demissa Hornem.	70	Br	Davies, 1955
102 serotina Mérat			
(a) serotina	35	Ga	Bidault, 1974
	68	Au	Dietrich, 1972
	70	No	Heilborn, 1924
(b) pulchella (Lonnr.) Van Ooststr.	70	Su	Heilborn, 1928
103 pallescens L.	62	Ge	Dietrich, 1972
	64	Br	Davies, 1956b
	66	Su	Lövqvist, 1963
104 hallerana Asso	50	It	Dietrich, 1964
	52	It	Dietrich, 1972
	54	Hs	Löve & Kjellqvist, 1973
105 rorulenta Porta	48-50	Bl	Dietrich, 1972
106 digitata L.	48 50 52	Br	Davies, 1956b
	54	Rm	Dietrich, 1972
107 ornithopoda Willd.			
(a) ornithopoda	c.46	Su	Heilborn, 1924
	52	No	Knaben & Engelskjon, 1967
	54	Su	Lövqvist, 1963
(b) ornithopodioides (Hausm.) Nyman	54	Au	Dietrich, 1972
	56	Ju	Druskević, ined.
108 pediformis C.A. Meyer			
(a) pediformis			
(b) rhizodes (Blytt) H. Lindb. fil.			
(c) macroura (Meinsh.) Podp.			
109 humilis Leysser	36 38	Ju	Druskević, ined.
110 glacialis Mackenzie	34	No	Knaben & Engelskjon, 1967
111 caryophyllea Latourr.	62 64 66 68	Su	Lövkvist, 1963
112 umbrosa Host			
(a) umbrosa	62	Ge	Dietrich, 1964
	66	Ge	Dietrich, 1972

(b) huetiana (Boiss.) Soó
(c) sabynensis (Less. ex Kunth) Kuk.
113 depressa Link
   (a) depressa
   (b) transsilvanica (Schur) Egorova
114 oedipostyla Duval-Jouve
115 illegitima Cesati
116 globularis L.

117 tomentosa L.	48	Ge	Dietrich, 1972
118 grioletii Roemer	48	It	Dietrich, 1964
119 ericetorum Pollich	30	Su	Lövqvist, 1963
120 melanocarpa Cham. ex Trautv.			
121 montana L.	38	Ge	Dietrich, 1972
122 fritschii Waisb.	30	He	Dietrich, 1972
123 markgrafii Kuk.			
124 pilulifera L.			
(a) pilulifera	18	Ge	Dietrich, 1972
(b) azorica (Gay) Franco & Rocha Afonso			
125 tricolor Velen.			
126 amgunensis Friedrich Schmidt Petrop.			
127 olbiensis Jordan	46	It	Dietrich, 1972
128 alba Scop.	54	Ge	Dietrich, 1964
129 supina Willd. ex Wahlenb.	38	It	Dietrich, 1964
130 liparocarpos Gaudin			
(a) liparocarpos	38	It	Dietrich, 1972
(b) bordzilowskii (V. Krecz.) Egorova			
131 sempervirens Vill.	30 31	Ge	Dietrich, 1967
	32	It	Dietrich, 1967
	34	Rm	Dietrich, 1967
	68	Ga	Lazare, 1976
132 firma Host	34	Ge	Dietrich, 1967
	34 68	Ju	Druskevic, ined.
133 kitaibeliana Degen ex Becherer	36	It	Dietrich, 1967
134 macrolepis	36 37	It	Dietrich, 1967
135 mucronata All.	34	Au	Dietrich, 1964
	36	Au Ge It	Dietrich, 1967
136 fuliginosa Schkuhr			
(a) fuliginosa	40	Ge Rm	Dietrich, 1967
(b) misandra (R. Br.) Nyman	40	No	Knaben & Engelskjon, 1967
137 ferruginea Scop.			
(a) ferruginea	39 40	Au Ge He It	Dietrich, 1967
(b) australpina (Becherer) W. Dietr.	38	It	Dietrich, 1964
	40	It	Dietrich, 1967
(c) caudata (Kuk.) Pereda & Laínz			
(d) macrostachys (Bertol.) Arcangeli	40	It	Dietrich, 1967
(e) tendae W. Dietr.	38	Ga It	Dietrich, 1967
138 brachystachys Schrank	40	Ge It Ju	Dietrich, 1967
139 fimbriata Schkuhr	42	He	Dietrich, 1967
140 atrofusca Schkuhr	38	No	Knaben & Engelskjon, 1967
	40	He	Dietrich, 1967
141 frigida All.	56	Au Ga He	Dietrich, 1967
	58	Ga	Dietrich, 1964
142 limosa L.	56 62	Su	Lövqvist, 1963
	64	Ge	Wulff, 1939
143 rariflora (Wahlenb.) Sm.	c.52	No	Knaben & Engelskjon, 1967
	54	Is	Löve & Löve, 1956
144 magellanica Lam.	58	He	Favarger, 1959

145 laxa Wahlenb.
146 atrata L.
    (a) atrata 52 54 Ju Druskević, ined.
    (b) aterrima (Hoppe) Čelak.
    (c) caucasica (Steven) Kuk.
147 parviflora Host 54 Au Dietrich, 1972
148 buxbaumii Wahlenb.
    (a) buxbaumii c.74 c.100 Su Lövqvist, 1963
    (b) alpina (Hartman) Liro 106 Is Löve & Löve, 1956
149 hartmanii Cajander
150 norvegica Retz. 56 Is Löve & Löve, 1956
    (a) norvegica
    (b) inferalpina (Wahlenb.) Hultén
    (c) pusterana (Kalela) Chater
151 holostoma Drejer
152 stylosa C.A. Meyer
153 bicolor All. 16 He Reese, 1953
c.48 Is Löve & Löve, 1956
50 He Davies, 1956
52 No Knaben & Engelskjon, 1967

154 rufina Drejer 60 Is Löve & Löve, 1956
155 paleacea Schreber ex Wahlenb. 72 73 No Faulkner, 1972
156 lyngbyei Hornem. 72 Is Löve & Löve, 1956
157 vacillans Drejer
158 recta Boott 73 c.74 75 Br Faulkner, 1972
159 halophila F. Nyl.
160 salina Wahlenb.
161 subspathacea Wormsk. 42 Is Löve & Löve, 1956
162 aquatilis Wahlenb. 76 Br Fe No Faulkner, 1972
77 Br Faulkner, 1972
84 Ge Dietrich, 1972
163 bigelowii Torrey ex Schweinitz 70 Is Löve & Löve, 1956
    (a) bigelowii 68-71 Br Hb Faulkner, 1972
    (b) rigida Schultze-Motel
    (c) arctisibirica (Jurtzev) A. & D. Löve
    (d) ensifolia (Turcz. ex Gorodkov)
        J. Holub
164 elata All. 74 Br Davies, 1956
75 76 Hb Faulkner, 1972
76 77 Br Faulkner, 1972
78 80 Su Lovqvist, 1963

    (a) elata
    (b) omskiana (Meinsh.) Jalas
165 cespitosa L. 78-80 Da Faulkner, 1972

166 buekii Wimmer
167 nigra (L.) Reichard 82 84 Su Lövqvist, 1963
83-85 Br Faulkner, 1972
    vulgaris ssp. juncella (Fries) Nyman 76 Is Löve & Löve, 1956
84 No Knaben & Engelskjon, 1967
168 acuta L. 72 74-76 78 Su Ehrenberg, 1945
82-85 Br Faulkner, 1972
169 trinervis Degl.
170 microglochin Wahlenb. 48 He Dietrich, 1972
58 Is Löve & Löve, 1956
171 pauciflora Lightf. 46 76 Ju Druskević, ined.
172 rupestris All. 50 He Favarger, 1959
52 No Knaben & Engelskjon, 1967

173 obtusata Liljeblad
174 nardina Fries   70   Is   Löve & Löve, 1956
175 capitata L.   50   Is   Löve & Löve, 1956
176 scirpoidea Michx.
177 pyrenaica Wahlenb.
178 pulicaris L.   58 60   Su   Lövqvist, 1963

SCITAMINEAE

CC MUSACEAE

1 Musa L.
   1 cavandishii Lamb. ex Paxton

CCI ZINGIBERACEAE

1 Hedychium Koenig
   1 gardneranum Sheppard ex Ker-Gawler

CCII CANNACEAE

1 Canna L.
   1 indica L.

MICROSPERMAE

CCIII ORCHIDACEAE

1 Cypripedium L.
   1 calceolus L.   20   Po   Skalińska et al.,
                                                                 1957

   2 macranthos Swartz
   3 guttatum Swartz

2 Epipactis Zinn
   1 palustris (L.) Crantz   40   Su   Löve & Löve, 1944
                                             40 44 46 48 Ga   Lévêque &
                                                                 Gorenflot, 1969
   2 helleborine (L.) Crantz   36   Ga   Lévêque &
                                                                 Gorenflot, 1969
                                                    38   Po   Skalińska et al.,
                                                                 1961
                                                    40   Cz   Uhríková &
                                                                 Feráková, 1978
                                                    44   Ga   Lévêque &
                                                                 Gorenflot, 1969
   3 leptochila (Godfery) Godfery   36   Da   Hagerup, 1947
   4 muelleri Godfery
   5 dunensis (T. & T.A. Stephenson) Godfery
   6 purpurata Sm.   40   He   Meili-Frei, 1966
   7 phyllanthes G.E. Sm.   36   Da   Young, 1953
   8 atrorubens (Hoffm.) Besser   40   Su   Löve & Löve, 1944
                                             40 + 0 - 7B   He   Meili-Frei, 1966
   9 microphylla (Ehrh.) Swartz   40   He   Meili-Frei, 1966

3 Cephalanthera L.C.M. Richard
   1 damasonium (Miller) Druce   32   Da   Hagerup, 1947
   2 longifolia (L.) Fritsch   32   Hs   Löve & Kjellqvist,
                                                                 1973

3 cucullata Boiss. & Heldr.
4 epipactoides Fischer & C.A. Meyer
5 rubra (L.) L.C.M. Richard   48   Au   Titz, 1966

4 Limodorum Boehmer
  1 abortivum (L.) Swartz   56   Lu   Coutinho, 1957
    ssp. trabutianum (Batt.) Rouy   60   Lu   Coutinho, 1957

5 Epipogium R. Br.
  1 aphyllum Swartz

6 Neottia Ludwig
  1 nidus-avis (L.) L.C.M. Richard   36   He   Meili-Frei, 1966

7 Listera R. Br.
  1 ovata (L.) R. Br.   34 + 0 - 2B He   Meili-Frei, 1966
                       34-38   Ho   Gadella & Kliphuis,
                                          1963
                       34 36   Su   Löve & Löve, 1944
                         38   Da   Löve & Löve, 1944
  2 cordata (L.) R. Br.   38   Po   Skalińska et al.,
                                          1961
                       40   Ho   Gadella & Kliphuis,
                                          1963

8 Spiranthes L.C.M. Richard
  1 spiralis (L.) Chevall.   30   Da   Hagerup, 1944
  2 aestivalis (Poiret) L.C.M. Richard
  3 romanzoffiana Cham.   60   Hb   Heslop-Harrison,
                                          1961

9 Goodyera R. Br.
  1 repens (L.) R. Br.   30   Su   Löve & Löve, 1942
                       40   Po   Skalińska et al.,
                                            1974

10 Gennaria Parl.
  1 diphylla (Link) Parl.

11 Herminium Guett.
  1 monorchis (L.) R. Br.   40   He   Heusser, 1938

12 Neottianthe Schlechter
  1 cucullata (L.) Schlechter

13 Platanthera L.C.M. Richard
  1 bifolia (L.) L.C.M. Richard   42   He   Heusser, 1938
  2 chlorantha (Custer) Reichenb. 42   Po   Skalińska et al.,
                                          1947
  3 micrantha (Hochst.) Schlechter
  4 hyperborea (L.) Lindley   84   Is   Löve & Löve, 1956
  5 obtusata (Pursh) Lindley   126   Su   Löve, 1954a

14 Chamorchis L.C.M. Richard
  1 alpina (L.) L.C.M. Richard   42   He   Heusser, 1938

15 Gymnadenia R. Br.
  1 conopsea (L.) R. Br.   40 80 c.97 Au   Groll, 1965
                      c.117 c.119
  2 odoratissima (L.) L.C.M. Richard 40   Ju   Sušnik & Lovka,
                                          1973

16 Pseudorchis Séguier
  1 albida (L.) Á. & D. Löve
  (a) albida
  (b) straminea (Fernald) Á. & D. Love   40   Rs(N)   Sokolovskaya &
                                                Strelkova, 1960
                                 42   Is   Löve & Löve, 1956
  2 frivaldii (Hampe ex Griseb.) P.F. Hunt

17 Nigritella L.C.M. Richard
  1 nigra (L.) Reichenb. fil.   64   Ju   Lovka et al.,
                                                   1972
  (a) nigra
  (b) rubra (Wettst.) Beauverd
  (c) corneliana Beauverd

18 Coeloglossum Hartman
  1 viride (L.) Hartman   40   Br   Richards, 1972

19 Dactylorhiza Necker ex Nevski
  1 iberica (Bieb.) Soó
  2 sambucina (L.) Soó   40   Ho   Vermeulen, 1947
                                 42   Fe   Sorsa, 1963
  (a) sambucina
  (b) insularis (Sommier) Soó   60   It   Scrugli, 1977
  3 sulphurea (Link) Franco
  (a) sulphurea
  (b) pseudosambucina (Ten.) Franco
  (c) siciliensis (Klinge) Franco
  4 incarnata (L.) Soó   40   Ga   Vaucher, 1966
  (a) incarnata
  (b) coccinea (Pugsley) Soó
  (c) pulchella (Druce) Soó
  (d) cruenta (O.F. Mueller) P.D. Sell   40   Su   Löve & Löve, 1944
  5 pseudocordigera (Neuman) Soó
  6 majalis (Reichenb.) P.F. Hunt &   80   Cz   Májovský et al.,
     Summerhayes   1974
  (a) majalis
  (b) occidentalis (Pugsley) P.D. Sell 80   Br   Vermeulen, 1938
  (c) alpestris (Pugsley) Senghas
  (d) purpurella (T. & T.A. Stephenson) 80   Br   Roberts, 1966
     D. Moresby Moore & Soó
  (e) praetermissa (Druce) D. Moresby 80   Br   Roberts, 1966
     Moore & Soó
  7 cordigera (Fries) Soó   80   Ho   Vermeulen, 1947
  (a) cordigera
  (b) bosniaca (G. Beck) Soó
  (c) siculorum (Soó) Soó
  8 traunsteineri (Sauter) Soó   80   Hs   Löve & Kjellqvist,
                                                   1973
                             122   Rs(B)   Vermeulen, 1938
  (a) traunsteineri
  (b) curvifolia (Nyl.) Soó
  (c) lapponica (Laest. ex Hartman) Soo
  9 russowii (Klinge) J. Holub   120   ?Rs(B) Löve & Löve, 1961a
10 elata (Poiret) Soó   80   Vermuelen, 1947
11 maculata (L.) Soó   40   Su   Löve & Löve, 1944
                         40 60 80 Au   Groll, 1965

  (a) maculata
  (b) elodes (Griseb.) Soó   40   Br   Maude, 1939
                                 80   Is   Löve & Löve, 1956

(c) schurii (Klinge) Soó
(d) islandica (Á. & D. Löve) Soó   80   Is   Löve & Löve, 1956
(e) transsilvanica (Schur) Soó
12 fuchsii (Druce) Soó
  (a) fuchsii   40   Ju   Lovka et al., 1971
  (b) psychrophila (Schlechter) J. Holub
  (c) sooana (Borsos) Borsos
13 saccifera (Brongn.) Soó   40
                              80   Europe   Vermeulen, 1947

20 Steveniella Schlechter
 1 satyrioides (Steven) Schlechter

21 Comperia C. Koch
 1 comperiana (Steven) Ascherson & Graebner

22 Neotinea Reichenb. fil.
 1 maculata (Desf.) Stearn   40   It   Scrugli et al.,
                                          1976

23 Traunsteinera Reichenb.
 1 globosa (L.) Reichenb.   42   Po   Skalińska et al.,
                                        1957

24 Orchis L.
 1 papilionacea L.   32   Hs   Löve & Kjellqvist,
                                 1973

 2 boryi Reichenb. fil
 3 morio L.   36   Hs   Löve & Kjellqvist,
                          1973

  (a) morio   36   It   Scrugli et al.,
                          1976

  (b) picta (Loisel.) Arcangeli
  (c) champagneuxii (Barn.) Camus
 4 longicornu Poiret   36   It   Scrugli et al.,
                                   1976

 5 coriophora L.   36   Ju   Sušnik & Lovka,
                                1973

  (a) coriophora
  (b) fragrans (Pollini) Sudre   20   Ga   Labadie, 1976
                                  38   It   Scrugli et al.,
                                              1976
  (c) martrinii (Timb.-Lagr.) Nyman
 6 sancta L.
 7 ustulata L.   42   Hs   Löve & Kjellqvist,
                             1973

 8 tridentata Scop.   42   It   Scrugli, 1977
  (a) tridentata
  (b) commutata (Tod.) Nyman
 9 lactea Poiret   42   It   Scrugli, 1977
10 italica Poiret   42   It   Del Prete, 1977
11 simia Lam.   42   He   Heusser, 1938
12 militaris L.   42   Cz   Májovský et al.,
                              1976

13 punctulata Steven ex Lindley
  (a) punctulata
  (b) sepulchralis (Boiss. & Heldr.) Soó
14 purpurea Hudson   42   Cz   Májovský et al.,
                                 1976

15 saccata Ten.   36   It   Scrugli et al.,
                              1976

16 patens Desf.			
17 spitzelii Sauter ex Koch			
(a) spitzelii	42	Hs	Löve & Kjellqvist, 1973
(b) nitidifolia (Teschner) Soó	40	Cr	Hautzinger, 1976
18 mascula (L.) L.	42	Hs	Löve & Kjellqvist, 1973
(a) mascula			
(b) wanjkowii (Wulf) Soó			
(c) signifera (Vest) Soó	42	Cz	Uhríková & Schwarzova, 1978
(d) olbiensis (Reuter ex Grenier) Ascherson & Graebner			
(e) hispanica (A. & C. Nieschalk) Soó			
19 pallens L.	40	He	Heusser, 1938
20 provincialis Balbis	42	It	Scrugli, 1977
(a) provincialis			
(b) pauciflora (Ten.) Camus			
21 anatolica Boiss.			
22 quadripunctata Cyr. ex Ten.			
23 laxiflora Lam.			
(a) laxiflora	36	It	Scrugli, 1977
(b) palustris (Jacq.) Bonnier & Layens	42	Cz	Májovský et al., 1976
(c) elegans (Heuffel) Soó			
25 Aceras R. Br.			
1 anthropophorum (L.) Aiton fil.	42	Ho	Kliphuis, 1963
26 Himantoglossum Koch			
1 hircinum (L.) Sprengel	24	He	Heusser, 1915
	36	Ju	Sušnik & Lovka, 1973
(a) hircinum			
(b) calcaratum (G. Beck) Soó			
(c) caprinum (Bieb.) Sundermann			
27 Barlia Parl.			
1 robertiana (Loisel.) W. Greuter	36	Ga	Raynaud, 1971
28 Anacamptis L.C.M. Richard			
1 pyramidalis (L.) L.C.M. Richard	36	Ho	Kliphuis, 1963
29 Serapias L.			
1 cordigera L.	36	It	Scrugli et al., 1976
2 neglecta De Not.			
3 vomeracea (Burm.) Briq.	36	Ju	Sušnik & Lovka, 1973
(a) vomeracea			
(b) orientalis W. Greuter			
4 lingua L.			
5 parviflora Parl.	36	It	Del Prete, 1977
30 Ophrys L.			
1 insectifera L.	36	Cz	Májovský et al., 1976
2 speculum Link	36	It	Scrugli, 1977
(a) speculum			
(b) lusitanica O. & A. Danesch			
3 lutea (Gouan) Cav.			

(a) lutea	36	Hs	Löve & Kjellqvist, 1973

(b) murbeckii (Fleischm.) Soó
(c) melena Renz

4 fusca Link	c.73	Bl	Dahlgren et al., 1971

(a) durieui (Reichenb. fil.) Soó

(b) fusca	72 76	Bl	Greilhuber & Ehrendorfer, 1975
(c) iricolor (Desf.) O. Schwarz	36	It	Scrugli, 1977

(d) omegaifera (Fleischm.) E. Nelson
5 pallida Rafin.
6 sphegodes Miller

(a) sphegodes	36	He	Heusser, 1938
(b) litigiosa (Camus) Becherer	36	He	Heusser, 1938

(c) tommasinii (Vis.) Soó

(d) atrata (Lindley) E. Mayer	36	Bl	Greilhuber & Ehrendorfer, 1975

(e) mammosa (Desf.) Soó ex E. Nelson
(f) aesculapii (Renz) Soó
(g) helenae (Renz) Soó & D. Moresby Moore
7 spruneri Nyman
(a) spruneri
(b) panormitana (Tod.) Soó
8 ferrum-equinum Desf.
(a) ferrum-equinum
(b) gottfriediana (Renz) E. Nelson

9 bertolonii Moretti	36	Ju	Sušnik & Lovka, 1973
promontorii	36-38	It	Greilhuber & Ehrendorfer, 1975

10 lunulata Parl.
11 argolica Fleischm.
12 reinholdii Spruner ex Fleischm.
13 cretica (Vierh.) E. Nelson
14 carmeli Fleischm. & Bornm.
15 scolopax Cav.

(a) scolopax	36	It	Scrugli, 1977
	38 40	Lu	Shimoya & Ferlan, 1952
(b) oestrifera (Bieb.) Soó			
(c) cornuta (Steven) Camus	36	Ju	Sušnik & Lovka, 1973

(d) heldreichii (Schlechter) E. Nelson

16 fuciflora (F.W. Schmidt) Moench	36	He	Heusser, 1938
(a) fuciflora	36 37 38	It	Greilhuber & Ehrendorfer, 1975
ssp. sundermannii	36	It	

(b) candica E. Nelson ex Soó
(c) oxyrrhynchos (Tod.) Soó
(d) exaltata (Ten.) E. Nelson

17 arachnitiformis Gren. & Philippe	36	It	Scrugli, 1977
	38		
18 tenthredinifera Willd.	36	It	Scrugli, 1977
19 apifera Hudson	36	Hs	Löve & Kjellqvist, 1973

(a) apifera
(b) jurana Ruppert

20 bombyliflora Link	36	It	Scrugli, 1977

31 Corallorhiza Chatel.
1 trifida Chatel.   42   Da   Hagerup, 1941

32 Calypso Salisb.
1 bulbosa (L.) Oakes

33 Liparis L.C.M. Richard
1 loeselii (L.) L.C.M. Richard

34 Microstylis (Nutt.) A. Eaton
1 monophyllos (L.) Lindley

35 Hammarbya O. Kuntze
1 paludosa (L.) O. Kuntze   28   Ho   Kliphuis, 1963

REFERENCES

Aakerberg, E., 1942. Hereditas, 28: 1-126.

Agapova, N.D., 1976. Taxon, 25: 635-636.

Ahti, T., 1962. Arch. Soc. Zool. Bot. Fenn. 'Vanamo', 16: 22-35.

Albers, A. and Albers, I., 1973. Österr. Bot. Zeits., 122: 293-298.

Albers, F., 1972. Ber. Deutsch. Bot. Gesell. 85 (5-6): 279-285.

Albers, F., 1973. Österr. Bot. Zeits., 121: 251-254.

Alden, B., 1976. Bot. Not., 129: 297-321.

Aleskoovskij, M.K., 1929. Izv. Saratovskogo Gosodarstvenn Instit. Selsk.

   Khozjajstva, 5: 375-382.

Almgaard, G., 1960. Kungl. Lantbrukshögsk. Annal., 26: 77-119.

Ančev, M.E., 1974. In: Ehrendorfer, F. and Ančev, M.E., Bot. Jour. Linn.

   Soc. London, 68: 268-272.

Ančev, M.E., 1975. In: Löve, Á., Taxon, 24: 514.

Ančev, M.E., 1976. Bulg. Acad. Sc., Phytol., 5: 51-56.

Arata, M., 1944. Nuovo Giorn. Bot. Ital., 51: 39-43.

Armenise, V., 1958. Nuovo Giorn. Bot. Ital., 64: 297-318.

Arnoldi, W., 1910. Flora, 100: 121-139.

Arrigoni, P.V. and Mori, B., 1971. Inf. Bot. Ital., 3: 226-233.

Arrigoni, P.V. and Mori, B., 1976. Inf. Bot. Ital., 8: 269-276.

Ashri, A. and Knowles, P.F., 1960. Agron. Jour., 52: 11-17.

Auquier, P., 1973. Candollea, 28: 15-19.

Auquier, P. and Kerguélen, M., 1977. Lejeunia, 89: 1-82.

Auquier, P. and Rammeloo, J., 1973. Bull. Soc. Roy. Bot. Belg., 106: 317-328.

Avdulov, N.P., 1928. Driev. Vsesojuz. Sezda Botanik. Leningrad, 1928: 65-66.

Azevedo Coutinho, L. de and Santos, A.C., 1943. Agron. Lusit., 5: 349-361.

Babcock, E.B., 1947. Univ. Calif. Publs. Bot., 22: 199-1030.

Babcock, E.B., Stebbins, G.L. and Jenkins, J.A., 1937. Cytologia, Fuji

   Jubil. Vol.: 188-210.

Badr, A. and Elkington, T.T., 1977. Pl. Syst. Evol., 128: 23-35.

Baez-Mayer, A.B., 1934. Cavanillesia, 6: 59-103.

Bagnoli, L., Guglielminetti, P. and Loreti, F., 1963. Nuovo Giorn. Bot. Ital.,
    70: 544-545.

Bajer, A., 1950. Acta Soc. Bot. Polon., 20: 635-646.

Bajon, R. and Greber, F., 1971. Actes Cell. Fl. Veg. Alp. Jur., Paris, 255-270.

Baker, H.G., 1948. New Phytol., 47: 131-145.

Baker, H.G., 1955. In: Darlington, C.D. and Wylie, A.P., Chromosome Atlas
    of Flowering Plants.

Baker, H.G., 1959. Proc. Bot. Soc. Brit. Is., 3: 288.

Baksay, L., 1951. In: Soó, R. and Javorka, S. A Magyar Növenyvilag
    Kezikönyve I-II. Budapest.

Baksay, L., 1954. Ann. Hist.-Nat. Mus. Natl. Hung., S.N., 5: 139-148.

Baksay, L., 1955. Ann. Hist.-Nat. Mus. Natl. Hung., S.N., 6: 167-176.

Baksay, L., 1956. Ann. Hist.-Nat. Mus. Natl. Hung., S.N., 7: 321-334.

Baksay, L., 1957b. Ann. Hist.-Nat. Mus. Natl. Hung., S.N., 8: 169-174.

Baksay, L., 1958. Ann. Hist.-Nat. Mus. Natl. Hung., S.N., 50: (N.S.9):
    121-125.

Baksay, L., 1961. In: Löve, Á. and Löve, D., Opera Bot., 5.

Ball, P.W., 1961. In: Löve, Á. and Löve, D., Opera Bot., 5.

Ball, P.W. and Tutin, T.G., 1959. Watsonia, 4: 193-205.

Bambacioni-Mezzetti, V., 1936. Anal. Bot., 21: 186-204.

Banach-Pogan, E., 1954. Acta Soc. Bot. Polon., 23: 375-382.

Banach-Pogan, E., 1955. Acta Soc. Bot. Polon., 24: 275-286.

Banach-Pogan, E., 1958. Acta Biol. Cracov., Bot., 1 (2): 91-101.

Barling, D.M., 1958. Proc. Bot. Soc. Brit. Is., 3: 85.

Barros Neves, J., 1939. Bol. Soc. Brot., Sér. 2, 13: 545-572.

Barros Neves, J., 1973. Bol. Soc. Brot., Sér. 2, 47: 157-212.

Battaglia, E., 1947. Atti Soc. Toscana Sci. Nat. Mem., Ser. B, 54: 70-78.

Battaglia, E., 1957. Caryologia, 9: 293-314.

Battaglia, E., 1957b. Caryologia, 10: 1-28.

Baumberger, H., 1970. Ber. Schweiz. Bot. Ges., 80: 17-95.

Beatus, R., 1936. Zeitschr. Vererb., 71: 353-381.

Beaudry, J.R., 1970. Natural. Canad., 97: 431-445.

Bedalov, M., 1969. Acta Bot. Croat., 28: 39-41.

Bedalov, M. and Sušnik, F., 1970. Caryologia, 23: 519-524.

Behre, K., 1929. Planta, 7: 208-306.

Bentzer, B., 1972. Bot. Not., 125: 406-418.

Bentzer, B., 1973. Bot. Not., 126: 69-132.

Bernstrüm, P., 1946. Hereditas, 32: 514-520.

Berudye, E.M. and Jandar, E., 1907. New Phytol., 6: 127-134, 167-174.

Beuret, E., 1972. Bull. Soc. Neuchât. Sci. Nat., 95: 35-41.

Bidault, M., 1968. Rév. Cytol. Biol. Veg., 31: 217-356.

Bidault, M., 1972. Ann. Sci. Univ. Besançon, Ser. 3., Bot., 12: 119-123.

Bidault, M., 1974. Compt. Rend. Séances Soc. Biol., 168: 466.

Bijok, K., 1959. Acta Bot. Soc. Polon., 28: 365-371.

Bijok, K., Krenska, B. and Pawlak, T., 1972. Acta Soc. Bot. Polon., 41:
    433-438.

Billeri, G., 1954. Caryologia, 6: 45-51.

Bir, S.S., 1960. Acta Bot. Neerl., 9: 224-234.

Björkman, S.O., 1954. Hereditas, 40: 254-258.

Björkman, S.O., 1960. Symb. Bot. Upsal., 17: 1-113.

Björkquist, I., 1961a. Bot. Not., 114: 365-367.

Björkquist, I., 1961b. Bot. Not., 114: 281-299.

Björkquist, I., Bothmer, R. von, Nilsson, O. and Nordenstam, B., 1969. Bot.
    Not., 122: 271-283.

Björse, S.A., 1961. In: Löve, Á. and Löve, D., Opera Bot., 5.

Blackburn, K.B., 1923. Nature (Lond.), 112: 687-688.

Blackburn, K.B., 1933. Proc. Univ. Durham Philos. Soc., 9 (2): 84-90.

Blackburn, K.B., 1939. Jour. Bot., 77: 67-71.

Blackburn, K.B., 1950. In: Tischler, G., Chromosomenzahlen der Gefässpflanzen
    Mitteleuropas; Junk. S-Gravenhage.

Blackburn, K.B. and Adams, A.W., 1955. Proc. Bot. Soc. Brit. Is., 1: 380.

Blackburn, K.B. and Heslop-Harrison, J.W., 1921. Ann. Bot., 35: 159-188.

Blackburn, K.B. and Heslop-Harrison, J.W., 1924. Ann. Bot., 38: 361-378.

Blackburn, K.B. and Morton, J.K., 1957. New Phytol., 56: 344-351.

Blaise, S., 1966. Compt. Rend. Acad. Sc. Paris, 262: 103-106.

Blaise, S., 1969. In: Löve, Á., Taxon, 18: 311.

Blaise, S., 1970. In: Löve, Á., Taxon, 19: 102-113.

Böcher, T.W., 1932. Bot. Tidsskr., 42: 183-206.

Böcher, T.W., 1938a. Dansk Bot. Ark., 9 (4): 1-33.

Böcher, T.W., 1938b. Svenska Bot. Tidskr., 32: 346-361.

Böcher, T.W., 1940a. Dansk Vid. Selsk. Biol. Medd., 15 (3): 1-64.

Böcher, T.W., 1947. Bot. Not., 1947: 353-360.

Böcher, T.W., 1954. Bot. Tidsskr., 51: 33-47.

Böcher, T.W. and Larsen, K., 1955. Bot. Tidsskr., 52: 125-132.

Böcher, T.W. and Larsen, K., 1958. Dansk Vid. Selsk. Biol. Medd., 10 (2):
   1-24.

Böcher, T.W. and Larsen, K., 1958b. New Phytol., 57: 311-317.

Böcher, T.W., Larsen, K. and Rahn, K., 1953. Hereditas, 39: 290-303.

Böcher, T.W., Larsen, K. and Rahn, K., 1955. Hereditas, 41: 423-453.

Bocquet, G., Favarger, C. and Zurcher, P.A., 1967. Zeits. Basler Bot. Ges.,
   3 (2): 229-242.

Bonnet, A.L.M., 1958. Nat. Monspel., Ser. Bot., 10: 3-6.

Bonnet, A.L.M., 1961. Nat. Monspel., Ser. Bot., 13: 7-13.

Bonnet, A.L.M., 1966. Nat. Monspel., Ser. Bot., 17: 21-29.

Borhidi, A., 1968. Acta Bot. Acad. Sc. Hung., 14 (3-4): 253-260.

Bosemark, N.O., 1956. Hereditas, 42: 443-467.

Bosemark, N.O., 1957. Hereditas, 43: 236-298.

Bothmer, R. von, 1967. Bot. Not., 120: 202-208.

Bothmer, R. von, 1969. Bot. Not., 122: 38-56.

Bothmer, R. von, 1970. Bot. Not. 123: 52-60.

Bothmer, R. von, 1975. Mitt. Bot. Staatssamm., München, 12: 267-288.

Bothmer, R. von and Bentzer, B., 1973. Ann. Mus. Goulandris, 1: 11-13.

Bowden, W.M., 1957. Can. Jour. Bot., 35: 951-993.

Bowden, W.M., 1964. Can. Jour. Bot., 42: 547-601.

Bozzini, A., 1959. Caryologia, 12: 459-470.

Bradshaw, M.E., 1963. Watsonia, 5: 321-326.

Bradshaw, M.E., 1964. Watsonia, 6: 76-81.

Brandt, J.P., 1952. Bull. Soc. Neuchat. Sci. Nat., 75: 179-188.

Brandt, J.P., 1961. Bull. Soc. Neuchat. Sci. Nat., 84: 35-87.

Bresinskya, A. and Grau, J., 1970. Ber. Bayer. Bot. Ges., 42: 101-108.

Breton-Sintes, S., 1971. These Univ. Clermont-Ferrand, E146: 1-70.

Brett, O.E., 1955. New Phytol., 54: 138-148.

Breviglieri, N. and Battaglia, E., 1954. Caryologia, 6: 271-283.

Brighton, C.A., 1976. Kew Bull., 31: 33-46.

Brighton, C.A., Mathew, B. and Marchant, C.J., 1973. Kew Bull., 28: 451-464.

Brügger, A., 1960. Blyttia, 18: 33-48.

Brummitt, R.K., 1973. Watsonia, 9: 369-373.

Brunsberg, K., 1965. Bot. Not., 118: 356-381.

Bruun, H.G., 1932. Symb. Bot. Upsal., 1: 1-239.

Burdet, H.M., 1967. Candollea, 22: 107-

Burdet, H.M. and Greuter, W., 1973. In: Greuter, W., Boissiera, 22:
    1-215.

Burdet, H.M. and Miege, J., 1968. Candollea, 23: 109-120.

Buttler, K.P., 1969. Rev. Roum. Biol., Ser. Bot., 14: 275-282.

Caldo, L. del, 1971. Inf. Bot. Ital., 3: 71-73.

Camara, A. and Coutinho, L.A., 1939. Agron. Lusit., 1: 268-314.

Campion-Bourget, F., 1967. Compt. Rend. Acad. Sc. Paris, 264: 2100-2102.

Cardona, M.A., 1973. Acta Phytotax. Barcinon., 14: 1-20.

Carolin, R.C., 1957. New Phytol., 56: 81-97.

Carpineri, R., 1971. Inf. Bot. Ital., 3: 75-79.

Carpio, M.D.A., 1954. Genet. Iber., 6: 101-111.

Carpio, M.D.A., 1957. Genet. Iber., 9: 163-185.

Cartier, D., 1965. Compt. Rend. Acad. Sc. Paris., 261: 4475-4478.

Cartier, D. and Lennoir, A., 1968. Compt. Rend. Acad. Sc. Paris, 266: 119-122.

Casas, J.F., 1973. Cuad. Cienc. Biol. (Granada), 2: 39-41.

Casas, J.F., Lainz, M. and Rejón, M.R., 1973. Cuad. Cienc. Biol. (Granada),
    2: 3-5.

Casper, S.J., 1962. Feddes Rep., 66: 1-148.

Casper, S.J., 1966. Biblioth. Bot., 127/128: 1-20.

Castro, D. de, 1941. Agron. Lusit., 3: 103-113.

Castro, D. de, 1943. Agron. Lusit., 5: 243-249.

Castro, D. de, 1945. Bol. Soc. Brot. Sér. 2, 19: 525-539.

Castro, D. de and Fontes, F.C., 1946. Broteria, 15: 38-46.

Cesca, G., 1961. Caryologia, 14: 79-86.

Cesca, G., 1961. Nuovo Giorn. Bot. Ital., 70: 542-543.

Cesca, G., 1972. Inf. Bot. Ital., 4: 45-66.

Ceschmedjiev, I.V., 1970. Bot. Zhurn., 55: 1100-1110.

Ceschmedjiev, I.V., 1973. Bot. Zhurn., 58: 864-875.

Ceschmedjiev, I.V., 1976. Taxon, 25: 642.

Chiappini, M., 1954. Nuovo Giorn. Bot. Ital., 61: 276-289.

Chiappini, M. and Scrugli, A., 1972. Inf. Bot. Ital., 4: 131-133.

Chiarugi, A., 1928. Nuovo Giorn. Bot. Ital., N.S., 34: 1452-1496.

Chiarugi, A., 1933. Nuovo Giorn. Bot. Ital., N.S., 40: 63-75.

Chiarugi, A., 1935. In: Tischler, G., Tab. Biol., 11: 281-304.

Chiarugi, A., 1949. Caryologia, 1: 362-376.

Chiarugi, A., 1950. Caryologia, 3: 148.

Choudhuri, H.C., 1942. Ann. Bot., N.S., 6: 184-217.

Chouksanova, N.A., Sveshnikova, L.I. and Alexandrova, T.V., 1968.
    Citologija. 10: 381-386.

Christoff, M., 1942. Zeitschr. Indukt. Abst. Vererb., 80: 103-125.

Christoff, M. and Christoff, M.A., 1948. Genetics, 33 (1): 36-42.

Christoff, M. and Popoff, A., 1933. Planta, 20: 440-447.

Christov, M. and Nikolov, T., 1962. Izv. Nauch. Inst. Rast (Sofia), 15:
    33-41.

Cinčura, F. and Hindakova, M., 1963. Biologia (Bratislava), 18: 184-194.

Clausen, J., 1931. Bot. Tidsskr., 41: 317-335.

Cole, M.J., 1962. Watsonia, 5: 113-116.

Constantinescu, G., Morlova, I., Cosmin, S. and Molea, I., 1964. Res.
    Roumania Biol. Ser. Bot., 9: 223-233.

Contandriopoulos, J., 1957. Bull. Soc. Bot. Fr., 104: 533-538.

Contandriopoulos, J., 1962. Thèse Fac. Sc. Montpellier, 1-354.

Contandriopoulos, J., 1964a. Rev. Gen. Bot., 71: 361-384.

Contandriopoulos, J., 1964b. Bull. Soc. Bot. France, 111: 222-234.

Contandriopoulos, J., 1966. Bull. Soc. Bot. France, 113: 453-474.

Contandriopoulos, J., 1967. 91e Congres des Soc. Sav. Rennes, 3: 271-280.

Contandriopoulos, J., 1968. Bull. Mus. Hist. Nat. Marseilles, 27: 45-52.

Contandriopoulos, J. and Cauwet, A.-M., 1968. Nat. Monspel., 19: 29-35.

Contandriopoulos, J. and Martin, D., 1967. Bull. Soc. Bot. Fr., 114 (7-8):
    257-275.

Cook, C.D.K., 1962. Watsonia, 5: 123-126.

Cook, C.D.K., 1968. In: Löve, A., Taxon, 17: 419-422.

Coombe, D.E., 1961. Watsonia, 5: 68-87.

Cope,T.A., 1976. Unpub. Ph.D. thesis, Univ. Leicester.

Cotton, R. and Stace, C.A., 1976. Genetica, 46: 235-255.

Coutinho, L. de A., 1957. Agron. Lusit., 19: 219-231.

Coutinho, L. de A. and Lorenzo-Andreu, A., 1948. An. Aula Dei, 1: 1-32.

Crane, M.B. and Gairdner, A.E.,11923. Jour. Genet., 13: 187-200.

Cugnac, A. de and Simonet, M., 1941a. Compt. Rend. Soc. Biol. (Paris),
    135: 728-731.

Cugnac, A. de and Simonet, M., 1941b. Bull. Soc. Bot. Fr., 88: 513-517.

Czapik, R., 1961. Acta Biol. Cracov. Ser. Bot., 4: 97-120.

Czapik, R., 1965. Acta Biol. Cracov., 8: 21-34.

Czapik, R., 1968. Watsonia, 6 (6): 345-349.

Dabrowska, L., 1971. Fragm. Flor. Geobot., 17: 383-386.

Dahlgren, K.V.O., 1952. Hereditas, 38: 314-320.

Dahlgren, R., Karlsson, T. and Lassen, P., 1971. Bot. Not., 124 (2): 249-269.

Daker, M.G., 1965. In: Löve, Á. and Solbrig, O.T., Taxon, 14: 88-89.

D'Amato, F., 1939. Nuovo Giorn. Bot. Ital., N.S., 46: 470-510.

D'Amato, F., 1945. Nuovo Giorn. Bot. Ital., N.S., 52: 86-87.

D'Amato, F., 1946. Nuovo Giorn. Bot. Ital., N.S., 53: 405-437.

D'Amato, F., 1957a. Caryologia, 10: 111-151.

D'Amato, F., 1957b. Caryologia, 9: 315-339.

D'Amato-Avanzi, M.G., 1957. Caryologia 9: 373-375.

Damboldt, J., 1964. Madroño, 17: 266, 268.

Danboldt, J., 1965. Öster. Bot. Zeit., 112: 392-406.

Damboldt, J., 1966. Ber. Deutsch. Bot. Ges., 78 (9): 373-376.

Damboldt, J., 1968. Taxon, 17: 91-104.

Damboldt, J., 1968b. Ber. Deutsch. Bot. Ges., 81: 43-52.

Damboldt, J., 1971. Taxon, 20: 787.

Damboldt, J. and Phitos, D., 1975. Pl. Syst. Evol., 123: 119-131.

Damboldt, J. and Podlech, D., 1963. Ber. Bay. Bot. Ges., 36: 29-32.

Damboldt, J., Graumann, G. and Matthäs, U., 1973. In: Löve, Á., Taxon, 22:
116.

Dangeard, P., 1941. Compt. Rend. Soc. Biol. Paris, 135: 581-583.

Dark, S.O.S., 1932. Ann. Bot., 46: 965-977.

Datta, R.M., 1932. Mem. Mchtr. Lit. Phil. Soc., 76: 85.

Davies, E.W., 1953. Nature (Lond.), 171: 659-660.

Davies, E.W., 1955. Watsonia, 3: 129-138.

Davies, E.W., 1956a. Watsonia, 3: 242-243.

Davies, E.W., 1956b. Hereditas, 42: 349-366.

Davies, E.W. and Willis, A.J., 1959. Proc. Bot. Soc. Brit. Is., 3: 328.

Delay, C., 1947. Rev. Cytol. Cytophys. Végét., 9: 169-223, 10: 103-229.

Delay, J., 1967. Inf. Ann. Caryosyst. Cytogénét., 1: 11-14.

Delay, J., 1968. Inf. Ann. Caryosyst. Cytogénét., 2: 17-22.

Delay, J., 1969. Inf. Ann. Caryosyst. Cytogenet., 3: 6, 7, 13, 17-27.

Delay, J., 1970. Inf. Ann. Caryosyst. Cytogenet., 4: 1-16.

Delay, J., 1971. Inf. Ann. Caryosyst. Cytogenet., 5: 37-40.

Del Prete, C., 1977. Inf. Bot. Ital., 9: 135-140.

Dersch, G., 1964. Ber. Deutsch. Bot. Ges., 76: 351-359.

Diannelidis, T., 1951. Portug. Acta Biol., Ser. A, 3: 151-170.

Dietrich, J., 1967b. Inf. Ann. Caryosyst. Cytogenet., 1: 23-28.

Dietrich, J., 1969. Inf. Ann. Caryosyst. Cytogenet., 3: 33-34.

Dietrich, J., 1970. Inf. Ann. Caryosyst. Cytogenet., 4: 29.

Dietrich, M., 1964. Madrono, 17: 266-267.

Dietrich, M., 1967. Feddes Rep., 75: 1-42.

Dietrich, M., 1972. In: Löve, Á., Taxon, 21: 333-346.

Dittrich, M., 1968. Ŏesterr. Bot. Zeit., 115: 379-390.

Dolcher, T. and Pignatti, S., 1971. Nuovo Giorn. Bot. Ital., 105: 95-107.

Dominguez, E. and Galiano, E.F., 1974. Lagascalia, 4: 61-84.

Donadille, P., 1967. Compt. Rend. Acad. Sc. Paris, 264: 813-816.

Doulat, E., 1943. Bull. Soc. Sci. Dauphine, 61: 1-232.

Dovaston, H.E., 1955. Notes Roy. Bot. Gard. Edinb., 21: 289-291.

Duckert, M.M. and Favarger, C., 1960. Bull. Soc. Neuchat. Sci. Nat., 83:
    109-119.

Durand, B., 1962. Rev. Cytol. Biol. Veg., 25: 337-341.

Eaton, R.D., 1968. Cytological Studies in the Genus Primula. Ph.D.
    thesis, University of Durham.

Edmonds, J.M., 1977. Bot. Jour. Linn. Soc., London, 75: 141-178.

Edmonds, J.M., Sell, P.D. and Walters, S.M., 1974. Watsonia, 10:
    159-161.

Egolf, D.R., 1962. Journ. Arn. Arbor., 43: 132-172.

Ehrenberg, L., 1945. Bot. Not., 1945: 430-437.

Ehrendorfer, F., 1952. Carnegie Inst. Yrbook, 51: 125-131.

Ehrendorfer, F., 1953. Öesterr. Bot. Zeits., 100: 670-672.

Ehrendorfer, F., 1955. Öesterr. Bot. Zeits., 102 (2-3): 195-234.

Ehrendorfer, F., 1959. Öesterr. Bot. Zeits., 106: 363-368.

Ehrendorfer, F., 1962. Ber. Deutsch. Bot. Ges., 75 (5): 137-152.

Ehrendorfer, F., 1962. Öesterr. Bot. Zeits., 109 (3): 276-343.

Ehrendorfer, F., 1963. In: Turrill, W.B. (Ed.) Vistas in Botany, 4: 158.

Ehrendorfer, F., 1964. Öesterr. Bot. Zeits., 111: 84-142.

Ehrendorfer, F., 1974. In: Ehrendorfer, F. and Krendl, F., Bot. Jour.
    Linn. Soc., London, 68: 268-272.

Ehrendorfer, F., 1975. In: Jordanov, D. et al. (Eds.) Problems of Balkan
    Flora and Vegetation. Sofia, Bulgarian Academy of Science, p. 178-185.

Ehrendorfer, F. and Ančev, M.E., 1975. Pl. Syst. Evol., 124: 1-6.

Eklundh-Enrenberg, C., 1946. Bot. Not.,1946: 529-536.

Elkington, T.T., 1974. Watsonia, 10 (1): 80.

Elkington, T.T. and Woodell, S.R.J., 1963. Jour. Ecol., 51: 769-781.

Ellerström, S. and Tjio, J.H., 1950. Bot. Not., 1950: 463-465.

Ellis, R.P. and Jones, B.M.G., 1970. Watsonia, 8: 45.

Engel, K., 1972. Taxon, 21: 495-500.

Engelskjon, T., 1967. Nytt Mag. Bot., 14: 125-139.

Erben, M., 1978. Mitt. Bot. Staatssamm., München, 14: 361-631.

Erdtman, G. and Nordborg, G., 1961. Bot. Not. 114: 19-21.

Ernet, D., 1972. Taxon, 21: 495-500.

Fabbri, F., 1957. Caryologia, 10: 388-390.

Fabergé, A.C., 1943. Jour. Genet., 45: 139-170.

Fagerlind, F., 1934. Hereditas, 19: 223-232.

Fagioli, A. and Fabbri, F., 1971. Inf. Bot. Ital., 3: 55-62.

Fagerlind, A., 1968. Blyttia, 26: 112-123.

Fahmy, T.Y., 1955. Rec. Trav. Lab. Bot. Geol. Zool. Fac. Sci. Univ.
    Montpellier, 7: 103-114.

Faulkner, J.S., 1972. Bot. Jour. Linn. Soc., London, 65: 271-301.

Faurg, J. and Pietrera, D., 1969. Ann. Fac. Sci. Marseilles, 42: 271-283.

Favarger, C., 1946. Ber. Schweiz. Bot. Ges., 56: 365-467.

Favarger, C., 1949a. Bull. Soc. Neuchât. Sci. Nat., 72: 15-22.

Favarger, C., 1949b. Ber. Schweiz Bot. Ges., 59: 62-86.

Favarger, C., 1952. Ber. Schweiz. Bot. Ges., 62: 244-257.

Favarger, C., 1953. Bull. Soc. Neuchât. Sci. Nat., 76: 133-169.

Favarger, C., 1954. VIII Congr. Internat. Bot. Compt. Rendus Rapp. and
    Comm., 9-10: 51-56.

Favarger, C., 1958. Veröffentl. Geobot. Inst. Rubel in Zurich, 33: 59-80.

Favarger, C., 1959. Bull. Soc. Neuchâtel Sci. Nat., 82: 255-285.

Favarger, C., 1960. Bull. Soc. Bot. France, 107: 94-98.

Favarger, C., 1961. Phyton, 9: 252-256.

Favarger, C., 1962. Bull. Soc. Neuchâtel. Sci. Nat., 85: 53-81.

Favarger, C., 1962b. In: Contandriopoulos, J., Ann. Fac. Sci. Marseilles, 32:
    1-354.

Favarger, C., 1965. Bot. Not., 118: 273-280.

Favarger, C., 1965b. Bull. Soc. Neuchâtel. Sci. Nat., 88: 5-60.

Favarger, C., 1965c. Monde Plantes, 348: 1-4.

Favarger, C., 1969a. Bull. Soc. Neuchâtel. Sci. Nat., 92: 13-30.

Favarger, C., 1969b. Taxon, 18: 434-435.

Favarger, C., 1969c. Acta Bot. Croat., 28: 63-74.

Favarger, C., 1971. Taxon, 20: 351.

Favarger, C., 1975. Anal. Inst. Bot. A.J. Cavanilles, 32: 1209-1243.

Favarger, C., 1975b. In: Löve, Á., Taxon, 24: 144.

Favarger, C. and Contandriopoulos, J., 1961. Ber. Schweiz. Bot. Ges., 71:
    384-408.

Favarger, C. and Huynh, K.L., 1964. Taxon, 13 (6): 201-209.

Favarger, C. and Küpfer, P., 1967. Compt. Rend. Acad. Sci. Paris, 264: 2463.

Favarger, C. and Küpfer, P., 1968. Collect. Bot., 7 (1): 325-357.

Favarger, C. and Küpfer, P., 1970. Bull. Soc. Bot. Suisse, 80: 269-288.

Favarger, C. and Villard, M., 1965. Bull. Soc. Bot. Suisse, 75: 57-79.

Fearn, G.M., 1977. Watsonia, 11: 254.

Felföldy, L.J.M., 1947. Arch. Biol. Hung., 17: 101-103.

Felföldy, L.J.M., 1947b. Rep. Hung. Agric. Exp. Sta., 47-49: 11-16.

Ferakova, V., 1968a. Biologia Bratislava, 23: 317.

Ferakova, V., 1968b. Folia Geobot. Phytotax., Praha, 3: 111-116.

Fernandes, A., 1931. Bol. Soc. Brot., Sér. 2, 7: 3-110.

Fernandes, A., 1937a. Bol. Soc. Brot., Sér. 2, 12: 159-219.

Fernandes, A., 1937b. Bol. Soc. Brot., Sér. 2, 12: 93-116.

Fernandes, A., 1939. Bol. Soc. Brot., Sér. 2, 13: 487-542.

Fernandes, A., 1950. Agron. Lusit., 12: 551-600.

Fernandes, A., 1950b. Rev. Fac. Ci. Univ. Coimbra, 19: 5-39.

Fernandes, A., 1959. Compt. Rend. Acad. Sci., Paris, 248: 3672-3675.

Fernandes, A., 1966. Bol. Soc. Brot., Sér. 2, 40: 207-261.

Fernandes, A., 1967. Portug. Acta Biol., Sér. B, 9: 1-44.

Fernandes, A., 1968. Collect. Bot. (Barcelona), 7: 381-392.

Fernandes, A. and Fernandes, R., 1945. Herbertia, 12: 85-96.

Fernandes, A. and Fernandes, R., 1946. Acta Univ. Conimbrig., 1: 1-33.

Fernandes, A. and Franca, F., 1972. Genet. Iber., 24: 31-36.

Fernandes, A. and Franca, F., 1973. Bol. Soc. Brot., Sér. 2, 46: 465-501.

Fernandes, A., Garcia, J. and Fernandes, R., 1948. Mem. Soc. Brot., 4:
    5-89.

Fernandes, A. and Queirós, M., 1969. Bol. Soc. Brot., Sér. 2, 43: 20-140.

Fernandes, A. and Queirós, M., 1970/1. Mem. Soc. Brot., 21: 343-385.

Fernandes, A. and Queirós, M., 1971. Bol. Soc. Brot., Sér. 2, 45: 5-121.

Fernandes, A. and Santos, M.F., 1971. Bol. Soc. Brot., Sér. 2, 45: 177-226.

Fernandes, R., 1970. Bol. Soc. Brot., Sér. 2, 44: 343-367.

Fernandes, R. and Leitao, M.T., 1969. An. Est. Exp. Aula Dei, 9 (2-4):
    210-222.

Fernandes, R.B., 1969. Bol. Soc. Brot., Ser. 2, 43: 145-158.

Fischer, M., 1967. Österr. Bot. Zeits., 144: 189-233.

Fischer, M., 1969. Österr. Bot. Zeits., 116: 430-443.

Fischer, M., 1970. Österr. Bot. Zeits., 118: 206-215.

Flovik, K., 1938. Hereditas, 24: 265-376.

Flovik, K., 1940. Hereditas, 26: 430-440.

Flovik, K., 1943. Nytt Mag. Naturv., 83: 77-78.

Fothergill, P.G., 1936. Proc. Univ. Durham Phil. Soc. 9: 205-216.

Fothergill, P.G., 1944. New Phytol., 43: 23-35.

Francini, E., 1927. Nuovo Giorn. Bot. Ital., 34 (2): 381-395.

Frey, L., 1969a. Fragm. Flor. Geobot., 15: 179-184.

Frey, L., 1969b. Fragm. Flor. Geobot., 15: 261-267.

Frey, L., 1970. Fragm. Flor. Geobot., 16: 391-394.

Frey, L., 1971. Fragm. Flor. Geobot., 17: 251-256.

Frisendahl, A., 1927. Acta Horti. Getob., 3: 99-142.

Fritsch, R.M., 1973. In: Löve, Á., Taxon, 22: 461.

Fuchs, A., 1938. Österr. Bot. Zeits., 87: 1-41.

Furnkranz, D., 1965. Österr. Bot. Zeits., 112: 421-423.

Furnkranz, D., 1967. Österr. Bot. Zeits., 114: 341-345.

Gadella, T.W.J., 1963. Acta Bot. Neerland., 12: 17-39.

Gadella, T.W.J., 1964. Wentia, 11: 1-104.

Gadella, T.W.J., 1966. Proc. Koni. Nederl. Akad. Wetenschappen., Ser. C, 69 (4): 502-521.

Gadella, T.W.J., 1972. Bot. Not., 125: 361-369.

Gadella, T.W.J. and Kliphuis, E., 1963. Acta Bot. Neerland., 12: 195-230.

Gadella, T.W.J. and Kliphuis, E., 1966. Proc. Kon. Nederl. Akad. Wetenschappen., Ser. C., 69: 541-556.

Gadella, T.W.J. and Kliphuis, E., 1967. Proc. Kon. Nederl. Akad. Wetenschappen., Ser. C., 70: 7-20.

Gadella, T.W.J. and Kliphuis, E., 1968. Taxon, 17: 199-204.

Gadella, T.W.J. and Kliphuis, E., 1968b. Proc. Kon. Nederl. Akad. Wetenschappen., Ser. C., 71: 168-183.

Gadella, T.W.J. and Kliphuis, E., 1970. Caryologia, 23: 363-379.

Gadella, T.W.J. and Kliphuis, E., 1972. Acta Bot. Croat., 31: 91-103.

Gadella, T.W.J. and Kliphuis, E., 1973. Kon. Nederl. Akad. Wetenschappen., Ser. C., 76: 303-311.

Gadella, T.W.J., Kliphuis, E. and Mennega, E.A., 1966. Acta Bot. Neerl., 15 (2): 484-489.

Gagnieu, A., Bouteiller, J. and Linder, R., 1959. Bull. Soc. Bot. Fr., 106: 148-154.

Gajewski, W., 1947. Acta Soc. Bot. Polon., 18: 33-44.

Gajewski, W., 1957. Monogr. Bot., 4: 1-416.

Gale, J.S. and Rees, H., 1970. Heredity, 25: 393-410.

Gamisans, J., 1973. Candollea, 28: 39-82.

Gamisans, J., 1974. Candollea, 29: 39-55.

Garbari, F., 1969. Nuovo Giorn. Bot. Ital., 103: 1-9.

Garbari, F., 1969. Nuovo Giorn. Bot. Ital., 103: 613.

Garbari, F., 1973. Webbia, 28: 57-80.

Garbari, F., Greuter, W. and Miceli, P., 1979. Webbia, 34: 459-480.

Garbari, F. and Martino, A. di, 1972. Webbia, 27: 289-297.

Garbari, F. and Tornadore, N., 1970. Inf. Bot. Ital., 2: 79-82.

Garbari, F. and Tornadore, N., 1971. Inf. Bot. Ital., 3: 153-154.

Garbari, F. and Tornadore, N., 1972. Nuovo Giorn. Bot. Ital., 106: 285.

Garbari, F. and Tornadore, N., 1972b. Inf. Bot. Ital., 4: 65.

Garbari, F., Tornadore, N. and Pecori, E., 1973. Inf. Bot. Ital., 5: 161-169.

Gardé, A., 1952. Genet. Iber., 3: 133-144.

Gardé, A. and Malheiros-Gardé, N., 1949. Agron. Lusit., 11: 91-140.

Gardé, A. and Malheiros-Gardé, N., 1953. Genet. Iber., 5: 115-124.

Gardé, A. and Malheiros-Gardé, N., 1954. Broteria, 23: 5-35.

Gardou, C., 1972. In: Löve, Á., Taxon, 21: 495-500.

Gauger, W., 1937. Planta, 26: 529-531.

Geitler, L., 1929. Züchter, 1: 243-247.

Gentscheff, G., 1937. Planta, 27: 165-195.

Gervais, C., 1965. Bull. Soc. Neuchât. Sci. Nat., 88: 61-64.

Gervais, C., 1966. Bull. Soc. Neuchât. Sci. Nat., 89: 87-100.

Gervais, C., 1968. Bull. Soc. Neuchât. Sci. Nat., 91: 105-117.

Gervais, C., 1972. Bull. Soc. Neuchât. Sci. Nat., 95: 57-61.

Gervais, C., 1973. Denkschr. Schweiz. Naturf. Ges., 88: 1-56.

Giorgi, B. and Bozzini, A., 1969. Caryologia, 22: 249-259.

Gill, J.J.B., 1965. Watsonia, 6: 188-189.

Glasau, F., 1939. Planta, 30: 507-550.

Glendinning, D.R., 1960. Nature (London), 188: 604-605.

Gorenflot, R., 1960. Rév. Cytol. Biol. Vég., 22: 79-108.

Gorenflot, R., Raicu, P. and Cartier, D., 1972. Compt. Rend. Acad. Sci.
    Paris, 274: 391-393.

Gorenflot, R., Raicu, P., Cartier, D., et al., 1973b. Compt. Rend. Acad.
    Sci., Paris, 274: 1501-1504.

Gori, C., 1954. Caryologia, 6: 241-246.

Gori, C., 1957. Caryologia, 10: 454-456.

Gori, C., 1958. Caryologia, 11: 28-33.

Grant, W.F., 1959. Canad. Jour. Genet. Cytol., 1: 313-328.

Grau, J., 1964a. Madroño, 17: 266-268.

Grau, J., 1964b. Österr. Bot. Zeits., 111: 561-617.

Grau, J., 1967a. Mitt. Bot. Staatssamm., München, 6: 517-530.

Grau, J., 1967b. Österr. Bot. Zeits., 114: 66-72.

Grau, J., 1968a. Mitt. Bot. Staatssamm., München, 7: 17-100.

Grau, J., 1968b. Mitt. Bot. Staatssamm., München, 7: 277-294.

Grau, J., 1970. Mitt. Bot. Staatssamm., München, 8: 127-136.

Grau, J., 1971. Mitt. Bot. Staatssamm., München, 9: 177-194.

Grau, J., 1976. Mitt. Bot. Staatssamm., München, 12: 609-654.

Graze, H., 1933. Jahrb. Wiss. Bot., 77: 507-559.

Greilhuber, J. and Ehrendorfer, F., 1975. Pl. Syst. Evol., 124: 125-138.

Griesinger, R., 1937. Ber. Deutsch. Bot. Ges., 55: 556-571.

Griff, V.G., 1965. Bot. Zhur., 50: 1133.

Groll, M., 1965. Ôesterr. Bot. Zeit., 112: 657-700.

Gross, R., 1966. Ber. Schweiz. Bot. Ges., 76: 306-351.

Guinochet, M., 1943. Rév. Cytol. Cytophysiol. Vég., 6: 209-220.

Guinochet, M., 1957. Bull. Soc. Hist. Nat. Afr. Nord., 48: 282-300.

Guinochet, M., 1964. Compt. Rend. Acad. Sci. Paris, 259: 3817-3819.

Guinochet, M., 1967. Compt. Rend. Acad. Sci. Paris, 264: 1623-1625.

Guinochet, M. and Foissac, J., 1962. Rév. Cytol. Biol. Végetal., 25 (3-4):
    373-387.

Guinochet, M. and Lefrane, M., 1972. In: Löve, Á., Taxon, 21: 495-500.

Guinochet, M. and Lemee, G., 1950. Rév. Gén. Bot., 57: 565-593.

Guinochet, M. and Logeois, A., 1962. Rév. Cytol. Biol. Vég., 25: 465-480.

Guittonneau, G., 1964. Bull. Soc. Bot. France, 111: 1-4.

Guittonneau, G., 1965. Bull. Soc. Bot. France, 112: 25.

Guittonneau, G., 1966. Bull. Soc. Bot. France, 111: 3-11.

Guittonneau, G., 1967. Bull. Soc. Bot. France, 114: 32-41.

Gustaffson, Å., 1932. Hereditas, 16: 41-62.

Gustaffson, Å., 1933. Bot. Gaz., 94: 512-533.

Gustaffson, Å., 1935. Hereditas, 20: 1-31.

Gustaffson, Å., 1939. Hereditas, 25: 33-47.

Gustaffson, Å., 1943. Acta Univ. Lund, II, 39 (2): 1-200.

Gustaffson, M. and Snogerup, S., 1972. Bot. Not., 125: 323.

Guşuleac, M. and Tărnavschi, I.T., 1935. Bull. Fac. Stiinţ. Cernauti, 9:
    387-400.

Gutermann, W., 1961. In: Löve, Á. and Löve, D., Opera Bot. 5.

Hadac, E. and Haskova, V., 1956. Biologia (Bratislava), 11: 717-723.

Häfliger, E., 1943. Ber. Schweiz. Bot. Gesell., 53: 317-382.

Hagerup, O., 1928. Dansk Bot. Arkiv., 6: 1-26.

Hagerup, O., 1932. Hereditas, 16: 19-40.

Hagerup, O., 1939. Hereditas, 25: 185-192.

Hagerup, O., 1940. Hereditas, 26: 399-410.

Hagerup, O., 1941a. Planta, 32: 6-14.

Hagerup, O., 1941b. Bot. Tidsskr., 45: 385-395.

Hagerup, O., 1944. Hereditas, 30: 152-160.

Hagerup, O., 1947. Kung. Dansk. Vidensk. Selsk. Biol. Medd., 20 (9): 1-22.

Hakansson, A., 1944. Hereditas, 30: 639-640.

Hakansson, A., 1953. Bot. Not., 1953: 301-307.

Hakansson, A., 1955. Hereditas, 41: 454-482.

Halkka, O., 1964. Hereditas, 52: 81-88.

Hall, A.D., 1937. Bot. Journ. Linn. Soc. London, 50: 481-489.

Halliday, G., 1960. Watsonia, 4: 207-210.

Halliday, G., 1961. In: Löve, A. and Löve, D., Opera Bot., 5.

Hambler, D.J., 1954. Nature (London), 174: 838.

Hambler, D.J., 1955. Proc. Bot. Soc. Brit. Isles, 1: 384-385.

Hambler, D.J., 1958. Bot. Journ. Linn. Soc. London, 55: 772-777.

Hamel, J.L., 1953. Rev. Cytol. et Cytophysiol., 14: 3-4.

Hamel, J.L., 1954. Mem. Soc. Bot. France, 34: 106-121.

Hämet-Ahti, L. and Virrankoski, V., 1970. Ann. Bot. Fenn., 7: 177-181.

Hamilton, A.P., 1968. Gladiolus Annual, 1968: 27-31.

Hanelt, P., 1963. Feddes Rep., 67: 41-180.

Hanelt, P. and Mettin, D., 1966. Kulturpflanze, 14: 137-161.

Harvey, B.L., 1966. Dissert. Abstr., Ser. B., 27: 365.

Hawkes, J.G., 1956. Proc. Linn. Soc., London, 166 (3): 97-144.

Harley, R.M., 1977. In: Harley, R.M. and Brighton, C.A., Bot. Jour. Linn.
    Soc., 74: 71-96.

Haskell, G., 1964. Jour. Roy. Hort. Soc., 89: 504.

Hedberg, O., 1958. Svensk. Bot. Tidskr., 52: 37-46.

Hautzinger, L., 1976. Pl. Syst. Evol., 124: 311-313.

Hedberg, I., 1970. Hereditas, 64: 153-176.

Hedberg, I. and Hedberg, O., 1961. Bot. Not.: 114: 397-99.

Hedberg, I. and Hedberg, O., 1964. Svensk. Bot. Tidskr., 58: 125-128.

Heilborn, O., 1918. Svensk. Bot. Tidskr., 12: 212-220.

Heilborn, O., 1924. Hereditas, 5: 129-216.

Heilborn, O., 1927. Hereditas, 9: 59-68.

Heilborn, O., 1928. Hereditas, 11: 182-192.

Heitz, B., 1967. Inf. Ann. Caryosyst. Cytogén., 1: 22.

Helms, A. and Jorgensen, C.A., 1925. Bot. Tidsskr., 39: 57-134.

Heneen, W.K., 1962. Hereditas, 48: 471-502.

Heneen, W.K. and Runemark, H., 1962. Hereditas, 48: 545-564.

Henin, W., 1960. In: Boulos, L., Bot. Not., 113: 400-420.

Heppell, R., 1945. In: Darlington, C.D. and Janaki-Ammal, E.K., Chromosome
    Atlas of Cultivated Plants, London, Allen and Unwin.

Heslop-Harrison, J., 1961. In: Löve, Á. and Löve, D., Opera Bot. 5.

Heslop-Harrison, Y., 1953. New Phytol., 52: 22-39.

Heslop-Harrison, Y., 1953b. Watsonia, 3: 7-25.

Heslop-Harrison, Y., 1955. Jour. Ecol., 43: 719-734.

Heslot, M.H., 1953. Compt. Rend. Acad. Sci. (Paris), 237: 2477-2479.

Heusser, C., 1915. Beih. Bot. Centralbl., 32: 218-277.

Heusser, C., 1938. Bull. Soc. Bot. Suisse, 48: 562-599.

Heywood, V.H., 1955. In: Turrill, W.B., Bot. Mag., 170: t.246.

Heywood, V.H. and Walker, S., 1961. Nature (London), 189: 604.

Hjelmquist, H., 1959. Bot. Not., 112: 17-64.

Hofelich, A., 1935. Jahrb. Wiss. Bot., 81: 541-572.

Holmen, K., 1961. In: Löve, Á. and Löve, D., Opera Bot. 5.

Holub, J. and Křisa, B., 1971. Folia Geobot. Phytotax. (Praha), 6: 82.

Holub, J., Měsíček, J. and Javůrková, V., 1970. Folia Geobot. Phytotax.,
    Praha, 5: 339-368.

Holub, J., Měsíček, J. and Javůrková, V., 1971. Folia Geobot. Phytotax.,

Praha, 6: 179-214.

Holub, J., Měsíček, J. and Javůrková, V., 1972. Folia Geobot. Phytotax.,

Praha, 7: 167-202.

Honsell, E., 1959. Ann. Bot. (Roma), 26: 177-188.

Hooper, M.D., 1961. In: Löve, A. and Löve, D., Opera Bot., 5.

Horánszky, A., Jankó, B. and Vida, G., 1972. Symp. Biol. Hung., 12: 177-182.

Horn, K., 1948. Blyttia, 6: 1-13.

Hou-Liu, P.Y., 1963. Acta Bot. Neerland., 12: 76-83.

Hovin, A.W. and Hill, H.D., 1966. Amer. Jour. Bot., 53: 702-708.

Howard, H.W., 1946. Nature (London), 158: 666.

Howard, H.W., 1947. Nature (London), 159: 66.

Hrubý, K., 1932. Preslia, 11: 40-44.

Hrubý, K., 1934. Bull. Intern. Acad. Czeck. Sci. Cl. Sci, Mat., Med.

Prague, 35: 124-132.

Hubbard, C.E., 1954. Grasses. Penguin Books, Harmondsworth.

Hultén, E., 1945. Bot. Not., 1945: 127-148.

Hylander, N., 1957. Bull. Hard. Bot. Bruxelles, 27: 591-604.

Iljin, M.M. and Trukhaleva, N.A., 1960. Doklady Akad. SSSR, 132: 217-219.

Ingram, R., 1967. Watsonia, 6: 283-289.

Jacobsen, P., 1954. Hereditas, 40: 252-254.

Jalas, J., 1948. Hereditas, 34: 414-434.

Jalas, J., 1950. Ann. Bot. Soc. Vanamo, 24: 1-362.

Jalas, J. and Kaleva, K., 1966. Ann. Bot. Fenn., 3: 123-127.

Jalas, J. and Kaleva, K., 1967. Ann. Bot. Fenn., 4: 74-80.

Jalas, J. and Pohjo, T., 1965. Ann. Bot. Fenn., 2: 169-170.

Jalas, J. and Rönkkö, M.T., 1960. Arch. Soc. Zoo. Bot. Fenn. 'Vanamo',

14 (2): 112-116.

Janaki-Ammal, E.K. and Saunders, B., 1952. Kew Bull., 1952: 539-542.

Jaretsky, R., 1928a. Jahrb. Wiss. Bot., 68: 1-45.

Jaretsky, R., 1928b. Jahrb. Wiss. Bot., 69: 357-490.

Jasiewicz, A. and Mizianty, M., 1975. Fragm. Fl. Geobot., 21: 277-288.

Johnson, M.A.T., 1982. Ann. Mus. Goulandris, 5: In Press.

Johnsson, H., 1941. Lunds Univ. Aarskr., Avd. 2, 37 (3): 1-43.

Johnsson, H., 1942. Hereditas, 28: 228-230.

Jones, K., 1955. In: Darlington, C.D. and Wylie, A.P., Chromosome Atlas
    of Flowering Plants.

Jones, K., 1958. New Phytol., 57: 191-210.

Jones, K., 1959. Welsh Pl. Breed. Stat. Rep. 1956-58: 68-71.

Jones, K. and Smith, J.B., 1966. Kew Bull., 20: 63-72.

Jonsell, B., Pálsson, J. and Portén, E.K., 1975. Svensk Bot. Tidskr., 69: 113-142.
    113-142.

Jörgensen, C.A., 1927. Jour. Genet., 18: 63-75.

Jörgensen, C.A., Sorensen, T. and Westergaard, M., 1958. Dansk Vid. Selsk.
    Biol. Skr., 9 (4): 1-172.

Juhl, H., 1952. Ber. Deutsch. Bot. Ges., 65: 330-337.

Kachidze, N., 1929. Planta, 7: 482-502.

Kaleva, K., 1969. Ann. Bot. Fennica, 6: 344-347.

Katznelson, J. and Zohary, D., 1967. Israel Jour. Bot., 16: 57-62.

Kay, Q.O.N., 1969. Watsonia, 7: 130-141.

Kerguélen, M., 1975. Lejeunia, 75: 9-16, 32-44, 145-182.

Kerguélen, M., 1976. Bull. Soc. Bot. Fr., 123: 317-324.

Khidir, M.O. and Knowles, P.F., 1970. Amer. Jour. Bot., 57: 123-129.

Khoshoo, T.N., 1959. Caryologia, 11: 297-318.

Kjellmark, S., 1934. Bot. Not. 1934: 136-149.

Kjellqvist, E., 1964. Bot. Not., 117: 389-396.

Kjellqvist, E. and Löve, Á., 1963. Bot. Not., 116: 241-248.

Kliphuis, E., 1962. Acta Bot. Neerl., 11: 90-92.

Kliphuis, E., 1962b. Proc. Kon. Akad. Wetenschap. Ser. C, 65 (3): 279-285.

Kliphuis, E., 1963. Med. Bot. Mus. Herb. Riksuniv. Utrecht, 195: 172-194.

Kliphuis, E., 1967. Acta Bot. Neerl., 15: 535-538.

Kliphuis, E., 1977. In: Löve, Á., Taxon, 26: 267-268.

Kliphuis, E. and Wieffering, J.H., 1972. Acta Bot. Neerl., 21: 598-604.

Knaben, G., 1948. Blyttia, 6: 17-32.

Knaben, G., 1950. Blyttia, 8: 129-155.

Knaben, G., 1954. Nytt Mag. Bot., 3: 117-138.

Knaben, G., 1958. Blyttia, 16: 61-80.

Knaben, G., 1959. Opera Bot., 2 (3): 1-74.

Knaben, G. and Engelskjön, T., 1967. Acta Boreali, 21: 1-57.

Knaben, G. and Engelskjon, T., 1968. Acta Univ. Berg. No. 4: 0-71.

Kollmann, F., 1973. Israel Jour. Bot., 22: 92-115.

Kollmann, F. and Feinbrun, N., 1968. Notes Roy. Bot. Gard. Edinb., 28 (2):
    173-186.

Kovanda, M., 1966. Preslia (Praha), 38: 48-52.

Kovanda, M., 1970a. In: Löve, Á., Taxon, 19: 106.

Kovanda, M., 1970b. Roz. Ceskosl. Akad. Ved., 80: 1-95.

Kožuharov, S. and Kuzmanov, B., 1962. Ann. Univ. Sofia, 57: 103-107.

Kožuharov, S. and Kuzmanov, B., 1964. Compt. Rend. Akad. Bulgar. Sc., 17:
    491-494.

Kožuharov, S. and Kuzmanov, B., 1964b. Ann. Univ. Sofia, 57: 103-107.

Kožuharov, S. and Kuzmanov, B., 1965a. Mitt. Bot. Inst. Sofia, 15: 255-258.

Kožuharov, S. and Kuzmanov, B., 1965b. Caryologia, 18 (2): 349-351.

Kožuharov, S. and Kuzmanov, B., 1967. Compt. Rend. Acad. Bulg. Sci., 20:
    469-472.

Kožuharov, S. and Kuzmanov, B., 1968. In: Löve, Á., Taxon, 17: 199-204.

Kožuharov, S. and Kuzmanov, B., 1969. Gen. Pl. Breeding, 2 (3): 181-185.

Kožuharov, S. and Kuzmanov, B., 1970. Mitt. Bot. Inst. Sofia, 20: 99-107.

Kožuharov, S. and Nicolova, T., 1975. In: Problems of Balkan Flora and

    Vegetation. Proc. 1st Internat. Symp. Balkan Flora and Veg. Varna,

    June, 1973- pp. 69-77. Sofia, Bulgarian Acad. Sciences.

Kožuharov, S. and Petrova, A.V., 1973. Compt. Rend. Acad. Bulg. Sci., 26 (6):

    829-832.

Kožuharov, S. and Petrova, A.V., 1974. In: Löve, Á., Taxon, 23: 377.

Kožuharov, S. and Petrova, A.V., 1975. In: Löve, Á., Taxon, 24: 369.

Kožuharov, S., Petrova, A.V. and Stoeva, M.P., 1974. Comp. Rend. Acad.

    Bulg. Sci., 27: 383-386.

Kožuharov, S., Popova, M. and Kuzmanov, B., 1968. Genetics Pl. Breeding,

    1 (3): 251-255.

Krendl, F., 1967. Öesterr. Bot. Zeits., 114: 508-549.

Krendl, F., 1974. In: Ehrendorfer, F. and Krendl, F., Bot. Jour. Linn.

    Soc., 68: 268-272.

Krendl, F., 1976. Ann. Natürhist. Mus. Wien, 80: 67-86.

Kress, A., 1963. Ber. Bay. Bot. Gesell., 36: 33-39.

Kress, A., 1963. Phyton, 10: 225-236.

Kubién, E., 1968. In: Skalińska, M., Pogan, E. and Jankun, A., Acta Biol.

    Cracov., Bot., 11: 199-224.

Kuhn, E., 1928. Jahrb. Wiss. Bot., 68: 382-430.

Küpfer, P., 1968. Bull. Soc. Neuchat. Sci. Nat., 91: 87-104.

Küpfer, P., 1969. In: Löve, A., Taxon, 18: 436-467.

Küpfer, P., 1969. Bull. Soc. Neuchat. Sci. Nat., 92: 31-48.

Küpfer, P., 1971. Acta Colloque Flore Veg. Chaines Alp. Jurass. Paris,

    167-185.

Küpfer, P. and Favarger, C., 1967. Compt. Rend. Acad. Sci. Paris, 264:

    2463-2465.

Kuzmanov, B., 1973. In: Löve, A., Taxon, 22: 651.

Kuzmanov, B. and Ančev, M.E., 1973. In: Löve, A., Taxon, 22: 461.

Kuzmanov, B., Georgieva, S., Nikolova, V. and Penceva, I., 1981. In: Löve,

    Á., Taxon, 30: 701-702.

Kuzmanov, B. and Jurukova, P., 1977. In: Löve, A., Taxon, 26: 557-565.

Kuzmanov, B. and Kožuharov, S., 1967. Compt. Rend. Acad. Bulg. Sci., 20 (5): 469-472.

Kuzmanov, B. and Kožuharov, S., 1970. In: Löve, Á., Taxon, 19: 265-266.

Kuzmanov, B., Popova, M. and Kožuharov, S., 1968. Genet. Pl. Breed. (Sofia), 2: 325-330.

Laane, M.M., 1966. Blyttia, 24: 270-276.

Laane, M.M., 1971. Blyttia, 29 (4): 229-234.

Labadie, J.P., 1976. In: Löve, Á., Taxon, 25: 636-639.

Ladizinski, G., 1971. Israel Jour. Bot., 20: 24.

Lambert, A.-M., 1969. Inf. Ann. Caryosyst. Cytogén., 3: 3.

Lambert, A.-M. and Giesi, J., 1967. Inf. Ann. Caryosyst. Cytogén., 1: 27-35.

Landolt, E., 1954. Ber. Schweiz. Bot. Ges., 64: 9-83.

Landolt, E., 1956. Ber. Schweiz. Bot. Ges., 66: 92-117.

Langlet, O.F.J., 1927. Svensk. Bot. Tidskr., 21: 1-17.

Langlet, O.F.J., 1932. Svensk. Bot. Tidskr., 26: 381-400.

Langlet, O.F.J., 1936. Svensk. Bot. Tidskr., 30: 288-294.

Langlet, O.F.J. and Söderberg, E., 1927. Acta Horti. Berg., 9: 85-104.

Larsen, K., 1954. Bot. Tidsskr., 50: 163-174.

Larsen, K., 1955a. Bot. Tidsskr., 52: 8-17.

Larsen, K., 1955b. Bot. Not., 108: 263-275.

Larsen, K., 1956a. Bot. Not., 109: 293-307.

Larsen, K., 1956b. Bot. Tidsskr., 53: 49-56.

Larsen, K., 1957a. Bot. Tidsskr., 53: 291-294.

Larsen, K., 1957b. Bot. Not., 110: 265-270.

Larsen, K., 1958a. Bot. Tidsskr., 54: 44-56.

Larsen, K., 1958b. Bot. Not., 111: 301-305.

Larsen, K., 1960. Bot. Tidsskr., 55: 313-315.

Larsen, K., 1963. Dansk. Bot. Ark., 20: 211-275.

Larsen, K., 1965. In: Löve, Á. and Solbrig, O.T., Taxon, 14: 91.

Larsen, K. and Laegaard, S., 1971. Bot. Tidsskr., 66: 249-268.

Larter, L.N.H., 1952. Jour. Genet., 26: 255-283.

Lazare, J.-J., 1976. In: Löve, Á., Taxon, 25: 483.

Lesins, K. and Lesins, J., 1961. Canad. Jour. Genet. Cytol., 3: 7-9.

Leute, G.-H., 1974. In: Löve, Á., Taxon, 23: 811-812.

Levan, A., 1940. Hereditas, 26: 317-320.

Lévêque, M. and Gorenflot,,R., 1969. Bull. Soc. Bot. Nord., France, 22: 27-58.

Liljefors, A., 1953. Act. Hort. Berg., 16: 277-329.

Lindqvist, K., 1960. Hereditas, 46: 75-151.

Litardière, R. de, 1943. Boissiera, 7: 155-165.

Litardière, R. de, 1945. Rev. Bot. Appl. Agric. Trop., 25: 16-18.

Litardière, R. de, 1947. Candollea, 11: 175-227.

Litardière, R. de, 1949. Mém. Soc. Hist. Nat. Afr. Nord., 2: 199-208.

Litardière, R. de, 1950. Bol. Soc. Brot., Sér. 2, 24: 79-87.

Lohammon, G., 1942. In: Löve, Á. and Löve, D., Bot. Not., 1942: 19-59.

Lopane, F., 1951. Caryologia, 4: 44-46.

Lorenzo-Andreu, A., 1951. An. Aula Dei., 2: 195-203.

Lorenzo-Andreu, A. and Garcia-Sanz, P., 1950. An. Aula Dei., 2: 12-20.

Löve, Á., 1940. Bot. Not., 1940: 157-169.

Löve, Á., 1941. Bot. Not., 1941: 155-172.

Löve, Á., 1942. Hereditas, 28: 289-296.

Löve, Á., 1954a. Vegetatio, 5: 212-224.

Löve, Á., 1954b. VIII Congr. Int. Bot. Rapp. & Comm., 9-10: 59-66.

Löve, Á., 1961. In: Löve, Á. and Löve, D., Opera Bot., 5: 1-581.

Löve, Á., 1967. In: Löve, Á., Taxon, 16: 445-461.

Löve, Á. and Kjellqvist, E., 1964. Portug. Acta Biol., Ser. A., 8: 69-80.

Löve, Á. and Kjellqvist, E., 1973. Lagascalia, 3 (2): 147-182.

Löve, Á. and Löve, D., 1942. Kungl. Fysiografiska Sallsk. Lund Forkadl., 12 (6): 1-19.

Löve, Á. and Löve, D., 1944. Ark. Bot., 31A, 12: 1-22.

Löve, Á. and Löve, D., 1948. Rep. Dep. Agric. Univ. Inst. Appl. Sci. (Iceland), Ser. B., 3: 9-131.

Löve, Á. and Löve, D., 1951. Svenska Bot. Tidskr., 45: 368-399.

Löve, Á. and Löve, D., 1956. Acta Hort. Gotob., 20 (4): 65-290.

Löve, Á. and Löve, D., 1958. Nucleus, 1: 1-10.

Löve, Á. and Löve, D., 1958b. Bot. Not., 111: 376-388.

Löve, Á. and Löve, D., 1961a. Bot. Not., 114: 33-47.

Löve, Á. and Löve, D., 1961b. Opera Bot. 5: 1-581.

Löve, Á. and Löve, D., 1961c. Amer. Fern. Jour., 51: 127-128.

Löve, Á. and Löve, D., 1961d. Bot. Not., 114: 48-56.

Löve, Á., Löve, D. and Raymond, M., 1957. Canad. Jour. Bot., 35: 715-761.

Löve, D., 1942. Svensk Bot. Tidskr., 36: 262-270.

Lovis, J.D., 1963. Brit. Fern. Gaz., 9: 110-113.

Lovka, M. and Sušnik, F., 1973. In: Löve, Á., Taxon, 22: 289.

Lovka, M., Sušnik, F., Löve, Á. and Löve, D., 1971. In: Löve, Á., Taxon, 20: 788-791.

Lovka, M., Sušnik, F., Löve, Á. and Löve, D., 1972. In: Löve, Á., Taxon, 21: 333-346.

Lövkvist, B., 1956. Symb. Bot. Upsal., 14 (2): 1-131.

Lövkvist, B., 1957. Bot. Not., 110: 237-250.

Lövkvist, B., 1962. Bot. Not., 115: 261-287.

Lövkvist, B., 1963. In: Weimarck, G., Skånes Flora.

Lundqvist, A., 1965. Hereditas, 54: 70-87.

Lungeanu, I., 1967. Acta Bot. Horti Bucurest, 1966: 37-59.

Lungeanu, I., 1971. In: Löve, Á., Taxon, 20: 792.

Lungeanu, I., 1972. In: Löve, Á., Taxon, 21: 679-684.

Lungeanu, I., 1973. In: Löve, Á., Taxon, 22: 651.

Lungeanu, I., 1975. In: Löve, Á., Taxon, 24: 503-504.

Majóvský, J. et al., 1970. Acta Fac. Rer. Nat. Univ. Com. Bot., 18:   45-60.

Majóvský, J. et al., 1974. Acta Fac. Rer. Nat. Univ. Com. Bot., 22:   1-20.

Majóvský, J. et al., 1976. Acta Fac. Rer. Nat. Univ. Com. Bot., 25:   1-18.

Majóvský, J. et al., 1978. Acta Fac. Rer. Nat. Univ. Com. Bot., 26:   1-42.

Malecka, J., 1962. Acta Biol. Cracov., Ser. Bot., 5:   117-136.

Malecka, J., 1967. Acta Biol. Cracov., Ser. Bot., 10:   195-206.

Malecka, J., 1969. Acta Biol. Cracov., Ser. Bot., 12:   57-70.

Malecka, J., 1970. Acta Biol. Cracov., Ser. Bot., 13:   155-168.

Malheiros, N., 1942. Agron. Lusit., 4:   231-236.

Malheiros, N., Castro, D. and Camara, A., 1947. Agron. Lusit., 9:   51-74.

Mansurova, V.V., 1948. Doklady Akad. SSSR, N.S. 61:   119-123.

Manton, I., 1932. Ann. Bot., 46:   509-556.

Manton, I., 1949. Watsonia, 1:   36.

Manton, I., 1950. Problems of Cytology and Evolution in the Pteridophyta,
    Cambridge.

Manton, I., 1955. In: J.E. Lousley (Ed.) Species Studies in the British
    Flora, Bot. Soc. Brit. Isles, London, pp. 90-103.

Manton, I. and Reichstein, T., 1961. Ber. Schweiz. Bot. Ges., 71:   370-383.

Marchant, C.J., 1968. Bot. Jour. Linn. Soc. (Lond.), 60:   411-417.

Marchant, C.J., 1972. Kew Bull., 26:   395-404.

Marchetti, D.B., 1962. Arch. Bot. Biografico, 38:   257-261.

Marchi, P., 1971. Inf. Bot. Ital., 3:   82-94.

Marchi, P., Capineri, R. and D'Amato, G., 1974. Inf. Bot. Ital., 6:   303-312.

Marklund, G., 1931. In: Holmberg, O.R., Skandinaviens Flora 1 (6, 1), 1-160.
    Stockholm.

Markova, M., 1970. Mitt. Bot. Inst. Sofia, 20:   81-92.

Markova, M. and Ančev, M.E., 1973. Compt. Rend. Acad. Bulg. Sci., 26:
    827-828.

Markova, M. and Ivanova, P., 1970. Mitt. Bot. Inst. Sofia, 20:   93-98.

Markova, M. and Ivanova, P., 1971. Mitt. Bot. Inst. Sofia, 21:   123-131.

Markova, M., Radenkova, J. and Ivanova, P., 1972. In: Löve, Å., Taxon, 21: 333-346.

Marks, G.E., 1956. New Phytol., 55: 120-129.

Marsden-Jones, E.M. and Turrill, W.B., 1957. The Bladder Campions. London, Ray Society.

Marsden-Jones, E.M. and Weiss, F.E., 1960. Proc. Linn. Soc. London, 171: 27-29.

Martinoli, G., 1943. Nuovo Giorn. Bot. Ital., 50: 1-23.

Martinoli, G., 1949. Caryologia, 1: 122-130.

Martinoli, G., 1950. Caryologia, 3: 72-78.

Martinoli, G., 1951. Caryologia, 4: 86-97.

Martinoli, G., 1953. Caryologia, 5: 253-281.

Martinoli, G., 1954. VIII Cong. Int. Bot. Compt. Rend. Séances of Rapp. Comm., 9-10: 78-79.

Mattick, G., 1950. In: Tischler, G., Die Chromosomenzahlen der Gefässpflanzen Mitteleuropas, Junk, S-Gravenhage.

Maude, P.F., 1939. New Phytol., 38: 1-31.

Maude, P.F., 1940. New Phytol., 39: 17-32.

McNaughton, I.H., 1960. Scottish Pl. Breed. Stat. Rep., 1960: 76-84.

Maugini, E., 1952. Caryologia, 5: 168.

Maugini, E., 1952b. Caryologia, 5: 101-112.

Maugini, E., 1953. Caryologia, 5: 249-252.

Meili-Frei, E., 1966. Ber. Schweiz. Bot. Ges., 75: 219-292.

Merxmüller, H., 1968. Mitt. Bot. Staatssamm., München, 7: 275.

Merxmüller, H., 1970. Taxon, 19: 140-145.

Merxmüller, H., 1974. Phyton (Austria), 16: 137-158.

Merxmüller, H., 1975. Anal. Inst. Bot. Cavanilles, 32 (2): 189-196.

Merxmüller, H. and Damboldt, J., 1962. Ber. Deutsch. Bot. Ges., 75 (7): 233-236.

Merxmüller, H. and Grau, J., 1963. Ber. Deutsch Bot. Ges., 76: 23-29.

Merxmüller, H. and Grau, J., 1969. Ber. Deutsch. Bot. Ges., 76: 23-29.

Měsíček, J. and Hrouda, L., 1974. Folia Geobot. Phytotax., 9: 359.

Mesquita, J.F., 1964. Bol. Soc. Brot., Sér. 2, 38: 119-136.

Messia, T.P. and Rejon, M.R., 1976. In: Löve, Á., Taxon, 25: 164.

Meusel, H. and Ohle, H., 1966. Österr. Bot. Zeitsch., 113: 191-210.

Meyer, D.E., 1957. Ber. Deutsch. Bot. Ges., 70: 57-66.

Meyer, D.E., 1958. Ber. Deutsch. Bot. Ges., 71: 11-20.

Meyer, D.E., 1959. Willdenowia, 2: 214-218.

Meyer, D.E., 1960. Amer. Fern Jour. 50: 138-145.

Meyer, D.E., 1961. Ber. Deutsch. Bot. Ges., 3: 393.

Meyer, D.E., 1962. Ber. Deutsch. Bot. Ges., 75: 24.

Meyer, D.E., 1967. Ber. Deutsch. Bot. Ges., 80: 28.

Miceli, P. and Garbari, F., 1976. Inf. Bot. Ital., 8: 207-216.

Miceli, P. and Garbari, F., 1979. Atti Soc. Tosc. Sci. Nat., Mem., Ser. B.,
   86: 37-51.

Michaelis, P., 1925. Ber. Deutsch. Bot. Ges., 43 (2): 61-67.

Mičieta, K., 1981. Acta Fac. Rer. Nat. Univ. Comen., Bot., 28: 95-103.

Milan, L., 1975. Biol. Vestn. (Ljubljana), 23: 25-40.

Mills, J.N. and Stace, C.A., 1974. Watsonia, 10: 167-168.

Mitra, J., 1956. Bot. Gaz., 117: 265-293.

Mirkovic, D., 1966. Acta Bot. Croat., 25: 137-152.

Mizianty, M. and Frey, L., 1973. Fragm. Flor. Geobot., 19 (3): 265-270.

Montgomery, F.H. and Yang, S.J., 1960. Can. Jour. Bot., 38: 381-386.

Monti, G. and Garbari, F., 1974. Nuovo Giorn. Bot. Ital., 108: 19-26.

Moore, D.M., 1958. Jour. Ecol., 46: 527-535.

Moore, D.M., 1973. In: Webb, D.A. and Halliday, G., Watsonia, 9: 333-344.

Moore, R.J., 1967. In: Löve, A. and Solbrig, O.T., Taxon, 16: 62-66.

Mori, M., 1957. Caryologia 9: 365-368.

Morisset, P., 1964. Proc. Bot. Soc. Brit. Is., 5: 378-379.

Morrison, J.W., 1959. Can. Jour. Bot., 37: 527-538.

Morton, J.K., 1956. Watsonia, 3: 244-252.

Morton, J.K., 1974. Watsonia, 10: 169.

Mosquin, T., 1967. Evolution, 21: 713-719.

Muntzing, A., 1929. Hereditas, 13: 185-341.

Muntzing, A., 1931b. Hereditas, 15: 166-178.

Muntzing, A., 1935. Hereditas, 20: 103-136.

Muntzing, A., 1958. Hereditas, 44: 280-329.

Muntzing, A., 1966. Bull. Bot. Soc. Bengal, 20: 1-15.

Muntzing, A. and Muntzing, G., 1941. Bot. Not., 1941: 237-278.

Muntzing, A. and Nygren, A., 1955. Hereditas, 41: 405-422.

Muntzing, A. and Prakken, R., 1940. Hereditas, 26: 463-500.

Murín, A., 1964. Caryologia, 17: 575-578.

Murín, A. and Májovský, J., 1976. In: Löve, Á., Taxon, 25: 487-489.

Murín, A. and Májovský, J., 1978. In: Löve, Á., Taxon, 27: 376-378.

Murín, A. and Váchová, M., 1970. In: Májovský, J. et al., Acta Fac. Rer. Nat.
  Univ. Comen., 16: 1-26.

Murray, B.G., 1975. In: I.B.K. Richardson, Bot. Jour. Linn. Soc., 71:
  211-234.

Nannfeldt, J.A., 1940. Symb. Bot. Upsal., 4 (4): 1-86.

Nannfeldt, J.A., 1942. In: Löve, Á. and Löve, D., Bot. Not., 1942: 19-59.

Natho, G., 1959. Feddes Rep., 61: 211-273.

Natividade, J.V., 1937. Bol. Soc. Brot., Sér. 2: 12: 21-85.

Negodi, G., 1937. Nuovo Giorn. Bot. Ital., 44: 667-672.

Netroufal, F., 1927. Österr. Bot. Zeitschr., 76: 101-115.

Neves, J. de Barros, 1944. Diss. Univ. Coimbra: 1-200.

Neves, J. de Barros, 1950. Agron. Lusit., 12: 601-610.

Nevling, L.I., 1962. Rhodora, 64: 277-282.

Nielson, N., 1924. Hereditas, 5: 378-382.

Nilsson, Ö. and Lassen, P., 1971. Bot. Not., 124 (2): 270-276.

Nordborg, G., 1958. Bot. Not., 111: 240-248.

Nordborg, G., 1966. Opera Bot., 11 (2): 1-103.

Nordborg, G., 1967. Opera Bot., 16: 1-166.

Nordenskjöld, H., 1937. Hereditas, 23: 304-316.

Nordenskjöld, H., 1951. Hereditas, 37: 325-355.

Noronha-Wagner, M. de, 1949. Genet. Iber., 1: 59-67.

Novotna, I., 1962. Preslia, 34: 249-254.

Nygren, A., 1946. Hereditas, 32: 131-262.

Nygren, A., 1948. In: Löve, Á. and Löve, D., Rep. Dep. Agric. Univ. Inst.
    Appl. Sci. (Iceland), Ser. B., 3.

Nygren, A., 1949. Hereditas, 35: 215-220.

Nygren, A., 1951. Hereditas, 37: 372-381.

Nygren, A., 1955. Ann. Roy. Agric. Cell. Sweden, 21: 179-191.

Nygren, A., 1956. Kungl. Lantbrukshogsk. Annal., 22: 179-191.

Nygren, A., 1957. Kungl. Lantbrukshogsk. Ann., 23: 488-495.

Nygren, A., 1962. Svensk. Vetenskap. Arsbok, 6: 1-29.

Ochlewska, M., 1966. Acta Biol. Cracov., Ser. Bot., 8: 135-145.

Ockendon, D.J., 1971. New Phytol., 70: 599-605.

Oehkers, F., 1926. Jahrb. Wiss. Bot., 65: 401-446.

Olsson, U., 1967. Bot. Not., 120: 255-267.

Ouweneel, W.J., 1968. Med. Bot. Mus. Herb. Riksuniv. Utrecht, 279: 184-188.

Packer, J.G., 1961. In: Löve, Á. and Löve, D., Opera Bot. 5.

Padmore, P.A., 1957. Watsonia, 4: 19-27.

Palmgren, O., 1939. Bot. Not., 139: 246-248.

Palmgren, O., 1942. IN: Löve, Á. and Löve, D., Kungl. Fysiogr. Sällsk. Lund.
    Forhändl., 12.

Palmgren, O., 1943. Bot. Not., 1963: 348-352.

Papeš, D., 1972. Acta Bot. Croat., 31: 71-79.

Papeš, D., 1975. Bot. Soc. Brit. Is. Conf. Rep., 15: 90-100.

Parreaux, M.-J., 1971. Ann. Litt. Univ. Besançon, 1971, 113-126.

Parreaux, M.-J., 1972. Annal. Sci. Univ. Besançon, Bot., 3 Sér., 13: 69-155.

Pauwels, L., 1959. Bull. Soc. Roy. Bot. Belg. 91: 291-297.

Pavone, P., Terrasi, C.M. and Zizza, A., 1981. In: Löve, Á., Taxon, 30: 695-696.

Pereira, A. de L., 1942. Bol. Soc. Brot., Sér. 2, 16: 5-40.

Perrenoud, R. and Favarger, C., 1971. Bull. Soc. Neuchât. Sci. Nat., 94: 21-27.

Persson, K., 1974. Opera Bot., 35: 1-188.

Peterson, D., 1933. Bot. Not., 1933: 500-504.

Peterson, D., 1935. Bot. Not., 1935: 409-410.

Peto, F.H., 1933. Jour. Genet., 28: 113-156.

Pettet, A., 1964. Watsonia, 6: 39-50.

Phillips, H.M., 1938. Chronica Bot., 4: 385-386.

Phitos, D., 1963. Mitt. Bot. Staatssamm. München, 5: 121-124.

Phitos, D., 1964. Ber. Deutsch. Bot. Ges., 77: 49-54.

Phitos, D., 1965. Österr. Bot. Zeits., 112: 449-498.

Phitos, D., 1966. Ber. Deutsch. Bot. Ges., 79: 246-249.

Phitos, D., 1970. Ber. Beutsch. Bot. Ges., 83: 69-73.

Phitos, D. and Damboldt, J., 1969. Ber. Deutsch. Bot. Ges., 82: 600.

Phitos, D. and Damboldt, J., 1971. Ann. Naturhist. Mus Wien, 75: 157-162.

Phitos, D. and Kamari, G., 1974. Bot. Not., 127: 302-308.

Pigott, C.D., 1954. New Phytol., 53: 470-495.

Piotrowicz, M., 1958. Acta Biol. Cracov., Ser. Bot., 1: 159-169.

Podlech, D., 1962. Ber. Deutsch. Bot. Ges., 75: 237-244.

Podlech, D., 1964. Madroño, 17: 266.

Podlech, D., 1965. Feddes Rep., 71: 50-187.

Podlech, D., 1970. Mitt. Bot. Staatssamm. München, 8: 211-217.

Podlech, D., 1975. XII Intern. Bot. Congr., Abstr.: 10.

Podlech, D. and Damboldt, J., 1964. Ber. Deutsch. Bot. Ges., 76 (9): 360-369.

Pogan, E., 1968. In: Löve, Á., Taxon, 17: 421.

Pogan, E., 1972. Acta Biol. Cracov, Bot., 15: 69-76.

Pogliani, M., 1964. Nuovo Giorn. Bot. Ital., 71: 558.

Polatschek, A., 1966. Österr. Bot. Zeits., 113: 1-46, 101-147.

Pólya, L., 1948. Arch. Biol. Hung. II, 18: 145-148.

Pólya, L., 1949. Acta Geobot. Hung., 6: 124-137.

Pólya, L., 1950. Ann. Biol. Univ. Debrecen, 1: 46-56.

Popova, M.T., 1972. In: Löve, Á., Taxon, 21: 164-165.

Popova, M. and Česchmedjiev, I.V., 1966. Naucni Trud. (Viss. Selsk. Inst. Vassil. Kolarov Plovdiv.), 15: 43-48.

Popova, M.T. and Česchmedjiev, I.V., 1975. In: Löve, Á., Taxon, 24: 143.

Popova, M.T. and Česchmedjiev, I.V., 1978. In: Löve, Á., Taxon, 27: 384-385.

Poucques, M.L. de, 1948. Bull. Soc. Sci. Nancy, N.S., 7: 33-39.

Priadcencu, A., Miclea, C. and Moisescu, L., 1967. Rev. Roum. Biol., Bot., 12: 421-425.

Prime, C.T., 1954. Jour. Ecol., 42: 241-248.

Pritchard, T.O., 1957. Jour. Ecol., 45: 966.

Proctor, M.C.F., 1955. Watsonia, 3: 154-159.

Przywara, L., 1970. Acta Biol. Cracov, 13: 133-142.

Puff, C., 1976. Canad. Jour. Bot., 54: 1911-1925.

Queirós, M., 1973. Bol. Soc. Brot., Sér. 2, 47: 77-103, 299-314.

Quézel, P., 1967. Taxon, 16: 240.

Quézel, P. and Contandriopoulos, J., 1965. Candollea, 20: 51-90.

Quézel, P. and Contandriopoulos, J., 1966. Bull. Soc. Bot. Fr., 113 (5-6): 351-353.

Rahn, K., 1957. Bot. Tidsskr., 53: 369-378.

Rahn, K., 1966. In: Löve, Á., Taxon, 15: 122-127.

Raitanen, P.R., 1967. Ann. Bot. Fenn. 4: 471-485.

Rao, Y.S., 1951. Current Sci., 20: 72.

Ratter, J.R., 1964. Notes Roy. Bot. Gard. Edinb., 25: 293-302

Ratter, J.A., 1967. Notes Roy. Bot. Gard. Edinb., 27: 300-333.

Ratter, J.A., 1976. Notes Roy. Bot. Gard. Edinb., 34: 411-428.

Raven, P.H., 1963. Quart. Rev. Biol., 38: 109-177.

Raven, P.H. and Moore, D.M., 1964. Watsonia, 6: 36-38.

Raynaud, C., 1971. In: Löve, Á., Taxon, 20: 795.

Reese, G., 1952. Ber. Deutsch. Bot. Ges., 64: 241-256.

Reese, G., 1953. Ber. Deutsch. Bot. Ges., 66: 66-73.

Reese, G., 1954. VIII Congr. Int. Bot. Rapp. & Comm., 9 and 10: 83-85.

Reese, G., 1961. In: Löve, Á. and Love, D., Opera Bot., 5.

Reese, G., 1963. Sch. Naturw. Ver. Schlesw. Holst., 34: 44-70.

Rejón, M.R., 1974. In: Löve, Á., Taxon, 23: 805-806.

Rejón, M.R., 1976. In: Löve, Á., Taxon, 25: 341-342.

Renner, O., 1951. Ber. Deutsch. Bot. Ges., 63 (5): 129-138.

Renzoni, G.C., 1963. Nuovo Giorn. Bot. Ital., 70: 493-504.

Renzoni, G.C. and Garbari, F., 1970. Nuovo Giorn. Bot. Ital., 104: 61-73.

Renzoni, G.C. and Garbari, F., 1971. Atti Soc. Tosc. Sci. Nat. Mem., Ser. B.
78: 99-118.

Ricci, I., 1961. Ann. Bot. (Roma), 27: 36-39.

Ricci, I. and Colasante, M.A., 1975. Ann. Bot. (Roma), 32: 218.

Richards, A.J., 1969. In: Löve, Á., Taxon, 18: 560-562.

Richards, A.J., 1972. In: Löve, Á., Taxon, 21: 165-166.

Ritter, J., 1973. Compt. Rend. Acad. (Paris), Sec. Biol., 167: 240.

Roberts, R.H., 1966. Watsonia, 6: 260-267.

Rodrigues, J.E. de, 1953. Diss., Univ. Coimbra.

Rodrigues, J.E. de, 1954. Bol. Soc. Brot. Sér. 2, 28: 117-129.

Rohner, P., 1954. Mitt. Naturf. Gess. Bern, 11: 43-107.

Rohweder, H., 1934. Bot. Jahrb., 66: 249-368.

Rohweder, H., 1939. Beih. Bot. Centralbl., 59: 1-58.

Rönning, O.I., 1962. Kungl. Norske Vid. Selsk. Skr., 1961 (4): 1-50.

Rosenberg, O., 1926. Arkiv. Bot., 20B (3): 1-5.

Rosenthal, C., 1936. Jahrb. Wiss. Bot., 83: 809-844.

Rosser, E.M., 1955. Watsonia, 3: 228-232.

Rottgardt, K., 1956. Beitr. Biol. Pflanzen, 32: 225-278.

Rousi, A., 1956. Ann. Bot. Soc. Vanamo, 29 (2): 1-64.

Rousi, A., 1961. Hereditas, 47: 81-110.

Rousi, A., 1965a. Ann. Bot. Fenn., 2: 1-18.

Rousi, A., 1965b. Ann. Bot. Fenn., 2: 47-112.

Rousi, A., 1967. Zuchter, 36 (8): 352-359.

Rousi, A., 1969. Hereditas, 62: 193-213.

Ruggeri, C., 1961. Caryologia, 14: 111-120.

Rune, O. and Ronning, O.I., 1956. Svensk. Bot. Tidskr., 50: 115-128.

Runemark, H., 1962. Bot. Not., 115: 8.

Runemark, H., 1967a. Bot. Not., 120: 84-94.

Runemark, H., 1967b. Bot. Not., 120: 9-16.

Runemark, H., 1967c. Bot. Not., 120: 161-176.

Runemark, H., 1970. Willdenowia, 6: 137-138.

Runemark, H. and Heneen, W.K., 1968. Bot. Not., 121: 51-79.

Rutishauser. A., 1936. Mitt. Naturf. Ges. Schaffhausen, 13: 25-47.

Rutland, J.P., 1941. New Phytol., 40: 210-214.

Ryberg, M., 1960. Acta Horti Berg., 19: 121-248.

Rychlewski, J., 1968. In: Skalińska, M., et al., Acta Biol. Cracov., 11:
    199-224.

Santos, A.C. dos, 1945. Bol. Soc. Brot., 19: 519-522.

Sañudo, A. and Rejón, M.R., 1975. Anal. Inst. Bot. Cav., 32 (2): 633-648.

Sauer, W., 1971. Carinthia II, 80: 79-87.

Sauer, W. and Chmelitschek, H., 1976. Mitt. Bot. Staatssamm. München, 12:
    513-608.

Saunte, L.H., 1955. Hereditas, 41: 499-515.

Saura, F., 1943. Rev. Fac. Agron. Vet. (Bs. Aires), 10: 344-353.

Scannerini, S., 1971. Inf. Bot. Ital., 3: 66-67.

Schafer, B. and La Cour, L., 1934. Ann. Bot., 48: 693-713.

Scheerer, H., 1939. Planta, 29: 636-642.

Scheerer, H., 1940. Planta, 30: 716-725.

Schmidt, A., 1961. Österr. Bot. Zeits., 108: 20-88.

Schmidt, A., 1962. Ber. Deutsch. Bot. Ges., 75: 78-84.

Schmidt, A., 1963. Österr. Bot. Zeits., 110: 285-293.

Schmidt, A., 1964. Flora, 154: 158-162.

Schmidt, A., 1964b. Ber. Deutsch. Bot. Ges., 77: 256-261.

Schmidt, A., 1965. Ber. Deutsch. Bot. Ges., 77 (1): 94-99.

Schneider, H., 1913. Flora, 106: 1-41.

Schneider, U., 1964. Feddes Rep., 69 (3): 180-195.

Schöfer, G., 1954. Planta, 43: 537-565.

Scholte, G., 1977. In: Löve, Á., Taxon, 26: 258-259.

Schönfelder, P., 1968. Oesterr. Bot. Zeits., 115: 363-371.

Schotsman, H.D., 1961a. Bol. Soc. Brot., Sér. 2, 35: 95-127.

Schotsman, H.D., 1961b. Bull. Soc. Bot. Suisse, 71: 5-17.

Schotsman, H.D., 1961c. Bull. Soc. Neuchât, Sci. Nat., 84: 89-101.

Schotsman, H.D., 1967. In: Jovet, P., Flora de France, 1, Paris.

Schotsman, H.D., 1969. Bull. Centre Etude Rech. Sci. Biarritz, 7: 869-872.

Schulle, H., 1933. Flora, 127: 140-184.

Schulz-Schaeffer, J. and Jura, P., 1967. Zeits. Pflanzenzucht, 57: 146-166.

Schürhoff, P.N., 1926. Die Zytologie der Pflanzen. Stuttgart, Verl. F. Enke,
    1-792.

Schwarz, O., 1964. Drudea, 1964: 5-16.

Schwarz, O. and Bässlar, M., 1964. Bot. Jahrb., 111: 193-207.

Schwarzenbach, F., 1922. Flora, 115: 393-514.

Scrugli, A., 1974. Inf. Bot. Ital., 6: 37-43.

Scrugli, A., De Martis, B. and Mulas, B., 1976. Inf. Bot. Ital., 8: 82-91.

Sealy, J.R. and Webb, D.A., 1950. Jour. Ecol., 38: 223-236.

Seitz, F.W., 1951. Zeits. Forstgenet. Forstpflanzenzücht., 1: 22-32.

Seitz, F.W., 1954. VIII Cong. Internat. Bot., Rapp. Comm., 13: 28-29.

Seitz, W., 1969. Ber. Deutsch. Bot. Ges., 82: 651-655.

Shimoya, C., 1952. Rec. Trav. Lab. Bot. Geol. Zool Fac. Sci. Univ.
    Montpellier, Ser. Bot., 5: 87-89.

Shimoya, C. and Ferlan, L., 1952. Broteria, 21 (4): 171-176.

Sibilio, E., 1961. Delpinoa, N. Ser., 3: 225-238.

Sikka, S.M., 1940. Jour. Genet., 40: 441-510.

Simola, L.K., 1964. Ann. Acad. Sci. Fenn. Ser. A., IV Bot., 78: 1-19.

Simonet, M., 1934. Ann. Sci. Nat., Bot., Ser. 10, 16: 229-338.

Simonet, M., 1955. Ann. Sci. Nat. Bot., Ser. 11, 16: 503-528.

Sinskaya, E.H. and Maleyeva, Z.P., 1959. Bot. Zhurn., 44: 1103-1113.

Sintes, S. and Cauderon, Y., 1965. Compt. Rend. Acad. Sci. Paris, 260:
    4249-4251.

Skalińska, M., 1950. Acta Soc. Bot. Polon., 20 (1): 45-68.

Skalińska, M., 1950b. Bull. Int. Acad. Polon. Sci. Lett. B, Sci. Nat., 1:
    149-175.

Skalińska, M., 1952. Bull. Int. Acad. Polon. Sci. Lett. 3BI: 119-136.

Skalińska, M., 1958. Acta Biol. Cracov., Ser. Bot., 1: 45-54.

Skalińska, M., 1959. Rep. IX Int. Bot. Congr., 2: 365.

Skalińska, M., 1967. Acta Biol. Cracov., Bot., 10: 127-141.

Skalińska, M., 1970. Acta Biol. Cracov., Bot., 13: 111-117.

Skalińska, M. and Kubień, E., 1972. Acta Biol. Cracov., Bot., 15: 39-50.

Skalińska, M., Malecka, J. and Izmailow, R., 1974. Acta Biol. Cracov., 17:
    133-163.

Skalińska, M., Piotrowicz, M., Czapik, R. et al., 1959. Acta Soc. Bot.

Polon., 28: 487-529.

Skalińska, M., Piotrowicz, M., Sokolowska-Kulczycka, A. et al., 1961. Acta Soc. Bot. Polon., 30: 463-489.

Skalińska, M. and Pogan, A., 1973. Acta Biol. Cracov., Bot., 16: 145-201.

Skalińska, M. et al., 1964. Acta Soc. Bot. Polon., 33: 45-76.

Skalińska, M., Pogan, E. and Pogan, A.L., 1966. Acta Biol. Cracov., Bot., 9: 31-58.

Skalińska, M., Pogan, E., Jankun, A. et al., 1968. Acta Biol. Cracov., Bot., 11 (2): 199-224.

Skalińska, M., Jankun, A., Wcislo, H. et al., 1971. Acta Biol. Cracov., Bot., 14: 55-102.

Skovsted, A., 1934. Dansk Bot. Ark., 9 (5): 1-52.

Smith, G.L., 1963. New Phytol., 62: 283-300.

Smith, P.M., 1967. Feddes Rep., 77: 61;64.

Snogerup, S., 1962. Bot. Not., 115: 357-375.

Snogerup, S., 1963. Bot. Not., 116: 142-156.

Snogerup, S., 1971. In: Nilsson, Ö. and Snogerup, S., Bot. Not., 124: 1-8, 179-186, 311-316, 435-441.

Snogerup, S., 1972. In: Nilsson, Ö. and Snogerup, S., Bot. Not., 125: 1-8, 131-138, 203-211.

Sokolovskaya, A.P., 1955. Bot. Zhurn., 40 (6): 850-853.

Sokolovskaya, A.P., 1958. Bot. Zhurn., 43: 1146-1155.

Sokolovskaya, A.P. and Strelkova, O.S., 1941. Doklady Akad. SSSR, N.S., 32: 145-147.

Sokolovskaya, A.P. and Strelkova, O.S., 1960. Bot. Zhurn., 45: 369-381.

Solbrig, O.T., Khyos, D.M., Powell, M. and Raven, P.H., 1972. Amer. Jour. Bot., 59: 869-878.

Söllner, R., 1954. Ber. Schweiz. Bot. Ges., 64: 221-354.

Šopova, M., 1966. God. Zborn. Biol. (Skopje), 16 (4): 107-144.

Šopova, M., 1972. God. Zborn. Biol. (Skopje), 24: 73-81.

Sørensen, T. and Christiansen, H., 1964. Bot. Tidsskr., 59: 311-314.

Sørensen, T. and Gudjonsson, G., 1946. K. Dansk. Vid. Selsk. Biol. Skr., 4 (2): 1-48.

Sorsa, M., 1963. Ann. Acad. Sci. Fennicae, Ser. A, IV, Biol., 68: 1-14.

Sorsa, M., 1966. Hereditas, 56: 395-396.

Sorsa, V., 1958. Hereditas, 44: 541-546.

Sorsa, V., 1962. Ann. Acad. Sci. Fenn. Ser. A, IV, Biol., 58: 1-14.

Speta, F., 1976. Linz. Biol. Beitr., 8: 239-322.

Staudt, G., 1962. Canad. Jour. Bot., 40: 869-886.

Stearn, W.T., 1978. Ann. Mus. Goulandris, 4: 83-198.

Stearn, W.T. and Özhatay, N., 1977. Ann. Mus. Goulandris, 3: 45-50.

Stebbins, G.L., Jenkins, J.A. and Walters, M.S., 1953. Univ. Calif. Publs. Bot. 26: 401-430.

Steiner, E., 1955. Bull. Torrey Bot. Club, 82: 292-297.

Stern, F.C., 1946. A study of the genus Paeonia. London.

Stern, F.C., 1956. Snowdrops and Snowflakes. London: Ballantyne & Co.

Stoeva, M.P., 1977. In: Löve, Á., Taxon, 26: 560.

Strandhede, S.-O., 1961. Bot. Not., 114: 417-434.

Strandhede, S.-O., 1963. Bot. Not., 116: 215-221.

Strandhede, S.-O., 1965. Opera Bot., 9 (2): 1-86.

Strandhede, S.-O. and Dahlgren, R., 1968. Bot. Not., 121: 305-311.

Strasburger, E., 1902. Jahrb. Wiss. Bot., 37: 477-526.

Strasburger, E., 1904. Jahrb. Wiss. Bot., 41: 88-164.

Strid, A., 1965. Bot. Not., 118: 104-122.

Strid, A., 1971. Bot. Not., 124: 490-496.

Stuart, D.C., 1970. Notes Roy. Bot. Gard. Edinb., 30: 189-196.

Styles, B.T., 1962. Watsonia, 5: 177-214.

Sugiura, T., 1938. Proc. Imp. Acad. Tokyo, 14: 391-392.

Sugiura, T., 1944. Cytologia, 13: 352-359.

Suomalainen, E., 1947. Ann. Acad. Sci. Fenn., Ser. A, 13: 1-65.

Sušnik, F., 1962. Biol. Vest., 10: 7-9.

Sušnik, F., 1967. Biol. Vest., 15: 63-66.

Sušnik, F., Druskovic, B., Löve, Á. and Löve, D., 1972. In: Löve, Á.,
    Taxon, 21: 333-346.

Sušnik, F. and Lovka, M., 1973. In: Löve, Á., Taxon, 22: 462-463.

Tⱥckholm, G., 1922. Acta Horti Berg., 7: 97-381.

Talavera, S., 1976. Lagascalia, 6: 275-296.

Talavera, S., 1978. Lagascalia, 7: 133-142.

Tamayo, A.I. and Grande, M.D., 1967. Bol. Real Soc. Espanola Hist. Nat.
    (Biol.), 65: 447-453.

Tarnavschi, I.T., 1935. Bull. Fac. Ştiint. Çernauti, 9: 47-122.

Tarnavschi, I.T., 1938. Bull. Fac. Ştiint. Çernauti, 12: 68-106.

Tarnavschi, I.T., 1948. Bull. Jard. Mus. Bot. Univ. Chij. 28 Suppl.:
    1-130.

Tarnavschi, I.T. and Lungeanu, I., 1970. In: Löve, Á., Taxon, 19: 609-610.

Teppner, H., 1971. Öesterr. Bot. Zeit., 119: 196-233.

Teppner, H., Ehrendorfer, F. and Puff, C., 1976. Taxon, 25: 95-97.

Thomas, P.T., 1940. Jour. Genet., 40: 141-155.

Tischler, G., 1934. Bot. Jahrb., 67: 1-36.

Titova, N.N., 1935. Soviet Bot., 2: 61-67.

Titz, W., 1964. Öesterr. Bot. Zeits., 111: 618-620.

Titz, W., 1966. Öesterr. Bot. Zeits., 113: 187-190.

Tombal, P., 1969. Inf. Ann. Caryosyst. Cytogénét., 3: 29-32.

Tournay, J.A., 1968. Bull. Jard. Bot. Bruxelles, 38: 295-381.

Tracey, R., 1977. Plant Syst. Evol., 128: 287-292.

Trela-Sawicka, Z., 1968. Acta Biol. Cracov. Bot., 11: 59-69.

Tschermak-Woess, E., 1947. Chromosoma, 3: 66-87.

Tschermak-Woess, E., 1967. Caryologia, 20: 135-152.

Turesson, B., 1972. Bot. Not., 125: 223-240.

Turesson, G., 1957. Bot. Not., 110: 413-422.

Turesson, G., 1958. Bot. Not., 111: 159-164.

Turesson, G. and Turesson, B., 1960. Hereditas, 46: 717-737.

Tutin, T.G., 1957. Watsonia, 4: 1-10.

Tutin, T.G., 1961. In: Löve, Á. and Löve, D., Opera Bot., 5.

Tzvelov, N.N. and Zhukova, P.G., 1974. Bot. Zhurn., 59: 265-269.

Uhl, C.H., 1948. Amer. Jour. Bot., 35: 695-706.

Uhl, C.H., 1961. In: Löve, Á. and Löve, D., Opera Bot., 5.

Uhríková, A., 1970. Acta Fac. Rer. Nac. Univ. Com. Bot., 18: 61-67.

Uhríková, A. and Feráková, V., 1978. In: Löve, Á., Taxon, 27: 379.

Uhríková, A. and Májovský, J., 1977. In: Löve, Á., Taxon, 26: 263.

Uhríková, A. and Schwarzová, T., 1978. In: Löve, Á., Taxon, 27: 379-380.

Ujhelyi, J., 1959. Feddes Rep., 62: 59-70.

Ujhelyi, J., 1960. Ann. Hist. Nat. Mus. Nat. Hung., Bot., 52: 185-200.

Ujhelyi, J., 1960b. Bot. Közlemenyek (Budapest), 48: 278-282.

Ujhelyi, J., 1961. Ann. Hist. Nat. Mus. Nat. Hung., Bot., 53: 208-224.

Ujhelyi, J., 1962. Ann. Hist. Nat. Mus. Nat. Hung., Bot., 54: 199-220.

Ujhelyi, J., 1970. Ann. Hist. Nat. Mus. Nat. Hung., Bot., 62: 175-195.

Ujhelyi, J., 1972. In: Vida, G. (Ed.) Symp. Biol. Hung., 12: 163-176.

Upadhya, M.D. and Swaminathan, M.S., 1963. Chromosoma, 14: 589-600.

Urbańska-Woytkiewicz, K., 1967. Acta Boreal., A, Scientia, 22: 1-14.

Urbańska-Worytkiewicz, K., 1968. Bull. Soc. Bot. Suisse, 78: 192-201.

Urbańska-Worytkiewicz, K., 1968b. In: Löve, Á., Taxon, 17: 419-422.

Urbańska-Worytkiewicz, K., 1970. Ber. Geobot. Inst. ETH, St: Rubel, 40:
    79-166.

Urbańska-Worytkiewicz, K., 1975. Aquatic Bot., 1: 377-394.

Vaarama, A., 1941. Ann. Soc. Zool. Bot. Fenn. Vanamo, 16 (2): 1-28.

Vaarama, A., 1947. Arch. Soc. Vanamo, 2: 55-59.

Vaarama, A., 1948. In: Löve, Á. and Löve, D., Rep. Univ. Inst. Appl. Sci.,

Dep. Agric., Iceland, Ser. B., 3.

Vaarama, A., 1949. Hereditas, 35: 251-252.

Vaarama, A. and Hiirsalmi, H., 1967. Hereditas, 58: 333-358.

Vaarama, A. and Leikas, R., 1970. In: Löve, Á., Taxon, 19 (2): 264-269.

Váchová, M. and Feráková, V., 1977. In: Löve, Á., Taxon, 26: 264.

Váchová, M. and Feráková, V., 1978. In: Love, Á., Taxon, 27: 382-383.

Vignoli, L., 1945. Nuovo Giorn. Bot. Ital., 52: 1-10.

Valdés, B., 1970. Bol. Roy. Soc. Esp. Hist. Not. (Biol.) 67: 243-256.

Valdés, B., 1970b. Bol. Roy. Soc. Esp. Hist. Not. (Biol.) 68: 193-197.

Valdés, B., 1973. Lagascalia, 3: 99-141.

Valentine, D.H., 1978. In: Valentine, D.H. and Lamond, J., Not. Roy. Bot.
    Gard. Edinb., 36: 39-42.

Van Loon, J.C. and Oudemans, J.J.M.H., 1976. Acta Bot. Neerl., 25 (5): 329-336.

Van Loon, J.C., Gadella, T.W.J. and Kliphuis, E., 1971. Acta Bot. Neerl.,
    20: 157-166.

Van Witsch, H., 1932. Österr. Bot. Zeits., 81: 108-141.

Vaucher, C., 1966. Bull. Soc. Neuchât. Nat., 89: 75-85.

Verma, S.C., 1958. Acta Bot. Neerl. 7: 629-634.

Vermeulen, P., 1938. Chron. Bot., 4: 107-108.

Vermeulen, P., 1947. Studies on Dactylorchids. Utrecht.F. Schotarius and
    Jens, pp. 1-180.

Viegli, L., Cela Renzoni, G., Corsi, G. and Garbari, F., 1972. Inf. Bot.
    Ital., 4 (3): 229-236.

Vignoli, L., 1933. Lav. R. Istit. Bot. Palermo, 4: 5-19.

Villard, M., 1970. Bull. Soc. Bot. Suisse, 80: 96-188.

Vladesco, A., 1940. Bull. Sect. Sci. Acad. Romaine, 23: 258-268.

Walter, R., 1973. Acta Biol. Cracov, 16: 227-233.

Walters, S.M., 1949. Jour. Ecol., 37: 192-206.

Walters, S.M., 1967. Acta Fac. Rer. Nat. Univ. Commen. Bot., 14: 7-11.

Walters, S.M. and Bozman, V.G., 1967. Proc. Bot. Soc. Brit. Is., 7: 83.

Walther, E., 1949. Mitt. Thuring. Bot. Ges. Beih., 1: 1-108.

Warburg, E.F., 1938. New Phytol., 37: 189-209.

Warburg, E.F., 1939. In: Maude, P.F., New Phytol., 38: 1-31.

Warburg, E.F., 1952. In: Clapham, A.R., Tutin, T.G. and Warburg, E.F., Flora of the British Isles. Cambridge University Press.

Wcislo, H., 1970. Acta Biol. Cracov., 13: 79-88.

Weimarck, G., 1971. Bot. Not., 124: 129-175.

Westergaard, M., 1938. Dansk Bot. Ark., 9 (5): 1-11.

Wettsteim-Westerheim, W. von, 1933. Züchter, 5: 280-281.

Whyte, R.O., 1930. Jour. Genet., 23: 93-121.

Wieffering, G.H., 1969. In: Löve, Á., Taxon, 18: 442.

Wilee, J., 1961. In: Löve, Á. and Löve, D., Opera Bot., 5.

Wilkinson, J., 1941. Ann. Bot. N.S., 5: 150-165.

Wilkinson, J., 1944. Ann. Bot., N.S., 8: 269-284.

Wilson, J.Y., 1956. Nature (London), 178: 195-196.

Wilson, J.Y., 1959. Cytologia, 23: 435-446.

Winge, Ö., 1917. Compt. Rend. Trav. Lab. Carlsberg, 11: 1-46.

Winge, O., 1940. Compt. Rend. Trav. Lab. Carlsberg, Ser. Physiol., 23: 41-74.

Witsch, H., 1932. Öesterr. Bot. Zeits., 81: 108-141.

Wolkinger, F., 1966. Ber. Deutsch. Bot. Ges., 79: 343-352.

Wolkinger, F., 1967. Phyton (Horn), 12: 91-95.

Wright, F.R.E., 1936. Jour. Bot., 74 (Suppl. 1): 1-8.

Wulff, H.D., 1936. Planta, 26: 275-290.

Wulff, H.D., 1937a. Jahrb. Wiss. Bot., 84: 812-840.

Wulff, H.D., 1937b. Ber. Deutsch. Bot. Ges., 55: 262-269.

Wulff, H.D., 1938. Ber. Deutsch. Bot. Ges., 56: 247-254.

Wulff, H.D., 1939a. Ber. Deutsch. Bot. Ges., 57: 84-91.

Wulff, H.D., 1939b. Ber. Deutsch. Bot. Ges., 57: 424-431.

Yeo, P.F., 1954. Watsonia, 3 (2): 101-108.

Yeo, P.F., 1956. Watsonia, 3 (5): 253-269.

Yeo, P.F., 1966. Watsonia, 6: 216-245.

Yeo, P.F., 1970. Candollea, 25 (1): 21-24.

Young, D.P., 1953. Bot. Not., 1953: 233-270.

Zakharyeva, O.I. and Makushenko, L.M., 1969. Bot. Zhurn., 54: 1213-1227.

Zeltner, L., 1961. Bull. Soc. Bot. Suisse, 71: 18-24.

Zeltner, L., 1962. Bull. Soc. Neuchât. Sci. Nat., 85: 83-95.

Zeltner, L., 1963. Bull. Soc. Neuchât. Sci. Nat., 86: 93-100.

Zeltner, L., 1966. Bull. Neuchât. Sci. Nat., 89: 61-73.

Zeltner, L., 1967. Bull. Soc. Neuchât. Sci. Nat., 90: 241-246.

Zeltner, L., 1970. Bull. Soc. Neuchât. Sci. Nat., 93: 1-164.

Žertová, A., 1962. Nov. Bot. Horti. Bot. Univ. Carol. Prag., 1962: 51.

Žertová, A., 1963. Biologia (Bratislava), 18: 697-700.

Zésiger, F., 1961. Ber. Schweiz. Bot. Ges., 71: 113-117.

Zhukova, P.G., 1965. Bot. Zhurn., 50: 1320-1322.

Zhukova, P.G. and Tikhonova, A.P., 1971. Bot. Zhurn., 56: 868-875.

Zickler, D., 1967. Inf. Ann. Caryosyst. Cytogénét., 1: 7.

Zickler, D., 1968. Inf. Ann. Caryosyst. Cytogénét., 2: 5-8.

Zickler, D. and Lambert, A.-M., 1967. Inf. Ann. Caryosyst. Cytogen., 1:
    3-6.

Zucconi, L., 1956. Caryologia, 8: 379-388.

413

414